MICHAEL FARADAY: SANDEMANIAN AND SCIENTIST

MICHAEL FARADAY: SANDEMANIAN AND SCIENTIST

A Study of Science and Religion in the Nineteenth Century

Geoffrey Cantor

First published 1991 by
THE MACMILLAN PRESS LTD
Houndmills, Basingstoke, Hampshire RG21 2XS
and London
Companies and representatives
throughout the world

ISBN 978-0-333-58802-4   ISBN 978-1-349-13131-0 (eBook)
DOI 10.1007/978-1-349-13131-0

A catalogue record for this book is available
from the British Library.

Reprinted 1993

Contents

List of Figures

Preface

In our culture the scientist occupies an ambiguous role and is represented by conflicting public images. On the one hand he (rarely she) is portrayed as benign – the hero who discovers truth; who reveals the knowledge that benefits the human condition through technology and medicine; who displays nature in all its colour and variety. On the other hand he is malign – he has sold his soul to make weapons of mass destruction and to destroy our environment. From these contrary perspectives the scientist emerges as hero and villain; as highly moral and as thoroughly immoral. There is also a third image which is sometimes linked to the preceding ones. This sets the scientist apart from the rest of society by enshrining his quest for higher truths within an acceptable amorality.

While such popular images have proved serviceable for a variety of purposes they prevent any sustained analysis of the complex components which link the personality and career of the individual scientist. One aim of this biography is to dispel such mythology and to paint a detailed picture of one particular scientist. In the case of Faraday there is no shortage of well-known stories that possess a wide currency partly because, unlike many of his present-day successors, he adopted a high public profile. This study is centrally concerned not only with the public Faraday – the celebrated lecturer and the discoverer of electromagnetic induction and the laws of electrochemistry – but also with the private Faraday and the relationship between them. Faraday's private life was dominated by his membership of a small Christian sect called the Glasites or Sandemanians who adopted and lived by a highly literal interpretation of the Bible. Thus we are confronted with the apparently surprising conjunction that one of the most successful practitioners in the history of science was also committed to fundamentalist religious principles. This book is principally concerned with exploring how Faraday's science and his religion were interrelated.

In popular images science is portrayed as separate from, if not sharply opposed to, religion. This certainly does not apply to Faraday, for whom science was an integral part of his religious experience, and whose science was, moreover, infused with a moral and religious vision. This constructive synthesis which we observe in

Faraday exemplifies how our understanding of the external world and our inner spiritual and moral selves can be interrelated. Two centuries after Faraday's birth we cannot expect to adopt wholesale either his approach to science (which was idiosyncratic even in his day) or his religion as pre-packaged solutions to the problems of the late twentieth century. However, a subsidiary aim of this study is to encourage readers to reassess their own views about the relation between science and religion.

This book was written for the non-Christian and the Christian, as well as for the non-scientist and the scientist. It must be emphasised that this study of Faraday should not be read as providing partisan support for any particular sect or denomination. The issues it raises are more general and more important. Finally, while he greatly respects the Sandemanians, the author is neither one of their number nor, indeed, a Christian.

GEOFFREY CANTOR

Acknowledgements

Writing can be a very lonely activity. Historical writing can also be congenial socially, since the historian must interact with many people, including historians, librarians and publishers. One of the greatest pleasures in writing this book has resulted from the generosity and encouragement of numerous individuals whom I have encountered during my search for Faraday and the Sandemanian connection.

I am deeply indebted to Gerard Sandeman, Mrs J.M. Ferguson, Jean Ferguson, Philip Embleton, Rosalind Brennand, and Mr and Mrs David Williamson for allowing me to consult material in their possession and to quote from it.

I am very pleased to acknowledge the intellectual stimulus I have received from two Faraday scholars, David Gooding and Frank James. Both of them have also commented on versions of my typescript, as have John Hedley Brooke and an anonymous referee. To all four I am very appreciative and feel sure that this book has benefited considerably from their comments. I would also like to thank the many other people who have contributed significantly towards the research for this book. Among these are James Alderson, Erica Blaikley, Brian Bowers, Malcolm Chase, Anne Connan, Elspeth Crawford, Phillip Eichman, Michael A. Faraday, Rosemary Faraday, Sophie Forgan, Mrs Gilfillan, Ivor Grattan-Guinness, Jean Hankins, Renata Hauser, Mark Jackson, David Knight, Lynn McMillon, Jack Morrell, Derek Murray, Janet Oppenheim, Adrian Sandeman, Simon Schaffer, Helen Smailes, Peter Swinbank, Cyril Vincent, Ken Weller, Gavin White and L. Pearce Williams.

During this research I have visited or contacted many libraries and I would like to express my gratitude to members of their staff. Among the libraries I have used extensively are those of the Universities of Dundee, Glasgow and Leeds; the libraries of the Royal Society of London and the Royal Institution; the Leeds Library, the Scott-Fanton Library, Danbury, and the Library of the Wellcome Institute for the History of Medicine, London. For permission to quote from manuscript material or to reproduce photographs I gratefully acknowledge the American Philosophical Society, the British Broadcasting Corporation, Dundee Libraries,

Dundee University Library, the Hunterian Art Gallery, the Institution of Electrical Engineers, the Director of the Royal Institution, the President and Council of the Royal Society, and the Scott-Fanton Museum and Historical Society.

For grants towards the cost of this research I would like to express my gratitude to the Royal Society of London and the Leeds University Research Fund. A substantial part of this book was written in the period January to September 1988 when I held the Gifford Fellowship at the University of Glasgow. I would like to thank the Gifford Trustees and the many people who helped make my stay in Glasgow both congenial and productive.

I am most grateful to Andrea Charters and Gay Lowe (Leeds) and Patricia Ferguson (Glasgow) for their friendly and prompt secretarial assistance. To Steven Gerrard, his colleagues at the publishers and Keith Povey I am indebted for their helpful advice and their conscientious supervision of the production of this book.

Lastly, I wish to express my appreciation to Barbara, David and Adrian who have been very supportive throughout the course of this research and who adopted Michael Faraday as one of the family.

GEOFFREY CANTOR

1 *Introduction*

On King Street, Dundee stands a small, attractive octagonal building with a sloping slate roof (Figure 1.1). The unusual shape first catches the eye, but the visitor is also soon aware that, despite its modern lines, brightly pebble-dashed walls and excellent state of

Figure 1.1 The Glasite meeting house at Dundee where Faraday delivered an exhortation on 9 August 1863.

Source: *The Dundee Yearbook*, 1904, 156.

repair, this structure long predates the neighbouring shopping centre and even the Victorian townhouses on the opposite side of the street. If you pause and look more closely, you soon become aware of several anachronistic features. Areas of wall that may once have contained windows and doors have been blocked in, indicating that the purpose of the building has changed since its construction. Moreover, one side abuts a thoroughly utilitarian and architecturally undistinguished, shoebox-shaped social hall, while an unsightly car park occupies much of the ground adjacent to the building. Stand further back and another contrast becomes apparent for on the townward side stands the sombre stonework of St Andrew's Parish Church (Church of Scotland), whose cemetery encroaches on one of the octagonal walls.

Less obvious to the casual visitor is a further contrast of a very different order. To the rear of the building is a small plaque which provides the only outward clue of the building's original function. It served, we are told, as the place of worship for a religious sect called the Glasites. A visit to the nearby local history library throws further light on the building, which was constructed by the Glasites in the early 1730s and served as their meeting house until the congregation dwindled early this century, after which it fulfilled other functions, including that of a furniture warehouse. Since this small sect originated near Dundee in dissension from the Church of Scotland, it is ironic that its meeting house should now be owned by the Parish Church whose tall, grey spires dwarf the more attractive building. Religious conformity and dissent stand here side by side but their contrasts and history are reified in these two opposing structures.

Few today remember the Glasites but in the eighteenth and nineteenth centuries they constituted a sizeable and influential segment of Dundee's population. At its peak, in the late eighteenth century, there were some 200 members in the Dundee church, which included several of the city's leading merchants and manufacturers as well as many skilled and unskilled workers. As in other Glasite churches, women outnumbered men. On Sundays congregants thronged to the meeting house (see Figure 1.2) which could hold about 700 worshippers and was, according to an account of 1772, 'almost constantly full'.[1] By the mid-nineteenth century the membership had declined to about 100, but on one Sunday in August 1863 it was particularly crowded, the regular congregation being swelled by an unusually large number of visitors. While this was a normal Sabbath in all other respects, the exhortation was delivered by an elder from the London church who was also one of the most outstanding scientists of the day. The visiting elder, who exhorted

Figure 1.2 Interior of the Dundee meeting house.
Source: Dundee Central Library.

on the gospel of St John, chapter 11, verses 25–26, was Michael Faraday, then in his 72nd year and in declining health.

Faraday had been associated with the Sandemanians – as the Glasites are called south of the Border – since his birth, and at the time of this visit to Dundee he had been a communicant for more than four decades. His exhortation and the few days he spent in Dundee passed off with characteristic lack of public notice. The *Dundee Courier and Argus* devoted one paragraph to the occasion while the rival *Dundee Advertiser* was even briefer.[2] Most Dundonians were doubtless far more preoccupied with the progress of the Prince and Princess of Wales from Perth to Aberdeen; or with the question whether the fountain donated by Mr Caird should be erected in the High Street; or with the controversy over whether Baxter Park (which bears the name of a leading Glasite family) should be used by itinerant preachers. Not that the inhabitants of Dundee were uninterested in science: the local press carried extensive reports later that month of the British Association meeting held in Newcastle and there was also a vocal lobby that sought (unsuccessfully) to attract the Association to Dundee in the following year.

It is perhaps not surprising that Faraday's visit attracted so little public attention, since he was not in Dundee for any purpose connected with science. Faraday's religious life has rarely attracted the attention of the press and his reason for visiting Dundee in 1863 was to meet friends and co-religionists from a sect that has never sought the public limelight. It appears, then, that Faraday travelled to Dundee (and also Glasgow) for purely social reasons and was asked by his hosts to deliver an exhortation at the meeting house. However, appearances are deceptive and a fuller appreciation of Faraday's visit to Scotland takes us beyond both his private views about religion and this circle of non-scientific friends who greeted him in Dundee. During the mid-1850s a schism resulted in the Edinburgh meeting house (which existed until 1989) being set apart from those in Dundee, Glasgow, Perth and London. In travelling to Scotland in 1863 Michael Faraday was acting on behalf of the London church and was engaged in shuttle diplomacy aimed at bringing accord among the different communities. He was therefore not only on a social call but also on Sandemanian business, and the exhortation he delivered made explicit reference to the topical issue of church unity.[3] Even though Faraday's attempt failed, his appearance at the Dundee meeting house provides us with one of the rare glimpses of this key figure in nineteenth century science outside the laboratory or the lecture theatre at the Royal Institution.

The elderly grey-haired man I have introduced you to is not the Faraday who features in most popular biographies, which usually stress the following points about his life and work. He was born into a poor family in 1791, received no schooling and was apprenticed to a bookbinder at 14 years of age. By studying the science books that he was required to bind and by attending lectures he became determined to pursue a career in science. In 1813 he was fortunate in being hired as Sir Humphry Davy's assistant at the Royal Institution, where he spent the next 54 years. In the early 1830s Faraday had established himself among the foremost British scientists, discovering benzene, electromagnetic rotation, electromagnetic induction and the laws of electrochemistry. In all he published nearly 400 scientific publications, most noteworthy among which was the series entitled 'Experimental researches in electricity', which laid the experimental and conceptual basis of that important subject. He was very successful as a lecturer, and many flocked to hear his courses (especially his juvenile lectures) and his Friday evening discourses at the Royal Institution. Although his life reads like a Smilesian success story, the standard biographies usually also refer in passing to his membership of an unusual religious sect.

In contrast to the conventional biography, this account of Faraday not only begins with him attending a Glasite meeting house but is centrally concerned with his membership of that small sect. Although Faraday is now remembered primarily as a scientist, his visit to Dundee indicates that, as an elder, he was one of the leading and most respected members of the Sandemanian community. More importantly, even a superficial acquaintance with Faraday the Sandemanian shows that his religion was not a peripheral aspect of his biography but instead it played a central role in his life. This last point cannot be too strongly emphasised. While many people wear their religion lightly, confining it perhaps to occasional or even frequent participation in some form of worship or membership of a religious group, Sandemanianism makes great demands on its members. It is not for the half-hearted or for those who wish to practise Christianity only on Sundays. Indeed, it is a way of life. In making his confession of faith, in 1821 at the age of 29, Faraday solemnly vowed to live according to the precepts laid down in the Bible and in imitation of Christ's perfect example. Sandemanians live strictly by the laws laid down in the Bible, and the sect's stern disciplinary code ensures that any backslider is either brought back into the fold or is excluded – 'put away' to use the conventional euphemism.

In joining this sect not only did Faraday commit himself to live in accordance with these doctrinal and moral principles but he thereby cut himself off from other branches of Christianity and from the practices of most of his contemporaries. As he noted in an often-quoted letter to the Countess of Lovelace, 'I am of a very small and despised sect of Christians, known, if known at all, as *Sandemanians*, and our hope is founded on the faith that is in Christ'.[4] Faraday thus turned his back on many of the religious and secular norms of early Victorian society and instead sought his religious and personal identity in this insular 'despised sect'.

While Faraday adopted through his Sandemanianism a distinctive doctrinal and moral code, he also entered fully into the life of the sect. To be a Sandemanian is to be a member of a closed and tight-knit spiritual brotherhood. Throughout his life, and particularly after 1821, he spent every Sunday and often many other occasions during the rest of the week with his co-religionists, very few of whom had the remotest connection with science. He prayed with them, exhorted them, was exhorted by them; they washed each other's feet and broke the communion bread together. But that is not all. They married into each other's families, supported each other in times of grief or during hardship, and they were even buried together. In life, even perhaps in death, Faraday inhabited this small, close-knit community numbering about 100 souls in London and perhaps 600 in the whole of Britain. The doctrinal and social norms of this community were also his. Thus there is an important sense in which Faraday was much closer to Thomas Boosey (a publisher and bookseller), George Leighton (a bookbinder) and Mary Straker (a poor but devout member of the congregation) than he was to the numerous scientists who passed through the doors of the Royal Institution or the Royal Society.

The exclusivity of the sect must be stressed, since the Sandemanians draw a sharp division between members and non-members. As one of the pamphlets defining church practices states, '*separation* is called for from the communion and worship of all such religious societies as appear in their public doctrine and practice to be setting aside or treating with indifference the plain commands of Christ and His apostles'. This position was based on such biblical passages as 1 John 4:1–6 and 2 Corinthians 6:14, where the righteous are divided from the unrighteous. The unrighteous included not only other religious groups but also those individuals who had not confessed their faith within the Sandemanian church. Frequently Sabbath services were attended by many of these non-Sandemanians, including the friends, children and even spouses of members, but they were excluded from the Love Feast, the kiss of charity and other acts which could be performed only among the faithful. To quote again from the same source, 'Christian charity cannot be extended beyond these limits'[5]. The Christian community, then, was clearly defined and sharply bounded.

Here we are dealing with a common sociological phenomenon, of a small group reacting against the prevailing cultural and religious institutions by differentiating itself from them. In its formation the Sandemanian church was a reaction to the corruption – both spiritual and institutional – of the Church of Scotland, and in its subsequent history it has maintained its distance from all other religious bodies. By maintaining a strict boundary the Sandemanians conform to the accepted definition of a sect[6]. The group's sectarianism was enforced through a rigid and highly literal interpretation of Scripture. This interpretation functioned to distinguish the righteous from the unrighteous and thus helped to police the boundary of this group. Even though they interpret and make sense of the social and political realms in biblical terms, the Sandemanians differ from many other Christian sects in their relative apoliticality and their attitude towards outsiders. Unlike, say, Jehovah's Witnesses or Plymouth Brethren, Sandemanians neither seek to extend their influence through missionary work nor do they view outsiders with pity and contempt. Instead, while accepting that outsiders cannot be given communion, they believe in following Christ's example and doing good to all men. Moreover, unlike many evangelical and fundamentalist sects, particularly in America, the Sandemanians are avowedly apolitical.

The other sociologically interesting aspect of the Sandemanians concerns their internal structure. Again in contrast to many other religious groups, especially religious cults, they do not have a leader or an extended social hierarchy; instead they insist on the spiritual equality of each member irrespective of wealth, age, sex or rank. There are, however, two offices within the church – the elders and deacons – but these do not constitute elevated ranks, since the bearers possess no authority or power over the other members; instead the elders, deacons and deaconesses[7] have certain biblically ordained offices to perform. Christian brotherhood is frequently emphasised by the Sandemanians on the basis of 1 Corinthians 1:10 which requires that 'ye all speak the same thing, and that there be no divisions among you; but that ye be perfectly joined together in the same mind and in the same judgment'. A very high degree of consensus has thus to be maintained within the church; for example, there has to be complete agreement over the election of elders and deacons. There must also be ways of preventing breakdown of that consensus; hence, should dissension occur on *any* issue, religious or secular, at the commencement of the Sabbath, the aggrieved parties must, according to the customs of the church, be excluded from the service. Although exclusions have occurred on numerous occasions, church discipline has in general been maintained effectively.

The foregoing discussion helps define the orientation of this study, for my main aim is to set Faraday firmly within the context of the Sandemanian church and then to investigate how the doctrines and norms of Sandemanianism relate to him, his views on a number of different issues and, most importantly, his science. Since the Sandemanians differentiate themselves from the rest of society and from other religious institutions, a study of the sect and its norms offers an appropriate and insightful framework for characterising Faraday's attitudes to a wide range of topics. Moreover, the internal homogeneity of the group suggests that if we can ascertain the norms of the Sandemanians we can more fully understand Faraday and his attitudes; this will be particularly helpful in areas where there is a paucity of evidence pertaining directly to Faraday.

Since his death in 1867 there have appeared more than a dozen book-length biographies of Faraday, numerous biographical sketches and a host of more academic articles. Many different themes have received attention within this extensive literature: Faraday's scientific discoveries, his public life, his moral attributes, his experimental procedures, his institutional situation, his finances, his metaphysics, his views on matter, forces and fields, to name but a few. Clearly no single approach to Faraday can offer a definitive account of his life and work and the present study is no more incomplete than its predecessors. Nevertheless, it is striking that Faraday scholars approaching their subject from a variety of different and even incompatible directions are unanimous in accepting that his religion played a very important role in his life and thought. Although James Riley's challenging assertion that 'Faraday was a Sandemanian first and a scientist second'[8] would not gain much support from historians of science, the consensus is that his religious views were closely connected with his scientific work. Let three recent examples stand for many.

In his extensive biography of Faraday, L. Pearce Williams argued that Faraday was motivated by the noble search for knowledge of God's creation, and his science was rooted in a religion which emphasised the intelligibility, beauty and symmetry of the divinely constructed universe. Likewise in several insightful articles David Gooding has shown that Faraday's views on a number of subjects (such as the use of mathematics in physical science and the role of physical forces) were intimately related to his metaphysics, and that his metaphysics was, in turn, underpinned by his theological beliefs. Finally in his innovative interpretation of the Royal Institution Morris Berman argued that Faraday was deeply involved in capitalist enterprises, but Berman opened his discussion of the subject by exploring the apparent contradiction between Faraday's religion and his activities supporting the capitalist economy.[9]

While these and other authors acknowledge the importance of religion for an adequate understanding of Faraday, his life and his science, it is surprising that few writers have paid any attention to the Sandemanian church, its history, membership, doctrines, practices and norms. Only Riley pursued the Sandemanian connection, but he published little on the subject, and the little that he published was concerned primarily with the previous two generations of the Faraday family.[10] The present study aims to fill this lacuna by showing that many different facets of Faraday's life and his science were closely linked to his Sandemanianism. Not only does this book explore a relatively uncharted area of Faraday scholarship, but it is also based on sources that have received little attention from historians of science. I have therefore immersed myself in the writings of Glas, Sandeman and the other leading Sandemanian authors and also the works of the few historians, principally John Hornsby, Derek Murray, Lynn McMillon and Jean Hankins,[11] who have pieced together parts of the history of this fascinating and now almost extinct Christian sect. Like them I have relied not only on published sources but also on the wealth of Sandemanian manuscripts that have accumulated in several different locations in Scotland, England and America. In the following pages I have used some of these manuscripts to shed light on crucial aspects of Faraday that are not adequately illuminated by published sources.

Since my aim is to interpret Faraday from the Sandemanian perspective, a brief but selective history of the sect is offered in chapters 2, 3 and 4. No attempt has been made to encompass all aspects of the sect's doctrines, practices and history, but those facets that contribute directly to our understanding of Faraday are emphasised. A full-length history of the sect is long overdue.

The structure of this book reflects three different, but interconnected, topics raised by the general problem of interpreting Faraday from a Sandemanian perspective. Firstly, in chapters 2 to 4 I show how the Faradays, especially Michael Faraday, fit into the history of the Sandemanian sect, particularly the small group of Sandemanians who lived in London. Taken together these chapters contain a biography of Faraday, but one which, although of limited scope, should provide the reader with a framework in which Faraday's life and work may be comprehended. Secondly, in chapters 5 and 6 I elaborate this framework by exploring Faraday's role in early Victorian society and his attitudes to contemporary religious and secular issues, particularly his views about science, the scientific community and its institutions. It will be argued that these attitudes derive primarily from the social practices and philosophy shared by the Sandemanians.

In chapters 7 to 9 I show that Faraday's religious beliefs underpinned his science. Taken individually these three chapters deal with Faraday's conception of the natural world, his views about scientific method, and (in chapter 9) his scientific practice. In this last chapter I offer two fairly detailed examples of scientific investigations that Faraday pursued, namely his researches into electromagnetism and into the relation between the gravitational and electrical forces. Although the level of technicality necessarily increases in these later chapters, I trust that the non-scientifically-trained reader will nevertheless be able to follow the argument, at least in outline, since in these chapters I establish a number of connections between Faraday's religion and his science.

Chapter 10, in which aspects of Faraday's character are explored, complements the earlier chapters. While researching this book I became increasingly aware of a number of themes common to both Faraday's religious experience and his science. These themes were, moreover, related to specific aspects of Faraday's personality, such as his strong need to order his experience and his inability to live with disorder. Faraday's psychological traits should be fully acknowledged if we are to understand not only his religion and his science but also the relation between them. This exploration into Faraday's personality is not intended as a reductive exercise but it is rather an attempt to show why he was so strongly attracted both to a peculiar religious sect and to science, or, more precisely, an idiosyncratic conception of science. This chapter also provides a way of linking several of the topics previously discussed, since the main features of Faraday's character can be related to his religion, his social experience and his science.

While this study is principally concerned with Faraday and his science it also raises the more general issue of the historical relation between science and religion. This is a widely discussed topic on which there is little consensus. However, a number of historians and others have repeatedly argued that science is necessarily compatible with rational, natural theology, but is consequently opposed to revealed theology. The example of Faraday effectively refutes this viewpoint, since he was not only a leading – possibly *the* leading – scientist of his generation, he was also a fully committed Christian who based his religion on a literal interpretation of the Bible. This conjunction between science and biblical Christianity lies at the heart of this study.

However, further analysis of this relation appears to be pre-empted by Faraday's oft-quoted remark to the Countess of Lovelace:

> There is no [natural] philosophy in my religion. . . . I do not think it at all necessary to tie the study of the natural sciences and religion together, and, in my intercourse with my fellow creatures, that which is religious and that which is philosophical [i.e. scientific] have ever been two distinct things.[12]

Here Faraday seems to be endorsing the view, frequently espoused by scientists both before and since, that science and religion inhabit separate domains that do not intersect. While this passage is open to various interpretations, it should not be used to foreclose discussion of the relation between Faraday's science and his religion, not least because here he is claiming that the strategy of separating science and religion was one he used in public. In private a very different relation might obtain. As we shall see, Faraday rarely linked science and religion in public, but when he did, he articulated fairly conventional views that barely betrayed his sectarian background. Yet, as I will argue in the following chapters, Faraday's science bore the imprint of his religion in many different ways.

This study of Faraday starts from the premise that Sandemanianism provided the central strand in his life to which other aspects, including his science, were directly related. Our first task, then, is to examine the sect through its history and doctrines.

2 *Anglicans, Dissenters and Sandemanians*

2.1 LOCATING FARADAY: THE REFORMS OF THE 1830s

Michael Faraday's life spanned almost eight decades. The England into which he was born in 1791 was bracing itself against the horrors perpetrated by the mob in France, while the state of the nation at the time of his death in 1867 was marked by mid-Victorian prosperity and security. Of the many manifest differences between these periods, the rate of change was most marked during Faraday's fifth decade; the 1830s saw the most concerted challenge to the old order in society and the partial emergence of the new. This age of reform left an indelible mark not only on politics but also on all aspects of British life, science included. In the biography of Faraday the year 1831 is particularly significant, since it marked his most celebrated discovery, that of electromagnetic induction. Of his many other scientific innovations, the other for which he is probably most widely known – the laws of electro-chemical decomposition – dates from just 3 years later. Indeed, while the 1820s witnessed Faraday's rise as a scientist, he was at the peak of his career in the 1830s, and it was during that decade that he wrote the first seventeen series of his *Experimental researches in electricity*, spanning more than 2,000 paragraphs and occupying some 660 pages in the final collected edition.

Yet the significance of the 1830s in the history of British science extends far beyond Faraday's research, since the decade saw the publication of such monumental works as Charles Lyell's *Principles of geology* (1830–3), John Herschel's *A preliminary discourse on the study of natural philosophy* (1830), and William Whewell's *History of the inductive sciences* (1837). It was also the decade of Darwin's voyage on the *Beagle* (1831–6) and John Herschel's observations at the Cape of Good Hope. Samuel Morse in America invented the electric telegraph and Henry Fox Talbot the photographic negative. The opening of the Liverpool–Manchester Railway in 1830, followed soon by other major routes, marked the impact of new technologies on the nation's life. Moreover, there was a new selfconsciousness in the scientific community characterised by the word 'scientist', coined in 1833, to describe those who were committed to the study of natural phenomena.[1] In the 1830s the organisational basis of British science also changed considerably, with the formation of the British Association for the Advancement of Science in 1831 and a manifest rise in the number of scientific societies throughout the country.[2] The new temper of British science can be seen not only in terms of innovations, both intellectual and institutional, but also in the clamour for reform of the country's oldest and most prestigious scientific society, the Royal Society of London. Although the process of reform occurred over a longer period, the main locus for the reforming lobby was the presidential election of 1830.

The conjunction of 'reform' and the 1830s strikes chords that take us far beyond science and into the political arena with the Reform Bills of 1831 and 1832 and the vast amount of related agitation that affected every facet of life in Britain. This Age of Reform resulted from the complex pressures that accrued from the country's rapid industrialisation, particularly during the previous few decades. The growth, concentration and overcrowding of population in London and the main industrial cities, such as Manchester, Glasgow and Birmingham, had altered the demographic picture and led to the widespread call for realignment in social structure. As E.P. Thompson has argued, the working class had fully emerged as a self-conscious group with a common interest by 1833. This interest was most forcibly expressed through demonstrations, riots and mob violence in the early 1830s, associated with the clamour for reform. Demonstrations by 100,000 artisans and working men were witnessed in the major cities and the cries 'Down with the Bishops!' and 'No Peers' were heard throughout the land. Particularly during the uncomfortable passage of the Reform Bill in 1832 Britain came to the brink of revolution and the old order was in serious danger of being swept away.[3] Revolution was averted by reforms that barely benefited the working-class, and working-class discontent thereafter became a recurrent feature of British political life, manifesting itself in the 1830s through trade unionism, Chartism and other Radical movements, and symbolised by the widespread support for the Tolpuddle Martyrs.

If the fear of revolution spurred political reform, the realities of the 1832 Reform Bill, particularly the £10 franchise, drove a deep wedge between the working class and the increasingly conservative middle class. Yet the manoeuvring for reform revealed other deep divisions in British society on matters both political and religious. The first of these divisions was over the question of Catholic emancipation in Ireland, which raised its head in the 1820s following Daniel O'Connell's formation of the Catholic Association, which aimed to remove their civil and political disabilities, and his election to Parliament in 1828 (although he was unable to take his seat). The pressure to prevent civil war by appeasing the Irish Catholics rapidly grew, as did the demand to refuse them emancipation, which was fuelled by the fear that Catholics would undermine the natural alliance between Church and state. Before Parliament in 1828 was the latest bill to grant Catholic emancipation, and although this passed the Commons by just six votes, it failed in the Lords, where the Duke of Wellington opposed it. A year later the bill – now modified to include securities against Catholic control – was passed, this time with the support of both Wellington and Peel. Soon Catholics were sitting in both Houses of Parliament and were holding other civil and political offices. However, Wellington's new readiness to uphold emancipation undermined his credibility with the rump of anti-Catholic Tories and helped seal the fate of his party.

In 1830 there was not only dissension among Tories but Wellington's refusal to countenance constitutional reform forced his opponents to rally under Lord Grey, whose new cabinet placed reform at the head of its agenda. Grey's aim was certainly not universal suffrage, but rather the need to find a political formula that would maintain power within the middle and upper classes while seeming to reform a manifestly corrupt system by extending the franchise. The Reform Bill's progress was slow and unsteady. In April 1831, after its second reading in the Commons, it faltered in committee and had to be abandoned and Parliament dissolved. Strengthened by a general election, the reformers managed to move the Bill through the Commons but it was torpedoed in the Lords on 8 October, since only two bishops voted for the Bill, while twenty-one stood firmly in opposition and six abstained. The Bill's defeat unleashed a wave of anger and violence against reactionary Tories and particularly against the Church since the bishops had so strongly resisted reform by their vote. Bishops and other clergy were attacked, both verbally and physically, the mob in Bristol sacking and burning the bishop's palace. The Anglican Church has rarely, if ever, been so widely abused as during the closing months of 1831.[4]

When the Bill was reintroduced in the spring of 1832, it passed through the Lords with the slender majority of nine, and the clerical vote had changed considerably, with twelve for reform and fifteen against. Yet the Bill's passage was further hindered on 7 May when the Lords passed an amendment which postponed the disfranchisement of pocket boroughs so that Grey requested the King to create new peers to ensure the Bill's acceptance. When the King refused, Grey and his cabinet resigned and Wellington tried vainly to form a government against the background of further public anger and violence. The tense 'days of May' were only ended when the King acceded to Grey's demand and the Bill passed through the House of Lords on 4 June, receiving the royal assent 3 days later. The Bill only extended the electorate by nearly 50 per cent but its main clauses disfranchised the pocket boroughs and somewhat corrected the imbalance in Parliamentary representation, although many anomalies still existed, especially in respect to Ireland.

Whatever the shortcomings of the Reform Bill, it stirred the deepest passions throughout the country and symbolised the mood for reform in other institutions, most particularly the Church. Indeed, constitutional and clerical reform were generally perceived as of a piece. During the 1820s there had been an increasingly vocal tide of opposition to the disabilities borne by dissenters. Moreover, both dissenters and reformers were prominent supporters of such secular educational enterprises as the Society for the Diffusion of Useful Knowledge and the new London University, which, unlike the ancient English universities, did not impose an Anglican religious test. Public dissatisfaction was also directed against the widespread (but by no means total) clerical opposition to Catholic emancipation and, more particularly, the bishops' response to the Reform Bill. If the Church was generally antipathetic to the changing temper of the age, there were also some blatant corruptions closer to home, particularly in its finance and structure. Grey and many other Whigs argued that the Church had to be seen to reform itself; however, while a couple of minor reforms were effected, a Bill to limit the number of livings an incumbent could hold failed to pass through Parliament in both 1831 and 1832. The failure of these mild Plurality Bills 'showed clearly how little the Church could be trusted to reform itself'.[5]

During Peel's short administration in 1835 a commission was formed to consider Church reform, showing that even a number of Tories perceived the need for change. However, when the Melbourne administration reformed the municipal corporations, dissenters were granted more power, including, in many cases, the power of ecclesiastical patronage. Moreover, Melbourne revived Peel's commission, whose recommendations were implemented in three Acts of Parliament. The first, the Established Church Act of 1836, reduced the considerable variation in the bishops' stipends and altered a few of the sees. The Pluralities Act, 2 years later, limited to two the number of benefices held by a clergyman and imposed other, related, ordinances, while the Dean and Chapter Act of 1840 eliminated all non-resident prebends, sinecure rectories and some resident canonries. With the considerable sum of money thus provided the commissioners were able to supplement poor livings and subsidise the formation of new parishes in the large cities. But although this tinkering with the Church corrected some of its most outstanding anomalies, the 'Whig reform of the Church of England was not a revolution'[6] and the changes that were enacted did not command widespread support. Instead they helped increase the popularity of the Oxford Movement, which sought to restore the Church to its role as designated in the Bible. Moreover, they still left many abuses within the Church and maintained the privileged relation between Church and state. It was the latter which rankled particularly with dissenters.

Dissenters played a major role in the reforms of the late 1820s and 1830s and benefited considerably during that period. On 9 May 1828 the 'great act', the Corporation and Test Repeal Act, was passed, relieving dissenters from political disabilities. The clamour to relieve their other disabilities soon increased: for example Parliament was petitioned to grant dissenters the right to celebrate births, marriages and deaths according to their own customs and not in conformity with the Anglican Church. As early as 1834 Lord John Russell introduced the Dissenters Marriage Bill, which would have required banns still to be read in a parish church. A year later Peel moved that dissenters could be granted a civil marriage but still require a clergyman as registrar. Neither of these half-hearted bills gained much support, even among dissenters. However, a bill passed in 1836 allowed dissenters to solemnise marriages in their own chapels (which could be licensed for that purpose) or to marry in a registry office. Although some dissenters, mainly Methodists, continued to marry in a parish church, many denominations availed themselves of the new law, which came into force on 1 March 1837.

While there was little opposition to this act from Anglicans, attempts to right a further grievance proved far more controversial. This was the liability which dissenters faced in having to pay church rates. In practice this liability varied considerably from parish to parish; in some, where there was adequate patronage, no rate was collected, while in others the property of dissenters was seized to pay for the upkeep of the church. Several attempts were made in Parliament in the mid-1830s to eliminate the church rate, but none of these succeeded, owing to the strenuous opposition from both clergy and Tories. Another grievance which came before Parliament at that time, but was likewise rejected, was the Test Act applied to the ancient English universities. Since incoming students were required to subscribe to the thirty-nine articles, Oxford was closed to dissenters, whereas Cambridge accepted them but required them to leave without a degree. This grievance was only righted in 1871.

Britain in the 1830s was thus deeply divided along complex social, economic, political and religious lines. Every institution was caught in the maelstrom of reform and counter-reform. The turmoil of change, and resistance to change, was everywhere apparent. The burning questions of the day were widely aired, discussed and written about. Everyone, we might assume, was caught up in matters of the moment. However, this generalisation is not entirely correct, since there were a few, Faraday included, who appear to have stood on the sidelines. We cannot reconstruct Faraday's conversations but his correspondence of the period is surprisingly lacking in references to political or ecclesiastical reform, the Corn Laws, the Poor Laws, the grievances of dissenters or any of the numerous pressing issues of the day.

While the Faraday correspondence that has survived is certainly far from complete, it is nevertheless voluminous, amounting to over 400 letters by Faraday in the 1830s. Moreover, it seems unlikely that the surviving letters are unrepresentative of his total correspondence in their almost total exclusion of contemporary social, political and ecclesiastical issues. Certainly many of Faraday's letters concern scientific matters, but, as other contemporary collections of scientific letters show, scientists were generally much concerned with just those issues on which Faraday was almost silent. This lacuna is all the more surprising since Faraday lived in close proximity to those issues and their proponents. He had been born into an artisan's family and had moved to a position which most people would associate with the professional middle classes. He was a dissenter who stood to gain from the relief of disabilities. He lived at the Royal Institution in the centre of London, in close proximity to men of influence – Lord Brougham, for example, was a near neighbour. The rich, the fashionable and the powerful passed through the doors of the Royal Institution and many of them were known to Faraday and even corresponded with him.

Yet there was one aspect of reform in the 1830s with which Faraday could identify; this was the (largely unsuccessful) attempt to reform the Royal Society, in which Faraday played a small but not insignificant role. That Faraday differentiated between the reform of science and reform in all other spheres is important for this study, since he viewed science as totally antithetical to politics. As he wrote to John Tyndall in 1850, 'When science is a republic, then it gains; and though I am no republican in other matters, I am in that'.[7] His denial that he was a republican in areas outside science should not make us rush to label him a Tory, since he stood if not above party politics, then at least tangential to it (See section 5.3).

How then are we to account for Faraday's unusual, even idiosyncratic, attitude towards the mainstream social, political and ecclesiastical issues of the day? This attitude does not derive from science, although it is clear that he took refuge in his basement laboratory to avoid the turmoil taking place above his head. Instead we should locate his attitude in the system of values to which he unswervingly adhered, and we should also understand those values as deriving from his religion. Sandemanianism offered both a religion and a social philosophy which distanced Faraday considerably from the turmoil of his day. Faraday's responses to contemporary events were not unique, but were shared by other members of the Sandemanian church. This perspective does not imply that Faraday was unaware or unaffected by the Age of Reform, but rather that he was able to control his interaction with that external world and to do so in a very different way from most of his contemporaries. To let one example stand for many: in February 1832, when Britain was rocked by civil strife and Parliament was torn over the issue of reform, the King proclaimed a day of fasting in an attempt to pull the country back from the brink. A few days later a further fast day was announced to try to halt the cholera epidemic. The question of fasts was discussed by the Sandemanians, who, although loyal to the Crown, refused to participate, since such fasts were not sanctioned by the Bible. Moreover, as one of their number complained, the King is 'head of that anti-christian hierarchy [the Church of England] which he has sworn to support in all its abominations'.[8]

While the Sandemanians usually accepted civic authority and the law of kings, in accord with their understanding of Romans 7:1–7 and 1 Peter 2:13–17, this example shows that they were totally unwilling to follow the decrees of William IV when they considered them to be opposed by the Bible. Indeed, the Sandemanians judged all issues by the touchstone of the Bible. This attitude necessarily placed them outside the main warring factions of the 1830s. They were not concerned about Catholic emancipation or even about relieving the disabilities suffered by (other) dissenters (except that they resented having to register marriages at Anglican churches). The corruptions within Anglicanism were also of no concern, since the Anglican church was corrupt, not owing to the poor pay or the absence of the clergy, but because the very concept of a national church was not sanctioned by Scripture. The calls for political reform likewise made little impact on the sect, since its members had nothing to gain or to lose by changes in the electoral roll or by the parliamentary machinations of the Whigs. Viewing the social and political turmoil of the 1830s, Sandemanians sometimes commented that they were witnessing the necessary destruction of the unrighteous. After all, the future was in the ! ands of God, and the prophecies in the books of Daniel and Revelation foretold how the present epoch would end. Nothing was changed by the violent clamour for or against reform, which was seen merely as the confirmation of biblical prophecy.

It is therefore no accident of selection that Faraday's extant correspondence of the 1830s contains so few references to contemporary social, political and ecclesiastical matters. His omission of these subjects is just what we would expect of a Sandemanian. Our first task is therefore to set Faraday firmly in the context of the Sandemanians in order to understand their (and his) peculiar responses to events in the social, political and ecclesiastical world. The remainder of this chapter will be concerned with the history of this insular sect.

2.2   THE SANDEMANIANS

2.2.1   *John Glas and the schism with the Church of Scotland*

Despite their proclaimed aversion to politics, the formation of the Sandemanians (otherwise known as Glasites) was a highly charged act in the context of Scottish ecclesiastical politics. The history of religion in Scotland in the early modern period was marked by inter-denominational strife and some of the issues raised by Glas in the 1720s had regularly reached boiling point after the mid-sixteenth century, when John Knox laid the basis of institutionalised Protestantism amid riot and bloodshed. Inseparable from the doctrinal and ecclesiastical reforms was the system of covenanting which the Protestants deployed in their ultimately successful bid for supremacy. First used in 1557, the Covenant was the outward, visible expression of solidarity among the reformers and the means of social control. During the subsequent slow evolution of reformed Protestantism there were many covenants. One of the most important was the National Covenant of 1638, which was anti-Catholic in intent, sanctioning only those decisions taken by free assemblies and parliaments and pledging the signatories to defend their ecclesiastical principles. The National Covenant soon became the popular political cause among a wide section of Scottish society when Charles I mounted an army to crush the dissidents. In 1640 the

Covenanters marched south, joining forces with the English Parliamentarians, and this cooperation led to the signing 3 years later of the Solemn League and Covenant. This covenant functioned not only to cement the bond between England and Scotland by an oath of mutual defence, but those who signed were pledged to eradicate all heresy, popery and prelacy. However, the subsequent confrontation between Cromwell and the Scottish Covenanters resulted in the most bloody scenes in Scottish history.

The history of covenanting cannot be dissociated from the question of church organisation and the rise of presbyterianism. As early as 1592 the Scottish Parliament recognised the legal existence of presbyteries and synods, but in the years following Cromwell's successes episcopalianism dominated and presbyterianism was largely driven underground. When secret open-air 'Conventicles' were held, particularly in the Western lowlands, government troops were sent in to suppress these seditious meetings. However, the presbyterian cause gained considerable support from the accession to the throne in 1685 of the Catholic James VII (James II in England) who was supported by the Scottish bishops. In the turmoil that followed, the country was pushed considerably in the direction of presbyterianism, and in the settlement of 1690 the Church of Scotland was given a constitution that abolished bishops in favour of a powerful, hierarchical structure consisting of synods, presbyteries and kirksessions, dominated by the General Assembly. Moreover, the government of each individual church was placed largely in the hands of the minister (who thus wielded considerable power) and the elders.

In the early eighteenth century covenanting had lost its previous association with seditious activities, becoming instead the badge of conformity. Nevertheless it remained a visible and popular practice symbolising endorsement of the Church's new form of organisation. Not only was this organisation presbyterian, but it also placed the Church of Scotland firmly under parliamentary control. Thus to accept the covenant amounted to a proclamation of allegiance to both Church and state, and a pledge to extirpate popery, prelacy and dissent.

Against this background the dissent of a young minister named John Glas (1695–1773) takes on political meaning.[9] Glas was educated at the Universities of St Andrews and Edinburgh and proceeded to the ministry at Tealing a few miles from Dundee. Thoughtful, well-read and popular with his congregation, his early career contained no outward signs of dissent. However, by the mid-1720s he found himself unable to reconcile the practice of covenanting with Scripture since the Bible allows no role to national churches or covenants. When he expressed this conviction to his congregation, nearly a 100 of its members joined together in July 1725 in Christian profession and agreed to follow Glas in founding their religion firmly on the Bible. Here was a clergyman within the established Church rejecting the covenant and challenging the structure of church government. Glas was questioning the very legitimacy of the Church of Scotland not in terms of the legal, parliamentary sanction that it had received in 1690 but in respect to the highest authority – God's word.

Glas's dissent gained wider significance when he preached at Strathmartine on 6 August 1726 on the subject of Christ's kingdom, arguing that His kingdom is spiritual, not mundane, and therefore totally unrelated to any nation or national church. This theme brought him into direct conflict with John Willison, a respected minister with liberal views on patronage, who preached at the same service and who not only declared his support for national covenants but also warned that 'Satan is in this matter transforming himself into an angel of light'.[10] A month later the issue came before the Presbytery of Dundee, at which Willison charged Glas with opposing 'the doctrine and *authority* of the church and the martyrs'.[11] The Presbytery ordered Glas to remain silent, but he refused, and the matter was referred to Synod. Although a committee was appointed in the autumn of 1726 to examine Glas's snub to authority, no report ensued. In the meantime Glas expounded his case in print, arguing in his *Testimony of the king of martyrs concerning his kingdom* (1727) that state religion was unscriptural. Later that year Glas was hauled before the Synod, which met at Montrose, and a committee examined his teachings but postponed further action until its next meeting. It was then that the Synod placed twenty-six written questions before Glas and studied his replies. It is clear from many of the questions that the Synod was deeply concerned that Glas was undermining the social order. For example, he was asked whether in his view magistrates had the right to quell heresy, and whether the Protestant religion should be defended by arms. Obviously dissatisfied with his answers, the Synod censured Glas.

The case next moved to the General Assembly, which confirmed the Presbytery's decision, despite protests from Glas's congregation. That was not the end of the matter, for Glas continued ministering to his flock at Tealing and preaching his own views. Exasperated by his insolence, the Synod then deposed Glas in October 1728. However, not only did many members of his congregation stand by him but four ministers openly supported him, claiming that the Synod had behaved unconstitutionally and immorally. Thus rumblings of rebellion were heard in the land. One of Glas's flock was close to the mark in claiming that, while he saw no grounds for censure, what was at issue was 'mere church authority'.[12] The final deposition was enacted on 12 March 1730. Glas was accused of preaching doctrines contrary to those of the established church, of continuing to preach after his suspension and of having refused to clear himself before the ecclesiastical authorities. However, the final and most telling accusation was that his actions 'manifestly tend to introduce the *greatest division, confusion, and disorder*, into this church'.[13]

Although his inspiration was spiritual, Glas's independency placed him in direct conflict with the Church. His father, also a minister, called him Ishmael – the outcast – and refused to help his son. He also warned Glas that his challenge to authority could only end in one way: 'His hand was against every man, and every man's hand against him'.[14] Glas's father-in-law considered that he was 'fighting in vain, for what he aimed at never would or could take place'.[15] Many viewed Glas as a dangerous heretic who not only espoused ridiculous, false doctrines but also undermined church authority and unity.

Historically, then, the Sandemanian sect dissented from the Protestant church, and it has subsequently been located in these terms both by its members and by many outsiders. However, Sandemanianism need not be studied solely in socio-historical terms; instead it can be examined through its doctrines. Interestingly, but not surprisingly, there is a high degree of convergence between these two approaches, since Sandemanian doctrines serve to legitimate the sect's perception of its position outside ecclesiastical politics.

In his early writings and sermons Glas not only strove to show that national covenants are contrary to Scripture, but also that the conception of the church contained in the Bible is incompatible with existing organised churches. In his *The testimony of the king of martyrs* Glas sought to undermine the authority of his opponents by showing that their conception of a church is the very antithesis of the true nature of the church as conveyed in the Bible. In Glas's account God made two covenants with man: the first, contained in the Old Testament, related solely to the mundane world and to the government of the ancient kingdoms; however, in the New Testament God offered a second covenant which related not to mundane, temporal worlds and kingdoms but to the eternal, pure and spiritual kingdom. 'My kingdom is not of this world', says Jesus (John 18:36); instead it is a 'spiritual kingdom'. The true Christian living in imitation of Jesus thus enters into this kingdom and participates in a church that is spiritual and unconnected with ecclesiastical organisations, presbyteries or civil kingdoms. On this, as on other topics, the Sandemanians drew a clear line between the spiritual and the mundane. Thus Glas, like Faraday a century later, deployed this distinction to set himself apart not only from other religious organisations but also from other aspects of society, particularly politics.

2.2.2 *Robert Sandeman and the eighteenth-century expansion of the church*

The early 1730s saw the flowering of Glas's stand against the Church of Scotland into a movement, but one which was rent by both internal and external strife. Having left the church at Tealing in the hands of a like-minded ordained minister, Glas moved to the neighbouring town of Dundee, where meetings in accord with his principles were already being held in a private house. Having been deposed and without a stipend, he and his wife struggled to provide for their growing family. He received many gifts from well-wishers and also opened a bookshop in Dundee, which now became the centre of his activities and of the embryonic Glasite church. With the exception of himself and the pastor remaining at Tealing, the church lacked ordained leaders, so that any further expansion was severely limited. However, soon after moving to Dundee he was persuaded to take a vacation during which a major crisis led to the resolution of this problem. On the first Sunday during his absence the meeting house was rent by a fierce dispute over whether communion could be observed without an ordained pastor. Although there were men present who had previously served as elders in the Church of Scotland, some held that these men could not dispense the Sacrament. At issue was the question whether 'ruling elders' were, or were not, identical to 'teaching elders'. Glas was rapidly recalled and a close study of the Bible resulted in the decision that the two terms referred to the same office, and therefore that non-ordained men could serve as elders, the only requirement being the person's moral integrity. A period of fasting followed, after which the meeting house elected Glas and a glover named James Cargill as its elders. This dispute, like so many that followed, ended in schism, and a number of brethren were put away.[16]

Although Glas's independency met with bitter opposition, disaffection with the Church of Scotland was widely felt, particularly in the rapidly expanding urban areas, where links with the local parish church were weak and where the parish structure was slow to respond to changing social conditions. Some drifted away from religious organisations while others were attracted to the increasing number of independent chapels, which offered a wide diversity of religious experience. Disaffection from the established church was most apparent among the labourers and semi-skilled workers, who, particularly in an urban environment, were frustrated by the traditional relationships of deference. Even in rural Tealing Glas's followers 'consisted mostly of the poorest sort'. Moreover, most of the better off soon deserted him not only because he preached against worldly wealth but also because his challenge to church authority compromised their social standing. Glas's reputation spread; he was often seen by his supporters as having taken a brave stand against the Church and as offering a purified and spiritual religion firmly based on the Bible. Moreover, Glas had not only dispensed with a ministry and had raised the status of the individual worshipper but had also founded a biblically sanctioned church which extolled fellowship. Despite, if not because of, the widespread condemnation of Glas, a few people – mostly semi-skilled workers and, as the movement developed, members of the urban middle classes – were attracted to his form of independency and sought him out, as is shown by his expanding correspondence.[17]

In 1733 Glas and a number of other members of his community were invited to Perth to attend the opening of a new Glasite meeting house. News of their impending visit caused alarm. A local clergy-man preached against them and applied to the magistrate to ban their meeting. They were attacked by a mob who pelted them with mud and other missiles, and threatened to burn the meeting house. The situation was brought under control by the sympathetic town clerk. By 1739 the Perth Glasites had built their own meeting house and for the rest of the century the community flourished, attracting many of the most influential local families.[18] In 1734 Glas was in Edinburgh supervising the formation of a further community of followers, and over the next few years several other Glasite meeting houses were established in the urban areas of Scotland. Since both the poor and the middling ranks of society were attracted to these meeting houses, it appears that in these areas the established Church failed to offer practical brotherhood and to heal social divisions.

One of those attracted to the Edinburgh meeting house was an energetic young student named Robert Sandeman (1717–73) who had intended either to enter the established Church or to pursue a career in medicine. The meeting with Glas convinced Sandeman to abandon these plans and he returned to his native Perth in 1735 to become an apprentice weaver. He married one of Glas's daughters in 1737 and by 1744 he had been elected to the elder's office in the Glasite church. The weaving business in which he held a partnership was thriving and he devoted much of his time to the church.[19]

While Glas laid the foundations of his church in Scotland in the 1730s and wrote extensively on theology – the 1782 edition of his collected works runs to five volumes – his son-in-law Robert Sandeman greatly assisted the sect's growth for a quarter of a century, beginning in the mid-1750s. At that time English Calvinists and Evangelicals were locked in battle, one major salvo being fired by the Northamptonshire Calvinist James Hervey in his book *Dialogues between Theron and Aspasio* (1755). Hervey was firmly committed to the doctrine of 'imputed righteousness', whereby God attributes Christ's righteous acts as our own. Thus a Christian who accepted this doctrine would look upon himself (or herself) as predestined and having a special relationship with God. Sandeman reacted strongly to this doctrine which he felt was thoroughly self-serving and, moreover, was devoid of basis in Scripture. Instead Sandeman insisted that the proper biblically ordained relationship with God did not presume any *a priori* standard of righteousness but instead had to be constructed on the basis of a reasoned faith. It is, he argued, only from this basis in faith that we can follow God's laws and attain salvation.[20]

Sandeman initially pursued his differences with Hervey in correspondence but in 1757 placed these before a wider public in his *Letters on Theron and Aspasio*, a work which was reprinted in 1759, 1762, 1768 and 1803 and also appeared in American editions in 1760 and 1838. Sandeman's book was the single most important event that raised the sect above its rather parochial Scottish beginnings. For example, it provided the crucial links which resulted in the formation of meeting houses both in the north-west Pennines and in London – meeting houses which were attended by members of the Faraday family and which will feature in section 4.1 and chapter 3 respectively. Yet Sandeman was not merely a controversial writer, since he travelled extensively and actively helped to create these and other communities.

A similar pattern occurred in America, where the *Letters* excited considerable interest in the early 1760s, and were followed by a visit from Sandeman. He reached Boston in 1764, accompanied by James Cargill, and was soon meeting several ministers and other sympathisers, including the President of Yale, Ezra Stiles. As in both Scotland and England, his presence excited controversy and a number of attempts were made to prevent Sandeman from preaching. Yet several congregations were founded, including the one at Danbury, which became the largest and longest-lasting in New England. By the time that Sandeman died in Danbury, Connecticut, in 1771 the small sect that bears his name had been implanted in American soil, where it remained until early this century.[21]

To provide an accurate account of the number of Sandemanian meeting houses proves surprisingly difficult, since not only is there often a paucity of reliable information, but it is sometimes unclear what should count as a Sandemanian meeting house. Having rebelled against the presbyterian structure of the established Church as unsanctioned by Scripture, Sandemanians eschewed not only the hierarchical church structure but also the circuits and assemblies that became central to denominations such as the Methodists. Each Sandemanian meeting house was run in a highly democratic manner with the elders discharging those basic organisational duties necessary for its day-to-day maintenance. The structure bequeathed by Glas was most vulnerable when a breakdown of unity occurred, for he insisted that communion could not proceed unless there was total unity among the congregants. In such cases of dissension one party would have to be severed from the church. Not surprisingly there were several schisms in the sect's history, which resulted in a substantial loss of members, although sometimes the dispossessed party proceeded to hold its own meetings in a neighbouring building. Another source of recurrent dispute concerned the relation between different meeting houses. The churches founded by Glas, Sandeman and their followers formed a loose association and there was no clear mechanism to govern inter-church relations. Indeed it was not clear whether there had to be complete agreement between the different churches or whether each was a separate entity. Disagreements sometimes occurred and these were often acrimonious and led to one or more meeting house being set apart from the others. Thus there are several cases of churches which called themselves Sandemanian but were not thus acknowledged by other such churches.

Figure 2.1 shows the locations of all the Glasite or Sandemanian churches that I have been able to trace, including some that were, for part of their history, not recognised by other Sandemanians. Some of these churches were short-lived, and not a few of them were attended by a mere handful of worshippers. Throughout the sect's history there were about forty meeting houses in Britain and a handful in America.

Information on the number of Sandemanians is also difficult to obtain. Such data as those provided by the 1851 religious census are unreliable since they omit some of the known meeting houses – for example, the one at Newcastle – and they fail to differentiate between those who had made their confessions of faith and those, children included, who merely attended service on the appointed day. Moreover, the meeting houses that did respond failed to follow the same convention in reporting the number of people attending in

·   Closed before 1821
■   Closed between 1821 and 1867
□   Closed after 1867

Figure 2.1 Locations of Sandemanian meeting houses in Britain.

the morning, afternoon and evening. By far the best data are the membership rolls of the various meeting houses, but the extant rolls provide only a very partial picture of the sect's history. Awaiting a more careful analysis, it appears that the sect started by Glas and extended by Sandeman at no time numbered more than 1,000 who had made their confession of faith. The peak probably occurred towards the end of the eighteenth century. Thereafter numbers declined slowly, with the fall in numbers periodically accelerated by schisms. An American visitor in 1834 counted 559 Sandemanians, including 36 on the far shores of the Atlantic.[22] By the beginning of the present century there were probably only a little over 100 Sandemanians remaining, mostly in London, Edinburgh and Glasgow. In 1989, the Edinburgh meeting house, with but one elder and a handful of congregants, was closed.

2.2.3   The church in Faraday's day

By the time Faraday made his confession of faith in 1821 the sect had begun its slow decline. Only seventeen meeting houses remained and by the time of his death in 1867 a further seven had disappeared. Similarly the number of members was also slowly declining; there were perhaps 600 members in 1821 and some 400 in 1867. Thus throughout his life he was surrounded by a few hundred co-religionists, about 100 of whom were in London. He was in weekly, if not daily, contact with these London Sandemanians. The London meeting house possessed a special relationship with a small group of poor Sandemanians at Old Buckenham, in rural Norfolk, and Faraday often visited them and helped maintain their congregation. Apart from these Norfolk brethren, the London meeting house was geographically isolated, the only other congregations in England being in Liverpool, Newcastle, Chesterfield, Nottingham and the north-west Pennines. Of these, the Newcastle congregation was doubtless the largest in Faraday's day, and it was also the one with which he was most closely associated, since several of his family settled in Newcastle and he visited them on a number of occasions.

The other Sandemanian communities were even further from London, the centre of gravity remaining in Scotland. Dundee, Perth, Edinburgh and Glasgow still flourished, and there were other smaller meeting houses – in such places as Aberdeen and Galashiels – spread across the country. The London community thus generally proceeded with little input from outside, but also with an awareness that the majority of its brethren were north of the Border. The Londoners were also in occasional communication with another far-flung outpost, the American Sandemanians, particularly those in Danbury.

If the map of Sandemanian meeting houses looks patchy, the relation between the meeting houses in Faraday's day was distinctly uneven. For much of that time he, like other Londoners, remained on close terms with those in other meeting houses, although sometimes schisms occurred. Indeed, a superficial reading of these sources might create the impression that the Sandemanians were frequently embroiled in controversy. Such an impression would, however, be false, since schisms were rare occurrences although ones which created a flurry of letters and copies of letters. Moreover, since this correspondence was preserved, letters relating to schisms figure prominently in extant Sandemanian collections.

There were only two major controversies during Faraday's 46-year membership. The year 1844 was particularly disastrous, owing to a rift which affected the whole sect. As one elder from Dundee bemoaned:

> not only have our many churches in this country but also our scattered breathren in America have been visited by terrible things in righteousness. The head of the church appearing among them as the Lord of the conscience searching his Jerusalem with candles and punishing the men settled in their lees [Zephaniah 1:12]. His chastening hand has been heavy upon you and upon us.[23]

Even the London community was affected by an inordinately large number of exclusions, Faraday's included. In 1855 the Edinburgh church was cut off from London and other churches on the question of eating game. The discord was considerable and during the last decade of his life Faraday tried to bring unity once more to a divided sect. These two divisions within the sect will be discussed further in sections 4.2 and 4.3.

While divisions within the church appeared very rarely on the horizon, the single most important feature of Faraday's perspective was the isolation of the London meeting house. This small fellowship comprising about 100 souls was not only isolated from other Sandemanians (geographically, but sometimes doctrinally) but it also functioned as an island which dissociated itself from the main currents of British life and strife. From the safety of the Sandemanian church Faraday viewed the rest of Victorian life almost as an irrelevant sideshow. To analyse this relationship further we need to examine the sect's self-perception and also how outsiders viewed the Sandemanians.

2.3 THE 'DESPISED SECT', FROM WITHIN AND WITHOUT

2.3.1 *From within*

One of the early Sandemanians' greatest concerns was to define what constitutes a church. They not only rejected the Church of Scotland's form of presbyterian government but also all other forms of church organisation not explicitly sanctioned by the Bible. Furthermore, they denied both that a church is constituted by any collection of people praying together and that it is a physical edifice constructed of bricks and mortar. Instead Glas turned to the Biblical description of the primitive church – which he equated with the kingdom of heaven – as God's blueprint for any properly constituted church. The subjects of this kingdom of heaven, although few in number, are *strangers and sojourners* on earth', and they are required by God to form churches and to perform their worship in accordance with biblical precept.[24] While national churches have no such divine legitimation in the New Testament, Glas and his followers argued that their conception of their own church was derived directly from God's word. Thus, after analysing the various uses of the words 'kingdom' and 'church' in the Bible, one of the early London Sandemanians, John Barnard, drew a firm distinction between the worldly kingdoms of the Old Testament and the heavenly kingdoms of the New. He then identified in the books of the Apostles descriptions of the primitive churches which, he argued, were the blueprints for the Churches of Christ. These Barnard defined as:

> a select society of CHRIST'S disciples united by the truth, and observing all the commands of CHRIST, in the closest fellowship with each other, and therefore of necessity separated from the world, whatever form it may assume, and exposed to its hatred; and especially so, to the hatred of that part of the world which shall take the lead in popular and pharisaical devotion.[25]

Since the difference between the church of Christ and conventional conceptions of a church is so great, Barnard warned that 'we are forbidden to think of an earthly establishment for his [Christ's] kingdom, or of temporal honour for his followers'.[26] It is thus clear that both Glas and Barnard considered that the true church had to be defined in transcendental terms, since its very nature contrasts with 'the state of an earthly nation', and it becomes 'not earthly, not of this world'.[27] Instead the legitimate church was constituted by all *true believers* who celebrate communion together. It was, as Barnard insisted, 'a society of saints and faithful brethren, so united one to another by the truth as to be capable of observing all ordinances, duties and discipline which CHRIST hath enjoined them'.[28] In an important sense it was therefore an invisible church, one which was expressed metaphorically in terms of Christ being 'the head of the body, the church' (1 Colossians 1:18).

Here, then, is the Sandemanians' conception of the true church and thus of themselves. A small, chosen, tightly knit group that practices the true, primitive and undefiled form of Christianity. While 'national churches flourish or decay by the smiles or frowns of princes',[29] God established the true church for eternity and its existence is unshaken by events in the mundane world. As a non-proselytising sect, Sandemanians took little interest in other churches and generally treated them with distant disdain. Other churches, especially national churches, had become corrupt and had forsaken the Scriptures. Barnard's reference to 'popular and pharisaical devotion' emphasises the Sandemanians' rejection of over-ritualised religions and those preachers who, currying favour with the public, forsook the word of God. Popular preachers, who set themselves up 'as a kind of factors, or rather ambassadors of God', were frequently denounced, as was the 'parson's shop'.[30]

If the church can thus be defined as the community of true believers, its internal structure is also completely specified in the Bible. Glas and his followers conceived the primitive church, and thus their own church, as a fellowship, where perfect union is achieved between its members. In this respect 1 Corinthians 1:10 was frequently cited by Sandemanians: 'Now I beseech you, brethren, by the name of our Lord Jesus Christ, that ye all speak the same thing, and that there be no divisions among you; but that ye be perfectly joined together in the same mind and in the same judgment'. Drawing on an analogy frequently used by political theorists, Barnard likened the church to a 'natural body' in which every part has its proper role and works for the good of the whole. However, the whole organism is destroyed if any part becomes diseased and ceases to function properly. The same considerations apply to the individual, since, in proper health, there is unity between the heart and the soul. Likewise in the individual's relation with the church, all work for the group's unity and any who give precedence to personal interest must be excluded. For Sandemanians unity is elevated to a necessary principle for the Church of Christ, as they often called themselves. Unity is the very antithesis of confusion, and confusion (cf. the confusion of tongues) is taken as the sign of chaos and not of a God-given world. As Barnard emphatically stated, 'God is not the author of confusion, but of peace'.[31]

Yet within this egalitarian conception of the church the Bible also prescribes certain office holders who are not superior to other communicants in terms of rank, power or spirituality. While Glas attacked the priesthood and the ministry as contrary to God's plan, he nevertheless considered that the Bible specified office holders to perform certain duties necessary for the maintenance of the church and its unity. From Glas's reading of Scripture it was clear that in the primitive church there were only two officers: elders (or bishops) and deacons. Since Acts 20:17 refers to 'the elders of the church', there must be at least two elders for a church to function and, in particular, a plurality of elders is required to celebrate the Lord's Supper. The elder's high moral character is specified in 1 Timothy 3:2–7: 'A bishop [= elder] then must be blameless, the husband of one wife, vigilant, sober, of good behaviour, given to hospitality, apt to teach; Not given to wine, no striker, not greedy of filthy lucre ...'. *The customs of the Churches of Christ*, a summary of Sandemanian practices, also states that a man is not debarred from becoming an elder by being 'engaged in trade or employment'.[32] The role of the elders is to teach and to lead the congregation. By contrast, the deacon (or deaconess), whose qualifications are specified in 1 Timothy 3:8–13, is required to meet the congregation's physical requirements and, in particular, to provide for the poor and the infirm.

In carefully protecting the church's unity the elders have to enforce a number of biblically ordained injunctions. If a person desires entry to the Sandemanian community, (s)he must convince first the elders and then the assembled members of the church that (s)he understands, accepts and is prepared to live according to God's precepts. Acceptance into the church is conditional only on the person's willingness to live by the Christian code, and both the 'rich and the poor' (Proverbs 22:2) and even 'the poorest and vilest of mankind' are not to be refused.[33] Those who have been accepted into the church must uphold its unity, and there are also a number of rules which are intended to prevent dissension. As 1 Corinthians 11:18–34 makes clear, a congregation may only celebrate the Lord's Supper if its members are 'visibly united in their profession and love one another, and there must be no strife, division, or ill-feeling among them'.[34] If division does exist – if any member of the church be opposed to any other on the Sabbath – the adversaries must be excluded. Furthermore, if any church member is found to have committed a sinful act, a scale of punishments ranging from admonition to excommunication is imposed. For the most flagrant acts of indiscipline, such as being 'a fornicator, or covetous, or an idolator, or a railer, or a drunkard, or an extortioner' (1 Corinthians 5:11) the offender must be separated from the church and its members may not keep his company or eat with him, unless they are close members of his family.

From the extant records of various Sandemanian meeting houses it would appear that a high proportion of members suffered exclusion: for example, nearly half the male Sandemanians listed in Appendix B were excluded. The reasons were not often recorded, but the following (probably typical) examples were included in a letter of 1797, in which the writer, from Dundee, mentioned three who had recently been 'put away'. The first was 'a young man who gave up with the Church after he had made a Confession of his Crimes'; the second was a man who 'was living in a State of Adultery for 2 years or more [and] was lying all this time' and tried to hide this 'till he could do no longer'; the third, a tailor, was put away for being drunk, presumably for being frequently drunk.[35] On many occasions an excluded member was readmitted to the church. Readmittance required a similar procedure to gaining initial membership, except that the applicant would have to show complete repentance for the act which had led to his or her exclusion. However if that person was excluded a second time (s)he could not be readmitted, since backsliding showed that the act of repentance was not genuine. Thus a second exclusion resulted in the person's final and irrevocable break with the Sandemanian church. Second exclusions were comparatively rare and only about 10 per cent of those listed in Appendix B suffered this fate.

Another type of separation occasionally occurred in the church's history. This was the separation between churches. All members accept that unity has to be preserved among members of a particular church and that the Sandemanians must be separated from all other religious organisations. It is also generally held that in matters of religion there is only one irrefragable truth. If, then, there is a difference of opinion between two Sandemanian meeting houses, the opposing parties must first attempt to reach consensus and thus unity. However, if agreement cannot be achieved, the opposing churches are required to withdraw from each other and their members cease communication. Although these were relatively rare events when set against the two and a half century history of the Sandemanian church, when they did occur they were invariably bitter and painful. Letters which begin 'Very dear brethren' end with words of a final parting. Men and women who had over decades been very close to one another and were often also joined through family ties found themselves trapped on opposite sides of an unbridgeable divide.

It may appear paradoxical but schisms and exclusions were enacted in order to preserve the unity of the Sandemanians and their conception of spiritual fellowship: a community of souls bound together by mutual love and by obedience to Christ's message. Yet it was deemed essential to police the boundary of this God-ordained community and when necessary to apply stringently the rules for membership and separation. Discipline was a key term for the Sandemanians, who accepted the Bible not only as the basis for all action but also as the rule-book for church organisation.

The picture that emerges is of a unified group of people who achieved an extraordinarily high degree of consensus. At the same time the periphery of the group was sharply bounded and there could be no intermediate position. Moreover, while those outside the church were invariably treated with considerable love and kindness, they were separated by an infinite divide. 'Be ye not unequally yoked together with unbelievers: for what fellowship hath right-eousness with unrighteousness: and what communion hath light with darkness?' (2 Corinthians 6:14). Thus Faraday could be on very agreeable terms with others, especially fellow scientists, yet remain totally severed from them on the most important issues. This divide will be explored further in chapter 6.

2.3.2 *From without*

If the Sandemanians considered themselves apart from the rest of contemporary society, outsiders acknowledged the sect's distance from the mainstream of social, political and religious life in Britain. Commentators, whether friend or foe, invariably noted not only the sect's exclusivity but also its peculiarites – the weekly Love Feast, the kiss of charity and, more trivially, the habit of taking Scotch broth at the Love Feast. However, beyond this emphasis on exclusivity and peculiarity there was no uniform response to the Sandemanians, and writers of different religious persuasions articulated a variety of views. From this range two frequently encountered positions deserve brief notice.

Generally the Sandemanians have received a harsh press, and the reasons are not hard to discover. Particularly for members of the established churches Sandemanianism represented stubborn inde-pendency and the challenge to authority – the response which first greeted Glas in the late 1720s. Nearly a century later he was described in the *London Christian Instructor* of 1819 as 'one of the most decided, religious cynics that the world has seen, whose sneers and censures were dealt without mercy and without discrimination on all that followed not with him'.[36] Again, in the eyes of a recent commentator concerned about present-day church unity in Scotland, Glas committed 'an act of folly, inexcusable provocation

and rebellion'.[37] Not surprisingly nineteenth-century Congregatio-
nalists such as David Bogue and James Bennett appreciated Glas's
stand against the established Church, but even they portrayed
Sandemanians as distant, conceited and argumentative, and unspar-
ingly dismissed Robert Sandeman as a bitter man who incorrectly
attacked such heroic nonconformists as Isaac Watts and Philip
Doddridge.[38]

A further point of contention concerned the Sandemanians'
rejection of an educated, trained ministry and their insistence on
using spiritually unblemished but often uneducated men as their
elders. Not surprisingly members of the established churches have
often looked with horror and contempt, mixed perhaps with fear, at
this aspect of the Sandemanians' snub to the authority of the church.
Such detractors have often portrayed the Sandemanians as ignorant
and uneducated, and they have often been likened to Quakers in this
respect. Moreover, since the Sandemanians proved fairly successful
in drawing workers and semi-skilled labourers away from the
established churches, the charge of ignorance has been mixed with
feelings of social contempt. Particularly at times of social tension,
such as the 1790s, when there was a great fear of secret societies, the
sizeable proportion of poor people in the sect and its commitment to
brotherhood further increased the suspicion in which it was held.
One London minister, for example, published a pamphlet in 1806
'intended as an antidote against the virulent poison of the
Sandemanian heresy', in which he charged the sect with harbouring
sinners and therefore fermenting rebellion.[39]

While many writers, especially clergymen, have poured scorn and
hatred on the sect, this was the response predicted by the Bible.
Since Jesus was despised by most of his contemporaries, those who
live their lives in imitation of Jesus would be treated similarly. 'If ye
were of the world, the world would love his own: but because ye are
not of the world, but I have chosen you out of the world, therefore
the world hateth you' (John 15:19).[40] This self-fulfilling prophecy
was clearly on Faraday's mind when Ada, Countess of Lovelace,
broached the question of his religion. 'I am', wrote Faraday, 'of a
very small and *despised* sect of Christians, known, if known at all, as
Sandemanians'.[41]

The other main reaction to the sect and its membership concentrated not on its danger but on a totally contrasting quality. One
term which frequently occurs in published accounts of Sandemanian
meeting houses is 'respectable'. For example, the well-informed and
sympathetic historian of Galashiels, the Rev. Robert Hall, claimed
that the local Glasite church contained some 'of the oldest and most
respected [names] in the town'.[42] Likewise a report on the London

meeting house in 1870 noted that the congregation consisted of 'people in respectable appearance'.[43] To study the social composition of the Sandemanians would take us too far from Faraday, but it is important to note that in urban areas the sect attracted not only labourers but also many from the middle classes, including a smattering of manufacturers and professional people. In Faraday's day there was considerable public concern that in urban areas the lower orders were alienated from the established Church, which they viewed as dominated by middle-class values. Moreover, while the dissenting chapels were growing at the expense of the Church of England, they too were becoming increasingly middle class and losing their earlier associations with the poor and the socially alienated.[44] The Sandemanians seem to have participated in this latter trend, becoming increasingly middle class and with a growing proportion of women. Yet throughout the sect's history it contained an unusual mixture of communicants drawn from different layers of society, except, significantly, the aristocracy. If the middling orders bore the badge of respectability, the sect stood for obedience to the rule of law and for opposition to revolution. Thus even the poorer members who adhered to the sect's strict discipline were, in this sense, respectable members of society.

What was common to both the working class and middle class communicants was a form of religious experience which turned away from the world and found peace and strength in a shared commitment to the kingdom of God. The strife between classes, conflicts in politics and in organised religion were not on the Sandemanians' agenda. The Age of Reform largely passed them by, and the excesses of the mob were interpreted as the accurate realisation of the prophetic drama. Not being of this world, they did not threaten existing institutions.

If at one level Faraday was a member of a 'despised sect', he was also viewed as a highly respectable member of British society. When lecturing at the Royal Institution or visiting the royal family, he was clearly accepted as part of polite society. At the same time he was not a member of that society but a sectary. We return to this apparent paradox in sections 5.5 and 5.6.

We must, however, be careful not to take the Sandemanians' self-image or the accounts written by their critics at face value. To understand the sect's proclaimed rejection of the mundane realm will assist us considerably in understanding Faraday's attitude towards social and political events in his day. It will also shed considerable light on his science, both his scientific practice and his views on the organisation of science. However, the Sandemanians were also greatly affected by pressures from without, especially social pressures. This is evident from such crude indicators as membership numbers; thus the London meeting house attracted many more new members in the turbulent 1820s and 1830s than during either of the previous two or the subsequent two decades. Other external events which appear to bear on the sect's history are the spate of secessions from the Church of Scotland in the 1720s, against which Glas's schism needs to be set, and the Disruption in 1843, which coincided with a major and painful dispute within the sect. Thus any account of the Sandemanians needs to take cognisance of both the sect's internal history and also the influence of external events. Similarly Faraday's life and work should be seen both within the context of the sect and in relation to other contemporary events. Indeed, while firmly locating him among the Sandemanians, a recurrent theme in this study is how Faraday mediated, though in a highly selective manner, between this small, isolated sect and metropolitan science, in which he played a very important, but ambiguous, role.

3 *The London Sandemanian Fellowship*

3.1 THE LONDON MEETING HOUSE

Sandemanianism gained its toe-hold in London around 1760, when the first community arose from the plethora of dissenting congregations that jostled with one another in the side streets of the City and competed for the rapidly expanding population. The staid established Church was often felt to be out of touch with people's lives and religious needs, whereas the dissenting chapels were more welcoming and offered many inspiring preachers and a range of religious experiences. In the mid-eighteenth century there were numerous Methodist, Quaker and Baptist meeting houses and also a number of Independent congregations that formed round individual preachers.[1] It was principally from these Independents that Sandemanianism gained its first influential converts, probably because it was seen as offering not just a form of Christianity based firmly on the Bible, but also a substantial organisational structure which Independent congregations often lacked. Paradoxically, Independent preachers were the very people whom Robert Sandeman had severely chastised in his influential and widely read *Letters on Theron and Aspasio* (1757).

One such Independent was Samuel Pike[2] (1717–73), who first encountered the Scottish sect's doctrines in Sandeman's book and was so impressed by Sandeman's arguments concerning the nature of man's justification before God that he initiated an assiduous correspondence in February 1758. In the course of their correspondence Pike was gradually persuaded by Sandeman, particularly on the question of popular preachers: initially he considered Sandeman too severe, contentious and uncharitable in wishing to dismiss such preachers, but he subsequently came to recognise that they stood in the way of the true Christian message based firmly on the Bible. As one of the popular preachers whom Sandeman had challenged, Pike was forced either to reject Sandeman's argument or to form his own Christian brotherhood in accordance with Sandemanian principles.

Over a period of some 2 years Pike moved ever closer to his mentor, but it also became increasingly clear that his relation to his congregation was becoming severely strained. Thus Sandeman warned Pike 'to be cautious, and count the cost well before you risque any thing on your connection with me'.[3] In the late summer of 1759 he publicly committed himself to Sandemanian principles, which resulted in a major split in his congregation, and during the ensuing months he was engaged in rescuing about 20 who were prepared to follow the same demanding path.

In the late 1750s and early 1760s several other Independent preachers followed similar paths. One was John Barnard[4] (1725–1804?) of Islington – subsequently the uncle of Edward Barnard, Faraday's father-in-law – who met both Glas and Sandeman and was, like Pike, soon engaged in correspondence which reorientated his theological views towards Sandemanianism. By July 1759 Sandeman was urging Barnard and his friends to form themselves into 'the Apostolic order', and he offered to send some communicants from Scotland to assist 'at their own expense'. He also warned Barnard not to accept into the London church those who were lukewarm and unwilling to commit themselves fully to primitive Christianity. Pike's name was explicitly linked to this warning, since he appeared to be 'filled with all the wisdom of a clergyman, by which you may easily understand I do not understand the wisdom that comes from above'.[5] Like Pike, Barnard also faced criticism from his congregants, and he saw that he had 'no prospect at present, as we cannot agree about the importance of the difference, or opposition rather, between popular Christianity and that of the Bible'. Barnard was rapidly losing his congregation and he had to confront the possibility of both losing his meeting house and returning to trade.[6] In July 1760 Barnard reported to Sandeman that a small fellowship, numbering but nine souls, had begun to perform regular Christian practices, but they felt very isolated and in need of support from Scotland. They were also clearly concerned about their uneasy relation with Pike and his followers.

One result of this correspondence was that a deputation consisting of Robert Sandeman, his brother William and John Handasyde, an elder from the Wooler (Northumberland) meeting house, visited London in April 1761 to offer advice. Their preaching excited considerable interest. It is clear that this deputation was not satisfied with the fledgling congregation which had gathered round Barnard. Nevertheless it appears that their visit clarified to Barnard and others the doctrinal and organisational basis on which Glas and Sandeman sought to base their Christianity. Following further soul-searching, Barnard's church, which met in Glover's Hall, became a legitimately constituted Sandemanian congregation on 23 March 1762. Several other dissenting ministers were by this time associated with the new community. One of the most enthusiatic converts was a Baptist minister from Chesham named Benjamin Vaughan, who was so impressed by the Sandemanians that he resigned from his church, moved to London and became a lacemaker.[7] Many other people, both rich and poor, joined the new church, so that its fellowship rapidly increased in size. In June 1766 Barnard reported to his Scottish mentor that 'Our number is now 106, besides several about the country', and 2 years later the number in London had risen to 149.[8] The latter number probably represents the maximum throughout the meeting house's existence of two and a quarter centuries.

During the first 8 years nearly 250 people joined the church, an average of about 30 per year. Thereafter the average dropped dramatically to four a year over the next century, with a mere one per year during the first two decades of the nineteenth century.[9] By early 1769 Barnard was writing to Robert Sandeman, who was planting the seeds of Sandemanianism in America, that the London church had 'put away nine in about five months, and received but one'.[10] The early flush of interest had almost abated. The church's rigid discipline took its toll, as did the following events, which were related by Robert Lyon, one of the elders, in 1781:

The Church of London became pretty numerous soon after its erection [in 1762]. Mr. Sandeman's works had made a great noise beforehand and the setting up of a Christian Church was so new a thing & uncommon that it drew the attention of all ranks and very much provoked the zeal of the religious. Among the Discenting ministers such as Mr. [Samuel] Pike, Mr. [Thomas] Prentice, Mr. Vaughn, Mr. [John] Boosy, Mr. Ivory[?] left the bulk of their congregation and brought a part of them with them and joined the Church. By those means numbers were added to the Church, many through the saving knowledge of the truth and many also having received it in word only, in this Church was to be seen in a very remarkable manner the fulfilment of the Apostle[']s word, 'Many walk of whom I tell you weeping that they are enemies of the cross of Christ whose God is their belly, who mind earthly things' [Philippians 3:18–9]. For as many rich people were added to the Church, so the poor who had joined it on an improper footing made a gain of their profession, endeavouring to live well at the expence of the former, supposing Gain was godliness. On the other hand we saw our words verified 'How hardly shall they who have the riches enter into the kingdom of Heaven'

[Mark 10:23]. For the rich became disgusted by degrees and when the proper time[?] came turned away again to this present evil world, these appearances called for the vigourous sacrifice of the Christian discipline and by that means our number is reduced from about 150 to about 90 and tho' we have a pretty numerous auditory, few in comparison have been added to us for several years.[11]

Thus after the initial flush of interest the London meeting house encountered a difficult period during which there were numerous exclusions. Even Barnard was later put away 'for not being sufficiently humble, or, as they thought, for entertaining too exalted notions of his preaching abilities'.[12] The stringent discipline and the extreme penalties that followed any disagreements helped ensure that the number of congregants remained limited.

It should be pointed out that the numbers given by Barnard and Lyon refer to those who had made their profession of faith and were thus subject to the church's discipline. Usually the public part of the service, although not the Love Feast nor the Lord's Supper, would be attended by a substantial number of other auditors – family, friends or just interested passers by. One such visitor soon after the turn of the century was Walter Wilson, who compiled information on dissenting churches in the metropolis. He noted the great neatness of the meeting house and also that the congregation was 'large and respectable'.[13] A later visitor, who heard Faraday exhort, commented that the Barbican meeting house was 'well filled by a middle-class congregation'.[14] More precise statistics of committed Sandemanians are provided by the rolls of the London church, which show that there were 106 in 1795 and 110 in 1842 (when Faraday was an elder). In response to the 1851 Religious Census it was claimed that all the 200 seats in the meeting house were filled, but this count would have comprised both members and a substantial number of non-members, children included.[15] Although the total membership remained fairly constant between 1795 and 1842, and then declined slowly, it is significant that, as in many other churches, the proportion of women increased considerably: 72 per cent in 1842 as compared with 55 per cent in 1795. It is also noticable that by about 1795 the membership had settled down in two related ways. First, unlike the earlier period, the majority of those who joined the community stayed with it for an extended period of time. Second, as will be discussed in the next section, the London Sandemanians had by that date become dominated by a few interlocking families.

Throughout Faraday's adult life not only did the membership of the London meeting house remain fairly constant but there were no major schisms within the congregation. There were, however, two events during this period which disturbed the tranquillity of the London meeting house. The first of these occurred in 1844 when Faraday and a number of other members were excluded for a short time, and the second in 1855 when the Edinburgh meeting house was separated over the question whether it is lawful to eat animals that had been strangled. Since Faraday played a part in both of these events, they will be discussed further in sections 4.2 and 4.3 respectively.

In 1885, nearly two decades after Faraday's death, the London meeting house was shaken by the severest schism in its history. The putative cause was an apparently trivial event: a member of the congregation, Mrs Agnes Young, was troubled about the 30-year separation from the Edinburgh meeting house and suggested to the elders in London and Edinburgh that the breach be healed. Matters escalated and in early September 1885 elders from several meeting houses assembled in London. The congregation was faced with the proposition that the church had acted in accordance with the divine law of discipline in separating from Edinburgh in 1855. There were only two responses possible: either the Bible's disciplinary code had been obeyed on that occasion (in which case no future accommodation with Edinburgh would be possible) or it had not. Faced with this stark choice, the London Sandemanians split into two factions, one comprising thirty-three members, the other forty. Out of these seventy-three members only fourteen were men, and they divided equally between the two camps![16] While no accurate information exists concerning the number of London congregants over the next few decades,[17] one of the congregations had been reduced to a mere eight members in 1936, and it contained only one man, Henry Young, the retired Librarian from the Royal Institution. Meetings were discontinued on Young's death in 1939. The other Sandemanian meeting house existed for almost a further half-century and closed in the autumn of 1984.

The history of the London congregation can also be traced through its buildings. As mentioned above, the congregation initially met at the Glover's Hall, but in the autumn of 1785 the meetings were transferred to a building, which could seat approximately 200 people, in Paul's Alley in the Barbican (Figure 3.1).[18] This new meeting house, which had been used by Baptists for about a century,[19] was the one Faraday attended as a child and where he prayed and preached for most of his life. The building has long since disappeared.

In September 1862, during Faraday's second period as an elder, the meeting house moved to new premises in Barnsbury Grove,

Figure 3.1 Interior of Paul's Alley meeting house which Faraday attended prior to 24 September 1862, when meetings were transfered to the new meeting house in Barnsbury Grove – see below.
Source: W. Jerrold, *Michael Faraday: man of science* (London, 1891), p.98.

Figure 3.2 Meeting house in Barnsbury Grove used by the Sandemanians 1862–99.
Source: Collection of Mrs J.M. and Miss J. Ferguson.

North London, an area where the majority of Sandemanians lived and where the Faradays bought a house. According to one visitor in the late 1860s the chapel (Figure 3.2) could seat 'three to four hundred people, and it is always nearly full', while a slightly later estimate put the seating capacity at 600.[20] Subsequently the London Sandemanians – two congregations after the schism of 1885 – met at other addresses in the same part of North London, while the building in Barnsbury Grove was sold in 1899 and used to house a telephone exchange. The connection with Faraday however, was considered so strong that in 1906 the aged Lord Kelvin unveiled a plaque commemorating the meeting house and Faraday's regular worship there. In his speech Kelvin drew attention to the text which appears above Christopher Wren's tomb in St Paul's Cathedral: 'Si monumentum requiris, circumspice' – 'If you want something to remember me by, look around you'.[21]

3.2 BLOOD RELATIONS

One striking feature of the nineteenth-century London Sandemanians is the dominance of a few families. Thus in the meeting house rolls such surnames as Sandeman, Barnard, Leighton, Baxter, Vincent and Faraday occur with surprising frequency. This recurrence of surnames was sometimes perceived as relating to the biblical description of the primitive church, on which the Sandemanians based their practices, since in Acts 1:15 it is recorded that the church's membership numbered about 5,000 men, but that their 'names together were about an hundred and twenty'.[22] On average, then, approximately forty men in the primitive church would have borne the same family name and were, presumably, related by blood. Although these figures for the primitive church do not correspond very closely to data on the Sandemanians, it is important to recognise that the church founded by John Glas was not only an invisible church – a group of committed believers – but also a network of families which over several generations intermarried and remained partially within the orbit of the church.

Since membership required the individual to make a personal act of faith, a child born into a Sandemanian household was not automatically a member of the church. Often families were large, and even if only two out of eight or ten children joined the church the family connection was maintained. However, the proportion varied considerably and in some instances none of the children turned to the church, while in other cases the majority of a generation followed that path. If the latter was a rare occurrence, it was the one which applied in Michael Faraday's case, since he, his brother Robert and one of his two sisters, Elizabeth, belonged to the London meeting house. All of Robert's four children became Sandemanians, as did at least four of Elizabeth's eight surviving children. Six of the twelve children of Faraday's other sister, Margaret, are also known to have made their confessions of faith (see Appendix A).

Similarly a high proportion of Sandemanians is to be found among the branches of the other leading families. In this respect a crucial document is Gerard Sandeman's edition of *The Sandeman genealogy* (1950) which contains numerous examples of intermarriage between members of these leading families. Moreover there are many instances of names that combine two such surnames, for example, John Glas Sandeman and George Handyside Baxter. The identification of the Sandeman family with the sect is further evidenced by the Sandeman coat of arms (Figure 3.3), which symbolises the family's commitment to the pure and undefiled form of Christianity advocated by Glas:

> The Crest is the Rock – the Truth as confessed by the Apostle Peter (Matthew 16:15–18). The naked female figure is the conventional symbol of Truth, but the Truth here represented conveys the Palm Branch of Peace to the earth upon which she stands; she has the 'Word' in her hand open, and the Light of the Gospel of the Sun of Righteousness shining in all its glory upon her breast (II Corinthians 4:3–6). Her white veil may represent the Garment of Righteousness (Revelation 19:9). It seems highly probable that the motto and arms were meant to embody verses 11–17 of the sixth chapter of Ephesians. The arms therefore stand as a Confession of Faith, while the motto may be freely translated 'The Truth formerly testified by the blood of Christ, is now testified by the sincerity of those who believe.'[23]

The ideals expressed here were shared by the Sandemanian fellowship, and Faraday would have accepted this coat of arms as encapsulating his religious commitments. Although he would have considered coats of arms as too closely associated with worldly political factions, this device was, in a sense, Faraday's coat of arms.

The interlocking families associated with the Sandemanian church would form a very complex genealogical chart. However, Appendix A contains but a small segment of this family network centred on the marriage between Michael Faraday and Sarah Barnard (1800–79). Both of these family trees have been extended back to the previous

Figure 3.3 Coat of Arms of the Sandeman family, first recorded in 1780.

Source: Gerard L. Sandeman, *The Sandeman Genealogy* (Edinburgh, 1950).

generation and forward to the next generation, so as to include all Faraday's nephews and nieces and their spouses. A few of this next generation may have been omitted for lack of adequate data, but I have eliminated those who died before reaching the age of 20, since few people confessed their faith before that age. The names of those who are known to have made their confession of faith are printed in italics. There were many Sandemanians among the three generations of these two intertwining family trees. In the first generation three out of the four parents (75 per cent) were Sandemanians, while nine out of thirteen (69 per cent) in Faraday's generation made their confession of faith. Turning to the third generation, the number is at least twenty-three out of fifty-nine (39 per cent). While some branches of the family drifted away from the Sandemanian church in that generation, others remained strongly represented. Thus fourteen (61 per cent) of the twenty three nieces and nephews on Faraday's side of the family retained their connection.

If we look at marriage patterns, we see that all the siblings of Michael and Sarah who married did so to people connected with the sect, at least six of whom had confessed their faith. As far as data are available for the next generation, the pattern continued, though to a lesser degree. At least seventeen marriages were made with Sandemanians and about twice that number to families connected with the sect. It should, however, be stressed that the above figures are probably underestimates, since, lacking a complete membership list for the whole country, I have almost certainly omitted a few individuals who joined meeting houses outside London. Most obviously some of Elizabeth Gray's (Faraday's elder sister's) children moved away from London, as did some branches of the Barnard family.

The above figures also doubtless underestimate the familial connections with the Sandemanian church, because many of those who did not confess their faith nevertheless retained long-term links with the sect. Most of the Faradays and Barnards whose names are not in italics would have attended the meeting house as children, often meeting their future husbands or wives through the sect. Later in life church families frequently met socially. Thus, for example, Faraday's younger sister, Margaret, married Sarah's brother John, although neither Margaret nor John confessed their faith. However, they, together with their children, often attended the meeting house on the Sabbath. A further example is provided by Faraday's nephew Walter Barnard, who did not join the sect but, according to his diary for 1860, nevertheless attended the meeting house on two Sundays out of three throughout that year.[24]

The above evidence indicates the close relationship between the leading families, such as the Faradays and Barnards, and the sect.

While some members of these families severed this religious connection, the majority retained some connection with the sect, and nearly half of those named in Appendix A confessed their faith. In visiting the meeting house on the Sabbath Faraday joined not only a group united in profession of the Christian faith but also an extended family circle. He was related to most members of the London meeting house. Likewise, when visiting the meeting houses at Newcastle or Glasgow, he met many co-religionists with whom he shared familial ties. In a strong sense, then, the Sandemanians were not just an invisible church but also a family church. Faraday – 'Uncle Faraday' as his nieces and nephews often called him – was not only tied to the other members by spiritual bonds but also by familial ties.

3.3 SOCIAL ANALYSIS OF MEMBERSHIP

Who joined the London Sandemanian community? Who were the people with whom Faraday worshipped and spent so much of his life? What were their backgrounds? Their social norms? In this section I shall offer tentative answers to these questions. However, before proceeding it is necessary to examine the limitations on the available primary sources. Appendix B lists the male Sandemanians who were contemporary with Faraday, i.e. those who belonged to the London meeting house between 1821, when Faraday made his confession of faith, and his death in 1867. The main source of names is the roll of the London meeting house. However, the roll is probably incomplete and there are also several minor difficulties in its interpretation.[25] I have omitted four men (Thomas Boosey, jnr. James Collett, James Parfit and William Todd) who are known to have belonged for less than 2 years. That leaves seventy-seven men on the list, but even in many of these cases I lack such data as the dates of birth and death, and whether they married.

For those whose dates of birth and of admission are known it appears that 29, Faraday's age in 1821, was a fairly typical age for admission, the majority being admitted in their twenties or very early thirties, although admission ages varied greatly. A far more significant feature of Appendix B is the high incidence of intermarriage between Sandemanians. The names of those wives who are known to have made their confession of faith are printed in italics, while those from Sandemanian families have their surnames picked out in capitals. The list of marriages has proved difficult to compile, since the Sandemanian rolls rarely contain information on this subject because the sect attached no great religious significance to marriage. Hence there are a number of uncertain entries, indicated by a question mark (?), in the final column. From the information available on fifty of the marriages made by the men on this list, it appears that thirty-eight (76 per cent) were to Sandemanians and in the vast majority of these cases (as well as several others) their wives were the daughters of established Sandemanian families. Thus Faraday's marriage to Sarah was a typical match for the son of a Sandemanian family. The data exhibited in this table further underscore the argument in the preceding section, that the Sandemanians were united not only by shared beliefs but also but strong familial ties.

If my list omits the many women who were members of the London community in Faraday's day – and there would be 163 by the criteria employed above – this should not be taken as impugning their significance. The list has been confined to men simply because biographical information is almost totally lacking for the women members, whereas a moderate amount of such information can be gleaned about the men. Moreover, since nearly all the members were interrelated, the list of men includes all the leading families. However, there are some important points to be made about the female Sandemanians. Since they outnumbered the men by more than two to one and were socially equal (except in respect to the elder's office), the ethos of the meeting house was probably influenced considerably by them. More easily ascertainable from the evidence is the high proportion who were single, e.g. Jane Barnard and Margery Ann Reid, who became companions to Faraday and Sarah. There were also quite a few elderly spinsters and widows in the community, including Sarah after Faraday's death.

Sarah Faraday, Jane Barnard and Margery Ann Reid are probably typical in that they, like many other middle-class Victorian women, devoted their lives to their families, homes and friends and did not pursue independent careers. Yet there were a few interesting exceptions among the London Sandemanians. For example, Charlotte Chater taught music, and Jemima Hornblower, for many years a deaconess in the community, conducted a school for girls in Stamford Hill. Three of Faraday's nieces attended the school and he also recommended it for the daughter of the Swiss chemist Christian Schoenbein, describing Miss Hornblower as 'a very dear friend of ours, and in her character and all that is about her, all that we could wish'.[26]

From trade directories, census returns and marriage and death certificates it has been possible to find the trades and occupations of nearly two-thirds of the men listed in Appendix B. However, surviving information is liable to favour those who owned their businesses or who made a mark in some field of endeavour. Thus information about Faraday and his work is far more readily available than, say, the occupations of the Deacon family about whom I have been able to find very little. Appendix B clearly shows that none of the London Sandemanian were from the gentry or aristocracy. On his death certificate Joseph Chater is described as a gentleman, probably indicating that he was retired and financially secure, though he had previously owned a stationers shop. A few of the male Sandemanians, including Faraday, Benjamin Vincent (Librarian) and Charles Vincent (Librarian and chemist), were not engaged in trade and belonged to the professionally skilled middle classes. Many more were businessmen or tradesmen and there were several small but thriving businesses run by London Sandemanians. For example, James Boyd, the smith and ironmonger from Dundee who had employed Faraday's father, expanded his business to become James Boyd and Son. The son later took over the concern, which by the 1860s appeared in trade directories as Alexander Boyd and Son of New Bond Street. Likewise Robert Faraday traded as a brass founder, and took advantage of the expansion of the gas companies by manufacturing and installing gas fittings. Robert's son, James, joined the family business, which later traded as Faraday and Son. There were also at least two cabinet makers in the meeting house, James Huddleston and Scott White-law.

The firm of Edward Barnard and Sons was a major employer of Sandemanians in London. Edward (1767–1855) had been apprenticed to a London goldsmith named Charles Wright in 1781 and had stayed with the firm through changes in ownership, becoming its foreman and later its manager. In 1808 he was taken into partnership, the business then being called Widow Emes and Barnard of Paternoster Row. The firm passed totally into the Barnard family in 1828, when Edward took three of his sons – Edward, John and William – into partnership. All three had been apprenticed to their father and ten subsequent members of the Barnard family also served their apprenticeships with the firm. It was very much a family business, and it was also highly successful until well into this century. According to the firm's historian, 'Edward Barnard and Sons was one of the largest and busiest' manufacturing silversmiths in nineteenth-century London.[27] Its ledgers show not only the considerable extent of its trade but also that it included nobility and royalty among its customers, perhaps the most noteworthy single item produced being the baptismal font used for Queen Victoria's first child in 1841. The firm moved into larger premises in Angel Street, St Martin's le Grand in 1838, and Edward brought up his family in a large house in the fashionable neighbourhood of Claremont Square.

There were strong familial and religious ties between Edward Barnard's firm and Reid, the Newcastle silversmith, whose family was prominent in the meeting house in that city. Two of Edward's daughters married into the Reid family, and one of his sons-in-law moved to London and worked in the silver trade. Again, the Barnards produced substantial amounts of silverware, which were sold in Reid's shop in Newcastle.

Edward Barnard and Sons was concerned both with the production and the design of silverware. The design role, which many members of the Barnard firm fulfilled, brought them into contact with the arts, in which Sandemanians were strongly represented. For example, Charles Blair Leighton (1823–55) was initially apprenticed to a silver-engraver and later studied lithography at the Royal Academy and exhibited there. He rapidly gained a reputation as a photographer and lithographer and joined his brother George Cargill Leighton in a photolithography business which was later particularly successful in producing the *Illustrated London News*. The extended career of Cornelius Varley (1781–1873) was even more diverse, for he was not only one of the leading water-colourists of his day but he was equally concerned with scientific and technological developments, having been awarded the Isis medal of the Society of Arts for improvements to optical instruments and a medal at the 1851 Exhibition for inventing the graphic telescope. His religious experiences were equally diverse, for he was a Sandemanian for a little over 2 years, although both his family and his wife's appear to have had long-term connections with the church.[28]

Among the accomplished Sandemanian artists was James Baynes (1766–1837). His family was initially connected with the Clapham and Kirkby Stephen meeting houses but he moved to London and joined the London church in 1792. The aim of this move was to allow him to study at the Royal Academy where he became a student of George Romney, the historical and portrait painter.[29] The Blaikley family from Scotland also provided several exhibitors at the Royal Academy, including the successful and accomplished portrait painter Alexander Blaikley (1816–1903), who made his confession of faith in Dundee in 1835 and moved to London 2 years later. He was warmly received by Edward Barnard – 'a fine-looking white-haired old gentleman in rather old-fashioned black dress, knee-breeches, ruffles, etc.' – and by Faraday, who invited him to breakfast at the Royal Institution. Blaikley shared a 'loving acquaintance' with Faraday, whose portrait he painted on at least

three occasions, including the group portrait, (Figure 3.4).[30] Another Scottish Sandemanian painter of note was John Zephaniah
Bell (1794–1885), who spent some time in London and painted a
portrait of Faraday which includes an electrophorus and the title
page of a manuscript bearing the date 1852 (Figure 3.5).[31]

Figure 3.4 Blaikley's group portrait (1855) of Faraday lecturing at the Royal
Institution before Prince Albert and other dignitaries.

Source: Royal Institution of Great Britain.

To this list must be added several other artists connected with the
London Sandemanians. Faraday's brother-in-law George Barnard
(1807–90) was a noted art teacher and writer who exhibited a number
of landscapes, including some drawn while he, his wife, Faraday, and
Sarah were touring in Switzerland in 1841.[32] Faraday's nephew
Frederick Barnard (1846–96) not only exhibited his work but also
gained a considerable reputation as an illustrator for magazines such
as *Punch* and *Illustrated London News* and contributed illustrations for
the household edition of Dickens's novels.[33] Charles Blair Leighton's
son, Edward, was also an artist, as was another of Faraday's nephews,
Frank Barnard, who joined the Sandemanians in 1877 (and therefore
does not appear in Appendix B). These connections with the world of
art are significant for Faraday's biography, since he took a keen
interest in art and also possessed a highly developed visual sense.

Figure 3.5 John Zephaniah Bell's portrait of Faraday (1852).
Source: Hunterian Art Gallery, Glasgow.

While the connections between Sandemanians and the world of art deserve further exploration, the other main area of employment was the book trade. Several members of the Leighton family were printers, Thomas Boosey specialised in importing sheet music and George Whitelaw worked as a printer and engraver, sometimes for Spottiswoode's, as did his son David. The firm of Grosvenor, Chater and

Company, wholesale stationers, brought together two of the Sandemanians on my list, while other members included another stationer, a newsagent and a bookbinder's foreman. An interesting example of cooperation among the Sandemanian brotherhood was the *Grand architectural panorama of London: Regent Street to Westminster Abbey, from original drawings made expressly for the work by R[obert] Sandeman, architect, and executed in wood by G[eorge] C[argill] Leighton* (J. Whitelaw, London, 1849). Faraday likewise used the professional services of several of his co-religionists in the printing trade. For example, we find him writing to one of the Leighton family in 1828 asking him to bind some issues of journals as soon as possible. Again he used 'an old friend Mr [Thomas] Boosey' to publish Gerrit Moll's 1831 defence of English science against the culumnies of Charles Babbage and others who argued that in England science was in decline.[34]

The professional men and proprietors of businesses were well represented among the deacons and elders but they only constituted approximately one third of the male Sandemanians listed in Appendix B. Some of the others were certainly employees: for example, David Sandeman, who was probably connected with the wine importers of the same name, is listed in directories as the Secretary of the Brighton and Hove Gas Company. Again, Arthur Young described himself in his 1871 census return as an average adjuster acting for a firm, while Thomas Barker worked as a merchant's clerk – probably a rather lowly occupation.

Several of the London Sandemanians were manual workers. Thus a letter of 1795 discussing the formation of a meeting house at Chesterfield names among its members William Stopard and Godfrey Smith, describing the first as a young farmer and the second as a servant.[35] Both these men were brought to London in the 1840s after the Chesterfied meeting house ceased to function. The community also included a saddler, a carrier and a coalminer from Nottingham. Of those whose occupations have not been traced, many would have been relatively unskilled employees whose names would not have occurred in the trade directories. That a significant proportion of the London Sandemanians were in this category is confirmed by Frank Barnard's claim in 1871 that out of not more than twenty men currently belonging to the London meeting house, most were 'quite poor, only seven or eight of them being masters of their own businesses'.[36] That the Sandemanians admitted such a diversity in social and economic background coheres with their profession of untainted Christianity, in which membership of the sect depends solely on faith.

Hence the community in which Faraday spent so much of his life contained a diverse mixture of women and men, – the (relatively) rich and the poor, professional men, employers and employees. The London Sandemanian meeting house thus contrasted sharply with the fashionable audience at the Royal Institution and the well-endowed scientists and patrons who attended the Royal Society of London and the Athenaeum.

4 *Faraday among the Sandemanians*

4.1 THE FARADAY CLAN

The initial connection between the Faraday family and the Sandemanian church was sealed by Michael Faraday's grandparents
more than half a century before his birth. To understand this
connection we must examine briefly the world of small independent
chapels in the northern Pennines. In that region, which includes
parts of Yorkshire, Lancashire and Cumbria, nonconformity was
slow to take hold, with only the Quakers attracting a significant
following in the late seventeenth century. However, in the early
decades of the eighteenth century a number of other nonconformist
denominations took root in this area. Congregationalist, Unitarian,
Baptist, and Methodist chapels were opened and each attracted its
own local following. While these nonconformist congregations often
formed part of larger, nationwide networks, there were other groups
which were peculiar to this region. Foremost among these were the
Inghamites, the followers of Benjamin Ingham[1] (1712–72), a York-
shireman who while a student at Oxford in the early 1730s had come
under the influence of John Wesley and his associates. Having served
alongside Charles Wesley in Georgia as a missionary to the Indians,
Ingham returned to Yorkshire in 1737 but soon found himself
ostracised by other clergy and, in 1739, he was prohibited from
teaching within the York diocese. Subsequently he became an
itinerant preacher in the northern and western parts of Yorkshire
and in neighbouring areas. An effective preacher, Ingham attracted
large audiences during his travels and formed numerous congregations. Estimates vary over the number of Inghamite chapels, one
source claiming sixty in 1743 and more than eighty by the end of the
following decade.[2] Some of these congregations were short-lived,
while others drifted in and out of the Inghamite fold. Despite a
sharp decline in the late eighteenth century, a few Inghamite chapels
have survived.

By the early 1760s many of Ingham's followers felt that, while he
offered an appealing form of Christianity based firmly on the Bible,
he had failed to provide his churches with a stable organisational

structure. The sprawling collection of Inghamite chapels lacked the discipline that Glas had imposed on his Church of Christ, and there was an ever-increasing danger that the movement would simply dissolve unless a firm, stable structure was rapidly found. Ingham had in 1760 read Glas's *Testimony of the king of martyrs concerning his kingdom* and Robert Sandeman's *Letters on Theron and Aspasio* and he was clearly impressed not only with the Glasites' theology but also with their church structure and discipline. It was therefore to the Glasites that Ingham turned in the hope of saving his independent sect, little realising the dangers involved. At his own expense he sent his two lieutenants, William Batty and James Allen (1734–1804), to Scotland to find out more about the Glasites at first hand.[3] These envoys were well received, and Allen was so enamoured with the Glasites that by the time he returned to Yorkshire he had become a convert to their cause and was openly opposing Ingham's half-hearted attempt to hold together his loose confederation of chapels. At a conference of leading Inghamites called in 1761 dissension was rife and, far from using the Sandemanian experience to heal wounds, Allen bitterly opposed Ingham. The conference resulted in schism, with Allen and his followers breaking with the Inghamite church and instead following Glas and Sandeman.

That 'horrid blast from the north', as Inghamites subsequently referred to the Sandemanian incursion, left Ingham's churches in disarray.[4] A number of congregants and even whole congregations turned to Allen and to Sandemanianism, so that by the late 1760s there were some six or seven Sandemanian churches in the region. Yet relations between these ex-Inghamites and the Glasites in Scotland soon became strained. While some commentators consider that Glas was less than generous in his dealings with Allen,[5] it is clear from the extant correspondence that the Sandemanians were highly suspicious of Allen's motives and complained that 'the greater part of them [the ex-Inghamites] were followers of Mr. Allen only, not of the apostles, and that the only principle himself was actuated by was the ambition of him being the head of a religious party in this World'.[6] Elders from both London and Scotland made several visits to Yorkshire to try to regularise matters and to establish churches that would follow Christ's teachings. In the long term the Sandemanians dissociated themselves from Allen, whom they described as 'once an elder in our churches, now an inveterate foe'.[7] However, the visiting Sandemanians experienced particular success with a congregation of 'about fifteen' who met at Clapham in Yorkshire under the guidance of Edward Gorrell, 'a man that has approved himself a faithful follower of Jesus Christ'.[8]

Not far from this Sandemanian meeting house stands Clapham Wood Hall, the small farmhouse inhabited by Faraday's grandparents, Robert (1728–86) and Elizabeth (née Dean, 1731–97). They had been Inghamites but associated themselves with the Allen faction and worshipped at Gorrell's chapel. It is not clear how the connection between Robert and Elizabeth and the Sandemanian chapel developed, since the only Faraday included in Gorrell's membership list dated 1774 was a Richard Faraday, probably one of Robert's elder brothers.[9] Their absence from this list may be ascribed to any of several causes: like many others, Faraday's grandparents may have worshipped at the Sandemanian meeting house without making their confession or they might have been excluded at the time Gorrell drew up his list or, again, they may have only joined at a later date.[10] Although the first of these seems the most likely possibility, it appears that Robert and Elizabeth were associated with the Clapham (often called the Hazelhall) meeting house, and probably closely associated. One other Faraday appears in the Sandemanian records for this area, and this was John (1759–1823), who was Robert's second son and a deacon at the Clapham meeting house.

John Faraday was one of ten children born to Elizabeth and Robert between 1757 and 1776. Such a large family proved a considerable strain on the father's limited earnings as a weaver or, as he is sometimes described, a slater. While in his teens their eldest son Richard moved from Clapham to the market town of Kirkby Stephen, where he became a respectable grocer and wool spinner and owned a significant amount of property. He married in 1777 Mary, third daughter of Michael and Betty Hastwell of Black Scarr Farm, Kaber, near Kirkby Stephen. Richard's movements are relevant to this account, since the third Faraday son, James (1761–1810), followed his eldest brother in at least two respects. Firstly, he moved in the direction of Kirkby Stephen, becoming an apprentice blacksmith in the small settlement of Outhgill. Secondly, he also married into the Hastwell family, being joined with the sixth child, Margaret (1764–1838), a maidservant, on 11 January 1786. Any connection between the Hastwell family and the Sandemanian meeting house in Kirkby Stephen remains unclear, as does James Faraday's precise relation to the church, although he joined it before he moved south in the harsh winter of 1790–1. His early connection with the Sandemanians is significant not only for the family's subsequent religious commitments but also for explaining more fully the circumstances under which James, Margaret and their first two children came to move to London.

On his blacksmith's meagre income James found it difficult to support the needs of his growing family. It appears that they were close to starvation and that James was forced to look for work in the London area. He moved first to Newington Butts, near the Elephant and Castle, but a few years later he was living in the West End of London, where he was hired by James Boyd, a Sandemanian who had moved south from Dundee many years earlier and who owned a flourishing smithy and ironworks. One source claims that Boyd later made Faraday his foreman.[11] Here, then, is one of many instances of Sandemanians helping one another in the area of employment. In moving to London the Faradays lived in rooms over a coach house in Jacob's Well Mews, Manchester Square, a mere stone's-throw from Boyd's smithy and ironworks in Welbeck Street. Late in 1809 the family moved the short distance to Weymouth Street.

James joined the London meeting house which was situated in Paul's Alley, Barbican, on 20 February 1791, very soon after he moved south. At that time there were two young children in the family – Elizabeth, born in 1787, and Robert, born in 1788. Margaret would also have been in the early stages of pregnancy with a third child. The couple's fourth child, also named Margaret, was born 11 years later, in 1802. It is, however, to the third child, Michael, born on 22 September 1791, to whom we now turn.

4.2 A PILGRIM'S PROGRESS

Little evidence remains about Michael Faraday's early religious experience. However, he may have been baptised since infant baptism was commonly practised by the Sandemanians in accordance with their understanding of Acts 2:38–9.[12] Moreover, like other children of Sandemanians, he would have attended the meeting house each Sabbath with his parents, brother and sisters and have become accustomed to the long, even tedious, services, the prayers and extensive quotation from the Bible. If the account of an eighteenth-century service in America is any guide, both the Sunday morning and evening services lasted for about 3 hours, with the Love Feast held during the intermission.[13] Children would have sat in the pews listening, or playing, but since they (and others who had not confessed their faith) were not admitted to the Love Feast, they 'remained in their pews and were there served with Scotch broth and sandwiches' when the communicants adjourned.[14] The connection with Scotch broth became immortalised in one of the names by which the Sandemanians became popularly known – the 'Kail Kirk'.

During Michael's childhood the whole family, including his mother, who did not become a church member, would have attended the meeting house in Paul's Alley. (see Figure 3.1, p.42). After his father's death in 1810 Margaret and her children continued to attend the meeting house regularly. Thus in a letter of 1817 to his friend Benjamin Abbott we find Faraday 'endeavouring to extricate myself from a meeting of some of our people respecting a singing school attached to our meeting house';[15] his participation in the singing school and his reference to 'our people' suggests that Faraday was at that time closely associated with the Sandemanians. Further, if indirect, evidence for Faraday's continued involvement with the London meeting house is derived from the choices of partner made by himself, his brother and his two sisters. Elizabeth had married a Sandemanian sadler named Adam Gray who was from Cupar but had moved to London. In 1821 Faraday married Sarah Barnard, daughter of Edward Barnard, the silversmith and elder in the London Sandemanians, while 5 years later his younger sister, Margaret, married one of Sarah's brothers. The Barnard family had included Sandemanians for three generations, Edward Barnard being the nephew of John Barnard, one of the Independent ministers who had joined with Robert Sandeman in the early 1760s. On her mother's side Sarah was descended from another leading Sandemanian family, the Booseys. Faraday's brother Robert likewise married into a prominent Sandemanian family, the Leightons. Such marriages within the Sandemanian network suggest that Faraday and his siblings remained in close contact with the London meeting house after their father's death. A further indication of this connection is the inclusion of two other sons of London Sandemanian families, George Barnard and Thomas Deacon, among the four young men who joined Faraday in 1818 in forming a study and improvement society.[16]

Until the Marriage and Registration Act of 1837 all marriages had to be licensed in church by an Anglican minister, and thus extant parish registers of the period sometimes contain information about Sandemanian marriages. The register for St Faith-in-the-Virgin, which is very close to St Paul's Cathederal, records the issue of a marriage licence on 12 June 1821 to Michael Faraday and Sarah Barnard. Edward Barnard, Sarah's father, was the witness.[17] Although prepared to obey the legal requirements concerning marriage, Sandemanians refused to participate in the Anglican service. As they explained in their account of the church's *Customs*, 'Marriage is regarded as a divine ordinance, honourable to all, and for which thanks may very properly be given, as is done for daily bread, or other earthly blessing, but there does not appear to be

Scripture warrant for making it a Church ordinance'.[18] Not surprisingly, they were keen to avail themselves of registry offices after 1837. On 19 September 1840 the first Sandemanian registry office wedding was held, after which Edward Barnard read a short paper expressing his delight that 'we can [now] legally unite in Marriage without wounding our consciences'.[19]

Since Sarah had made her confession of faith more than 2 years before their marriage, it was Faraday who followed her into the Sandemanian fold. Although on many previous occasions he must have considered his religious position, his marriage to Sarah may have provided a further impetus to join the sect, which he did a little over a month later. Yet Faraday would have been well aware that marriage, or indeed any other change in one's worldly situation, is irrelevant to matters of religious belief. Hence his oft-cited retort to Sarah when she asked him why he had not consulted her: 'That is between me and my God'.[20] His father-in-law and brother may also have played some role in his decision, since Edward Barnard had recently become an elder and brother Robert made his confession of faith less than 4 months later. For his admission to the church on 15 July 1821 Faraday would have been required to demonstrate before the assembled congregation his faith in the saving grace of God and his commitment to live in imitation of Jesus Christ. According to the sect's published *Customs*, the candidate has to 'appear to understand and believe the TRUTH, and express a readiness to do whatever Christ has commanded'. He would then have responded to questions first from the elders (Eliezar Chater, George Leighton and Edward Barnard, his father-in-law) and then from any other members. When everyone was satisfied, 'prayer is offered up, with the laying on of hands, as a Scriptural token of blessing and receiving, in the name of the Lord; the members saluting with the holy kiss, to testify hearty welcome into the fellowship, and love for the sake of the truth professed'.[21]

Since Faraday's religious practices were both personal and confined to his weekly, if not daily, private intercourse with other members of this small socially isolated community, there are few visible benchmarks to chart his path during the next 46 years. The roll of the London meeting house is suitably succinct. Against his name and date of admission the following notes are inserted: 'Deacon 1 July 1832; Elder 15 Oct 1840; Ex[cluded]. 31 March 1844; Restored 5 May 1844 [;] to Elder 21 Oct 1860; *d* 25 August 1867'.[22] Yet each of these dates marks an important event in his connection with the Sandemanian church. We shall therefore comment on them as far as extant evidence permits.

In respect to the first of these events there is a letter of the young John James Waterston, the scion of a leading Edinburgh Sandemanian family, who was destined for an impressive but inadequately acknowledged career in science. On a visit to London in July 1832 he sent home news of his travels and of the meeting house which he visited: 'Michael and Robert Faraday were made deacons the other week', he reported. In the same letter Waterston also mentions a mutual friend's enthusiastic assessment of Faraday's scientific abilities and his indifference to honours.[23] While turning his back on such worldly rewards, Faraday accepted with the deacon's office the duty of ministering to the community's physical needs, such as visiting the sick and poor and organising the running of the meeting house. Moreover, the position of deacon, like that of elder, reflects the community's unanimous assessment of Faraday's high moral character. The service of ordination would have been very similar to the ordination of an elder, described below, but without fasting and the right hand of fellowship.

The next benchmark in Faraday's participation in the London meeting house was his appointment to the elder's office in October 1840. His election, by the unanimous agreement of the congregation, was one of the most important events in his life. His ordination took place on the morning of Thursday, 15 October, and was preceded by a church fast. Several members of the congregation would have read relevant passages from the Bible, particularly those dealing with the office of elder and the expected qualities of its incumbents.[24] Following the laying on of hands, the other elders would also have given him the 'right hand of fellowship'. The reading of 1 Timothy 6:10–16 would have committed him to 'follow after righteousness, godliness, faith, love, patience, [and] meekness' and to 'fight the good fight of faith'. At the conclusion of this passage the church's agreement would have been signified by a rousing 'Amen'.[25] Now sitting with the elders, Faraday would have commenced his ministry with a prayer, and at the end of the service the congregation would have adjourned for breakfast.

For the next 3½ years he played a leading role not only at the London meeting house but also in the national and international community of Sandemanians. However, on 31 March 1844 he was excluded from the sect. The reason usually given for his exclusion is that he was invited to visit the Queen one Sunday early in 1844. Hence arose a problematic situation for a Sandemanian, who is required by the word of God both to worship and partake communion at the meeting house on the Sabbath and also to be a loyal subject. In choosing between the calls of religious and secular duties Faraday would thus have been faced with a dilemma. However, by responding to the summons he failed to appear at the meeting house that Sunday. When called to explain his action, he was 'ready to justify his own conduct in obeying her [the Queen's] commands'. The published source for this incident is the first edition of John H. Gladstone's biography, published in 1872, which seems to imply that the reason why Faraday was excluded was not so much that he accepted the Queen's command, as that he was not repentant but insisted on defending his action.[26] This is consistent with the Sandemanian attitude towards offences, since those who have transgressed are not automatically excluded. To the contrary, 'if a man be overtaken in a fault, ye who are spiritual, restore such an one in the spirit of meekness' (Galatians 6:1). What Faraday may have lacked on this occasion was the 'spirit of meekness'.

Superficially this standard account appears acceptable. However, the records of the London Sandemanian church indicate a further aspect of this incident that casts doubt on the above account.[27] Not only was Faraday excluded on 31 March 1844 but eighteen others suffered similarly at about the same time and among these were several of his close friends and relatives, including his brother (Robert), sister-in-law (Margaret Faraday) and father-in-law (Edward Barnard). Since communion cannot be held if there is dissension in the church, any disagreements which are not healed before the Sabbath must lead to the exclusion of the dissenting party. Could Faraday's visit to the Queen have created such controversy, with some 20 per cent of the membership being excluded at that time? This seems unlikely but not impossible. The standard account is, however, problematic for four further reasons. Firstly, neither Frank James (who is editing Faraday's correspondence) nor I have been able to find any evidence that Faraday visited the Queen at that time. The Queen's diary and the 'Court Circular' in *The Times* are silent and no contemporary letters mention the incident.[28] Secondly, Faraday was not the first of those nineteen excluded, as one might expect if his actions precipitated this controversy. His father-in-law, Edward Barnard, who had been an elder for more than 20 years, was excluded on 3 March. Faraday, together with fourteen others, was excluded on 31 March and three others were excluded a few weeks later.[29] Thirdly, the second edition (1873) of Gladstone's biography was emended, and in place of Faraday's supposed visit to the Queen his exclusion was attributed to an unspecified cause, known only to 'the parties immediately concerned'.[30]

Fourthly, and most importantly, 1844 was a particularly bad year for the sect, with a major dispute effecting most, if not all, of the meeting houses. A letter by George Baxter, one of the Dundee elders, indicates the nature of the dispute. The letter, addressed to the church in London, opens by noting that the churches in both Britain and America 'have been visited by terrible things in righteousness. The head of the church appearing among them as the Lord of the conscience searching his Jerusalem with candles and punishing the men settled in their lees [Zephaniah 1:12]. His chastening hand has been heavy upon you and upon us'. The main point at issue, it appears, was the question whether it was lawful, according to Matthew, chapter 18, for the elders to make decisions on behalf of the church. For the Dundonians it was clear that Scripture required the involvement of the whole church and not just the elders. Moreover, all decisions had to be arrived at through the honest study of the Bible. The letter also hints at other disciplinary disagreements that had 'given this church much trouble about two years ago and since'. In relating the dispute with Edinburgh Baxter also 'hinted that something of the same error had been practised among you' in London, presumably referring to the exclusions earlier that year.[31] Most of the above points are confirmed by a subsequent letter by Robert Cree, a Glasgow elder, who also complained that 'these were truly awfull times' dominated by 'the hidden things of darkness . . . when many appeared to be given up to walk in their own ways'.[32]

From this broader perspective Faraday's exclusion on 31 March 1844 looks very different from the reason offered in the standard account. From the foregoing discussion it would appear that his exclusion resulted from this searing dispute over discipline that reverberated throughout the Sandemanian churches and not from the lapse of an individual who, supposedly, visited the Queen one Sabbath. While we cannot completely eliminate the standard account, as either the cause of Faraday's exclusion or as a partial causative factor, the lack of contemporary confirmatory evidence shifts the probability firmly in favour of the account offered here.

The exclusion affected Faraday deeply and brought him 'low in health and spirits'.[33] He was, however, restored to the community on 5 May 1844, having expressed his sincere repentance, and most of the other eighteen were restored at about the same time. Although Faraday's exclusion from the church lasted for only five weeks, a further 16 years were to pass before he was re-elected to the elder's office.

Many years later George Sims published in his column in *The Referee* an account of Faraday's connection with the Sandemanian church (which Sims had occasionally attended as a child) in which he claimed that in 1856 Faraday had been put away because his 'scientific researches, he confessed, [had] unsettled his simple faith as a Sandemanian'.[34] No evidence has so far come to light concerning any rift between Faraday and his church in 1856. Moreover, there are two sources which categorically deny that Faraday experienced this supposed conflict between his science and his religion.[35] Although we should be careful not to endorse such claims uncritically, Faraday's science does not seem to have been reponsible for his exclusion in 1844 or any disruption in 1856. Nevertheless, as will be explored in section 10.3, Faraday did come close to his second and final exclusion in 1850.

For the 16 years following the events of 1844 Faraday was relegated to the ranks of the London Sandemanians but he was again elected elder in 1860, following the resignation of Thomas Boosey, and served alongside George Whitelaw, Stephen Leighton and Benjamin Vincent, the Librarian at the Royal Institution.[36] What the entry in the London meeting house roll fails to mention is that Faraday laid down the elder's office on 5 June 1864. The circumstances surrounding his resignation, which will be explored further in section 10.3 were directly related to Faraday's science – not to any supposed conflict with his research but to a personal crisis brought about by the offer of the Presidency of the Royal Institution. Faraday subsequently remained in the church for the final 3 years of his life.

4.3 FARADAY'S ACTIVITIES IN THE CHURCH

In this section I want to discuss briefly several aspects of Faraday's membership of the Sandemanian church. Perhaps the most important aspect is the most difficult to document. In his daily life he sought to embody the beliefs, precepts and practices of primitive Christianity. He encompassed the

> faith of the Gospel, that *Jesus is the Christ, the Son of the living God,* and, as such, the only *Prophet, Priest,* and *King* of God's church: so that all the divine knowledge we can possibly acquire is contained in His revealed Word, as He is the *Prophet*; all our hope of salvation is derived from the complete atonement He has made, and the perfect righteousness He has wrought out for sinners, as He is the *Priest*; and we are bound to do whatsoever He hath commanded us, as He is the *King* and Head of the Church.[37]

Faraday lived by the Bible and by the demanding discipline imposed by the Sandemanians. His Christianity was not limited to Sunday observance but infused all aspects of his life – his social intercourse, his views on social and political issues and his science.

Every Sunday morning and Wednesday evening he would leave the Royal Institution and travel to the meeting house in the Barbican (later Barnsbury). The Sabbath morning service, which lasted about 3 hours, consisted of prayers, readings from the Bible, an exhortation by one of the elders and the singing of several psalms. At about noon the members would adjourn to a neighbouring room for the Love Feast, which included a simple meal. The Love Feast was intended 'to cultivate mutual knowledge and friendship, to testify that we are all brethren of the same family; that the poor and rich may partake together'.[38] In the early afternoon the service continued for a further period of about 3 hours, with more prayers, psalms and blessings, including the collection. The Sabbath service concluded with the kiss of charity and the celebration of the Lord's Supper, at which the bread was broken.

On the Wednesday evenings Faraday would join the rest of the congregation for prayer, Bible reading and public exhortation, every man being encouraged to exhort in order to edify the proceedings. At the Love Feast and also sometimes at the weekday meeting various matters relating to the congregation would be discussed, e.g. the admission of a new member, the reason for someone's absence or the need to repair the meeting house roof.

If the assumption of the elder's office in 1840 weighed heavily on Faraday, there is no direct evidence to suggest that these problems continued, although his 'breakdown' during the early 1840s covers much of this period. What little evidence there is points to his normal practice of the elder's duties, most obviously his participation in the Sabbath services, including the exhortations that he was expected to deliver. He would have regularly exhorted the congregation on themes taken from the Bible. There are a few extant copies of his exhortations and also a number of cards on which he outlined their contents.[39] The Sandemanian exhortation consists of numerous passages from Scripture moulded together into a discourse by a minimum of connecting text. This form of exhortation is intended to direct the auditor to the unadorned word of God and to prevent His message from becoming diluted or from being corrupted by merely human interpretations and concerns. Sandemanians thus seek to avoid the pitfalls of popular preachers who, they considered, perverted God's word for their own dubious ends. The significance for Faraday's science of this literalist and 'anti-hypothetical' approach to the Bible will be discussed further in section 8.1. Yet, by the same token, their distaste for fancy rhetoric also made their exhortations unpalatable to non-Sandemanians, who enjoyed listening to fiery or intellectually demanding sermons.

Following a visit to the meeting house a few years after Faraday's death, the broad churchman Charles Maurice Davies claimed that during the exhortation the elder wept and almost lost 'his voice from emotion – nay, more than that, he made many of his congregation cry too, out of mere sympathy, for the discourse was rather critical than pathetic'. He added, 'I never remember undergoing such a protracted process of depression in my life'.[40] Clearly the Sandemanian exhortation is an esoteric form of rhetoric; it is highly stylised, lacks rhetorical flourishes and is thus totally different from the sermon of a Wesley or a Whitfield. Not surprisingly, some commentators have compared Faraday's exhortations unfavourably with his brilliantly delivered science lectures.[41] A few accounts of Faraday's style of delivery at the meeting house exist. According to one auditor, he 'read a long portion of one of the Gospels slowly, reverently, and with such an intelligent and sympathising appreciation of the meaning, that I thought I had never heard before so excellent a reader'. Another noted the 'devoutness of his manner' and a third was greatly impressed by the 'amazing sense of the power and beauty of the whole filled one's thoughts at the close of the discourse'.[42]

Faraday performed numerous pastoral duties among the London brethren, such as visiting those in need and tending to them, both materially and spiritually. Extant sources contain a few references to these duties. For example, he informed Benjamin Vincent (1819–99) that he had seen the aged and infirm 'Mr. [William] Clarke in bed'.[43] He comforted the sick, the dying and the bereaved. On at least one occasion, but probably on several, he baptised an infant.[44] Faraday was an integral member of the Sandemanian community and he spent a considerable amount of his time ministering to his fellow Sandemanians and performing his duties in the church.

The London meeting house also took under its wing a congregation at Old Buckenham, a tiny village some 20 miles south-west of Norwich. In the 1760s a small church had flourished in the nearby village of Banham, but it had been cut off due to 'lack of brotherly love and discipline'. A few of its members had, however, formed the core of the Old Buckenham community, which was founded about 1790.[45] Its members, who numbered thirty-two in 1838, were mainly tenant farmers and most of them were poor. Moreover, it was a relatively unstable community, lacking the experienced leadership that had been present in the London church since its foundation. Hence the deacons and elders from London took a fraternal interest in the Old Buckenham church, which they visited regularly, and they were prepared to intervene if a breakdown in discipline occurred. These tasks sometimes fell to Faraday, and on a number of occasions he, usually accompanied by Sarah, took the train to Norfolk in order to conduct the Sabbath service and resolve any disciplinary problems.[46] His niece Caroline Reid has left a detailed account of the Faradays' visit in the summer of 1842:

> [2nd July] The meeting [house] is a very pleasant cheerful room, could hold about a hundred, sloping roof, seats comfortable – sort of forms with a little raised desk. I was much struck by the Brethren's prayers. It is the first time of visiting any church but our own. My uncle read and Mr. [Samuel] Lancaster preached . . . After noon Mr. [David] Fisher read and my uncle preached on charity and the bond of perfectness.

> [9th July] Pleasant walk over the green and along the street[,] Sweet Sackford Lane[,] to a meeting with uncle William [Barnard], Mr. Lancaster, my uncle and aunt Faraday. Uncle Faraday read and Mr. Fisher preached in the morning from 17 John 1, 2, 3, Etc. I thought a very excellent and sober discourse, just what I need to be reminded of. At dinner 25 besides ourselves which with several babies quite filled the two tables. The children take their dinner in the little kitchen at the same time. Walked in the lane after dinner. Uncle Barnard read and Uncle Faraday preached from Romans 15:13. at the meeting house quite full. Tea at John Loveday's on the green.

> [Monday 10th July] Mr. Fisher came to breakfast with us to say farewell. Seemed to feel our leaving very much. No doubt my uncle Faraday has been a great help and support to the elders and this time after the painful separation of Thomas Loveday and his son Foulger.[47]

From this report and a number of extant letters it is clear that the Old Buckenham congregation was highly dependent on visits from its London brethren and that Faraday was held in great love and respect. For his part Faraday, one of the most eminent scientists of his day, freely ministered to this small, poor and sometimes wayward congregation in rural Norfolk, supporting them and taking a brotherly interest in their well-being.

There is also some evidence of his frequent interaction with individuals in the London community. The letters he sent to other Sandemanians contain numerous references to the health and movements of other members. Miss Jemima Hornblower, a teacher from Stamford Hill, has had a limb removed; Mary Straker is

seriously ill; someone has suffered an accident; another is travelling.[48] He even offered advice to the elders at Dundee on how to
heat their meeting house.[49] Likewise letters between other Sandemanians sometimes contained information about Faraday's activities or his health.[50] These items of news are, in a sense, trivial, but
they are also indicative of Faraday's daily concern with the
congregation and evidence of practical fellowship. As he wrote to
William Buchanan in Edinburgh, 'I find myself talking more about
our friends here [he was visiting Brighton at the time with several
other Sandemanians] than friends with you but I thought you would
like to hear a few words of them and there is much mingling &
ought to be such community of feeling & spirit as to make us one
body[,] so I do not think I am far wrong'.[51]

When he visited other Sandemanian communities, Faraday
conveyed the love and good wishes of the London fellowship. Thus
he made several visits to Newcastle, where a number of his relatives
– bearing such names as Deacon, Reid, Proctor and Spence – had
settled, and they were also joined by William Paradise, one of the
elders at the London meeting house who transferred to Newcastle in
1843. Faraday spent several weeks there in the summer of 1842 and
he was in Newcastle again in August 1861 when the local Sandemanians moved to a new and improved meeting house in Picton
Terrace, off New Bridge Street. Faraday reported to Vincent that
after the Wednesday meeting 'we had a very merry party . . . which
indeed went off so well, that they have since called it the house
warming'.[52] As a Newcastle newspaper later noted, while on such
visits Faraday took the opportunity to meet local scientists and to
inspect engineering works which utilised scientific principles.[53] Less
frequently he also visited the churches at Edinburgh, Glasgow and
Dundee.

While these visits were partly social, thus instantiating the Sandemanian conception of a Christian fraternity, they were also
diplomatic. As already noted, schisms between the various meeting
houses sometimes occurred and when this happened whole congregations were set apart. As discussed in the preceding section, 1844
was a particularly painful year for the Scottish churches, and it was
also the year in which Faraday was excluded for several weeks.
Another dark cloud appeared on the horizon towards the end of
1854, and this brought the London, Dundee and Edinburgh
communities into a conflict that deserves further analysis, since it
shows Faraday active in the sect's decision-making and disciplinary
procedures. The matter at issue was whether the killing of game was
subject to the requirement in Acts 15:29: 'That ye abstain from
meats offered to idols, and from blood, and from things strangled'.

The major question discussed in the London meeting house was whether the blood had to be drained immediately after the animal's death or whether time could elapse. One concern was whether, if the blood was not completely drained at the time of death, some of it might seep into the animal's flesh, from where it could not be effectively removed. The eating of such blood-laden meat was, of course, contrary to divine command. It appears that there was considerable difficulty obtaining consensus on this matter among the members of the London meeting house. One member, Thomas Bassett, resigned on this issue, since he considered that the elders had acted inconsistently and they had also tried to impose their view on the rest of the congregation. From Bassett's letter of resignation it appears that Faraday played a prominent role in this discussion, and that he had expressed some disagreement with the elders on this question. Faraday appears to have suggested that, since Scripture does not specify when and how the blood should be drained, a test should be held to find out whether blood could be cleaned effectively some time after the animal's death. Despite his initial reservations, Faraday must finally have joined the consensus led by the elders. His own position was precarious, since he had suffered one exclusion and a second would have placed him outside the church. Instead he concurred and was one of the signatures to the letters withdrawing from Edinburgh. It was Bassett, however, who withdrew from the sect.[54]

Despite these internal disagreements, the London and Dundee meeting houses soon agreed the clear meaning of the Scriptural prohibition on the eating of blood, declaring that it is contrary to divine law to eat an animal unless its blood has been drained at the time of its death, for otherwise some of the blood is bound to remain in the animal's flesh. Edinburgh adopted the opposing viewpoint. The position of the London church was forcefully stated in a letter of 7 November signed by the four elders, the four deacons and four other members, including Faraday.[55] Their arguments were reaffirmed in letters from the Dundee congregation.

A four-man deputation from Dundee therefore visited Edinburgh during the following month in order to resolve the matter. The Dundonian contingent was pleased to be supplemented and supported by two elders from London, Stephen Leighton and Benjamin Vincent. By contrast, the three Edinburgh elders were annoyed by the presence of the Londoners, whom they had neither invited nor expected. Far from achieving unity, the meeting ended in discord. 'We then went home to our Lodging & felt much shocked at the arbitrary manner in which the discipline had been conducted & grieved to see a whole Church resisting the Divine Word for the gratification of their own lusts', reported Alexander Moir, one of the elders from Dundee.[56] During the ensuing weeks letters passed between the churches, and the issue was concluded in a letter from London of 31 January 1855, which was signed by Faraday and others, terminating relations with the Edinburgh church.[57] This was a distressing period for Faraday, since he had to sever his connections with such close friends as William Buchanan and also with members of his own family. Nevertheless, for Sandemanians, Faraday included, the discipline of the Church of Christ was considered of greater importance than personal friendships.

The rift with Edinburgh in 1854–5 left a deep impression on Faraday and his companions. Writing to Benjamin Vincent on 12 October 1860, Faraday referred to the 'great matter':

> As to the *great matter* I say little of it here[.] It is constantly in my thoughts but I cannot write much about [it]. Nor is that needed – it is in better hands than mine and in his working & guidance in whom I hope to trust for surely he made us and not we ourselves and he guides his own as a shepherd his sheep.[58]

The 'great matter' was his reappointment to the elder's office. However, he was soon participating in another matter of great importance. In mid-1861 a meeting was held at Old Buckenham between Vincent and George Baxter, one of the Dundee elders. Faraday was clearly delighted by this meeting, since 'Such events as you have had the privilege of being engaged with are sources of great happiness & comfort to the people of Christ':[59] the constituent churches were working together towards unity and this, in turn, must be good in the sight of God. At the end of August he was in Newcastle, walking along the cliff-top with William Paradise, who, he reported to Vincent, 'joins in, most heartily, with the view taken by the Elders of London, Dundee and Glasgow, [and] says he could never see how the contrary conclusion could be sustained by the Scriptures'.[60] The internal politics of the Sandemanian churches was once again on the agenda and, with the omission of Edinburgh from this list, it appears that the two elders were discussing the schism with Buchanan's church.

In August 1863 Faraday was in Scotland again visiting elders in Glasgow and Dundee, and also meeting representatives from the Aberdeen meeting house. While in Glasgow, he and his friends were awaiting the arrival of George Baxter (of Dundee), but since he 'has not made his appearance here . . . it is concluded (& indeed known) that the case is compromised and that he has returned to Dundee'.[61]

Clearly Baxter had been on some mission, and if that had been successful, he would have proceeded to Glasgow; evidently, however, success had eluded him. It is most likely that the mission was an attempt at reconciliation with the Edinburgh meeting house by bringing it into line with the London, Glasgow and other communities. While Faraday was meeting several elders in Glasgow, related activities were taking place in other parts of the country, and several other elders were also shuttling between the meeting houses. Benjamin Vincent was away from London for an extended period and Faraday subsequently travelled to Dundee (where, as we saw in chapter 1, he preached on the Sabbath), returned to Glasgow and later proceeded to London. All this activity was aimed at uniting the churches, especially bringing an accord with the Edinburgh meeting house. Moreover, we can now appreciate that his presence in Dundee was as much due to church politics as to personal friendships. The words with which he opened his Dundee exhortation now adopt an urgent relevance in respect to the state of the churches: 'On this occasion, my dear brethren, when through the mercy of God we are assembled together, enjoying the great privilege of *dwelling together in unity*, I would ask your attention' to John 11:25-6'.[62]

Writing from Hampton Court on 25 August Faraday expressed to Vincent his heartfelt wish

> that the Father of Mercies will have given one heart & one way to us, as a people whom he hath chosen in his great mercy & long suffering to be with him in his beloved Son Jesus Christ. How great is the love he hath manifested in him – it ought indeed to move us to love one another.

While we may read into this Faraday's concern with the schism with Edinburgh there was at that time another discord of direct concern to Faraday, but much closer to home. On the previous Sabbath some members were unable to sit down together for the Lord's Supper, since 'Discipline was in hand & all were not at peace'. It is not clear whether the problem that afflicted George Leighton, Euphemia Conacher and Sarah Faraday was the disagreement with Edinburgh or some other issue.[63] Presumably the matter was rapidly resolved, or at least there is no extant evidence that it continued to fester.

Yet Faraday's responses both to this local controversy and to the schism with Edinburgh show that he conceived his actions (and those of other Sandemanians) in terms of God's laws and His mercy. God's protection was assured only if we acted in accordance with His law but we might expect punishment if we transgressed those laws. This was the disciplined, lawlike universe that Faraday inhabited.

In these last two sections Faraday's biography has been presented from an unusual standpoint. In place of the traditional picture, with Faraday making discoveries in his laboratory or teaching science in the Royal Institution's lecture theatre, I have presented his life as viewed from the London Sandemanian meeting house. The rest of the world, including both science and Victorian society, was on the outside and did not directly impinge on the proceedings of this small sect. However brief this biography, its aim has been to convey to the reader this crucial, but often neglected, part of Faraday's life. But this was a central part of his life since he viewed his membership of the Sandemanians – its Christian beliefs, practices and fellowship – as more important than his career in science.

4.4 FARADAY'S INFLUENCE: SANDEMANIANS AT THE ROYAL INSTITUTION

Being a small sect, numbering about 100 souls, and sharply different-iated from the rest of Victorian society, the London Sandemanians were extremely close-knit. Faraday moved in this small group. He not only prayed with its members but interacted socially with them both at the meeting house and in each other's homes. There are numerous references in his letters to social calls on fellow Sandeman-ians, perhaps to take tea, and, less frequently, to other Sand-emanians visiting him at the Royal Institution. Moreover, he was in the almost continual company of Sarah and, during his later years, also in the company of his Sandemanian nieces Jane Barnard and Margery Ann Reid. Yet his contact with his co-religionists was not confined to prayer and social activities, since he assisted them in their trades and in the furtherance of their careers.

As noted earlier, he sometimes used the services of Sandemanian printers and bookbinders. He also on several occasions tried to help the members of Sandemanian families find employment. One young man, a 'Mr. David', presumably David Buchanan, scion of a leading Edinburgh Sandemanian family, stayed with the Faradays for an extended period while seeking employment in London. He was, however, unsuccessful since, Faraday complained, he did 'not appear to know anything of bookkeeping or any other branch of [useful] knowledge'.[64] Faraday also tried to obtain an army post for one of Benjamin Vincent's brothers, but failed as he lacked appropriate contacts.[65] Likewise he proposed Joseph Chater, junior, to the post of collector for the Society of Arts and supported the appointment of

Robert Faraday to install gas lighting at the Athenaeum, although in his letter of support he judiciously sought to avoid appearing nepotistic.[66] Faraday also commissioned the Sandemanian artist Alexander Blaikley to paint a portrait of William Wyon, the chief engraver at the Royal Mint.[67] In an age when personal influence was all-important in obtaining employment and extending trade Faraday was a man of public standing who was often well-placed to assist his fellow Sandemanians in various ways. That he did so relatively infrequently suggests both that his public influence should not be overrated and that, for reasons to be discussed in chapter 5, he used what influence he possessed sparingly.

However, he used his influence more assertively on several occasions to assist Sandemanians gain employment at the Royal Institution, and thereby helped to cement a long-term connection between the sect and the Institution. In one sense, Faraday can be charged with nepotism, but this charge needs to be set against the Royal Institution's need to retain efficient, honest employees and contractors. As their minutes show, the managers were continually hiring and firing; if a doorman was found to be drunk and rude to the members, a more responsible, congenial replacement had to be found. When a new member of staff was hired, the Managers tried to ensure that the appointee was diligent, honest and responsible. There was no better recommendation than that which came from one of the Managers or their trusted servants, such as Faraday; he could vouch for the high moral character of his fellow Sandemanians and their families.

In May 1848 a new Assistant Secretary was sought by the Managers, the previous incumbent having been caught embezzling the Institution's money. Four applications were received and considered, but the Managers unanimously favoured the candidate 'to whose character and fitness Mr Faraday, who had known him many years, freely testified'.[68] Benjamin Vincent, the appointee, was from one of the leading families at the London meeting house and made his confession of faith in 1832, subsequently being elected to both the deacon's and the elder's office. This was not the first occasion that Faraday had spoken on Vincent's behalf. Five years earlier he had recommended Vincent to translate into English the travel diary of the Swiss chemist Christian Schoenbein, which had been published in German in 1842. Writing to Schoenbein, Faraday described Vincent as 'a young man of my acquaintance who is a corrector of the press and acquainted with many languages, more or less'.[69] At that time Vincent was employed by the publisher Gilbert and Rivington and was responsible for their books on philology; hence his knowledge of languages, which other sources describe as 'remarkable linguistic abilities' and 'one of the profoundest classical scholars of his age'.[70] After corresponding with Schoenbein and offering him a portion of the work in translation, Vincent was charged with finding a publisher. Although he was unsuccessful, he subsequently translated another of Schoenbein's works into English, again experiencing difficulty in interesting publishers in the project.[71]

Obviously a talented man, Vincent made rapid progress at the Royal Institution. At the end of 1848, only 8 months after his appointment as Assistant Secretary, he also accepted the post of Keeper of the Library, and his annual salary increased by £50 to £200.[72] In his new position Vincent was required to provide two sureties of £200, for which he named a distant relative and Michael Faraday. This was one of only two occasions when Faraday is know to have discharged this role; the other was in respect to a member of another Sandemanian family, George Whitelaw, who was apprenticed to Spottiswoode the printer.[73] Following Faraday's death two members of the Barnard family, Edward Barnard, junior, and John Barnard – Faraday's nephews – were appointed Vincent's sureties.[74]

Vincent's career at the Royal Institution spanned over 40 years. During that time he performed the numerous routine duties connected with both the Assistant Secretaryship and the post of Librarian. In connection with the first of these positions, he was responsible for writing many of the reports of lectures and discourses that appeared in the national press. As his obituary notice in *The Times* states, he was thus 'instrumental in impressing upon public attention the importance of scientific research to the commercial prosperity of the community, and doubtless also contributed greatly to the prosperity of the Royal Institution'.[75] In respect to the latter post, he particularly distinguished himself by preparing the impressive two-volume catalogue of the Royal Institution Library (1857 and 1882). Vincent is now more generally known for his editions of *Haydn's dictionary of dates and universal information*, a major work of reference which is still available in facsimile, and for his *A dictionary of biography* (1877). He also compiled an index to the persons, places and subjects cited in the Bible, an index which has since appeared in numerous editions of the Bible.[76]

While Vincent ministered to the Managers and members, he also performed an important and less conspicuous function at the Royal Institution by assisting Faraday. As Sophie Forgan has pointed out, Faraday's activities at the Royal Institution were not confined to its public space; he also occupied the private suite of rooms where he lived but where access was restricted to his family and a few close friends.[77] Vincent, as friend, colleague and co-religionist, was an intimate of Faraday and one of the chosen few who moved freely between the public and private domains at the Royal Institution. It is clear from the surviving letters that Vincent often acted as Faraday's personal secretary, particularly when Faraday was out of town. Like Faraday's nieces, Jane Barnard and Margery Ann Reid, Vincent formed part of Faraday's personal entourage, but unlike these nieces he maintained a formal and highly visible position within the Royal Institution.

Vincent retired from the Royal Institution in 1889, owing to 'failing eyesight', and was rewarded for his long service by being granted all the privileges of the Institution, the title of 'Honorary Librarian' and a pension of £150 per annum, which was half his current salary.[78]

The Sandemanian connection does not end with Faraday and Vincent. Shortly after his appointment as Keeper of the Library in January 1849, Vincent approached the Managers to ask whether one of his sons could accompany him throughout part of the day. Since the son, Charles Wilson Vincent (1837–1905), 'bore a good character and was well disposed', the Managers consented to this request.[79] In May 1851 Vincent again wrote to the Managers, but this time to point out the need for an assistant, since he was forced to close the Library when absent from the Institution, and it was also desirable that the Library should be opened in the early evening between 6pm and 8pm. Vincent also noted that he felt able to make this request in the light of the Institution's improved finances. The Managers consented to the new appointment and immediately offered it to Charles Vincent, at a salary of £35 per annum, since he had during the previous three years 'conducted himself well, learnt the duties of the office, and made himself useful to the members and others frequenting the Library'.[80] Charles Vincent remained as Assistant Librarian for a period of 6 years, after which another brother, Robert, held the post for a short time before having to resign owing to ill-health. Charles's name reappears in the Managers' minutes in the 1870s when he resumed the post of Assistant Librarian for a further period of some 6 years, during which time he received a small gratuity but no salary. Presumably this was a part-time position to fit in with his other activities.

The first clue to Charles Vincent's other interests appears in a letter his father addressed to the Managers in May 1854 requesting permission for Charles to be absent from the Institution for 3 days a week in order to attend lectures at the Royal College of Chemistry. Possibly owing to a recommendation from Faraday, Charles had been offered free admission by August Hofmann, 'subject to consent by Sir Henry de la Beche'. Benjamin Vincent promised that the

Members would not be inconvenienced, since another son (presumably Robert) would be present in the Library. The Managers consented to this request and a similar one made in the following spring.[81] Charles's chemical studies are pertinent to his future career, since in November 1857 Faraday reported that Charles had resigned from the Institution and was now employed by a chemical manufacturer.[82] Little is known of Charles's activities in the chemical industry, but in the 1870s he was engaged in literary ventures when he edited *The Year-Book of Facts in Science and Art* and published a work on Burton brewing water. His most impressive achievement was the massive and informative *Chemistry, theoretical, practical, and analytical, as applied to the arts and manufactures*, which appeared under his editorship in 1877-9. The title page of that work shows that he was a Fellow of the Royal Society of Edinburgh and of the Chemical Societies of London and Berlin.[83] He moved still further away from industrial chemistry when, in 1879, after his second period as Assistant Librarian of the Royal Institution, he was appointed Librarian at the Reform Club in Pall Mall. Following in the family tradition, he compiled a printed catalogue of the Reform Club Library.

However, according to the historian of the Reform Club, 'Vincent was respected, for the work on the catalogue, but he was not liked as much as [his predecessor Henry] Campkin'. A more ominous tone is conveyed by the Club's minutes for 1901, when, in response to a request from Vincent that his salary be raised, the committee awarded him £325 per annum together with a gratuity of £25 'in consideration of your present distress'.[84] Charles had been a Sandemanian for only 5 years but had been excluded in 1864, long before his (unspecified) troubles at the Reform Club.

When Robert was forced to resign as Assistant Librarian in 1858, his place was taken by John Handasyde Buchanan. Although the name Buchanan is not uncommon, it is one which is frequently encountered in Sandemanian circles, especially in the Edinburgh meeting house. More conclusive of Buchanan's background is his unusual middle name which belongs to a Sandemanian family; indeed an earlier John Handasyde was the elder from Wooler (Northumberland) who had been sent to London in 1762 to help form that congregation. Buchanan departed early in 1860 to take up employment in a merchant's counting house, and his position was accepted by Henry Clayton Hughes, son of the Royal Institution's Clerk of Accounts, who was not a Sandemanian.[85] Towards the end of 1879 – after Charles Vincent had joined the staff of the Reform Club – Henry Young (1859-1939) was appointed temporary library assistant, and it was Young who became Assistant Secretary and

Keeper of the Library on Benjamin Vincent's retirement in 1889. Young was from another Sandemanian family, and at the time of his appointment had been a member of the church for over a decade. He named as one of his sureties the Sandemanian printer and lithographer George Cargill Leighton, MRI, and his bond was subsequently accepted by two other Sandemanians, Charles P. Moir in 1893 and R. H. Sandeman in 1899. Young became an elder in 1891 and at the time of his death in 1939 he was the last male member of one of the two Sandemanian meeting houses in London.

In her recently published account of the Royal Institution Gwendy Caroe paints a rather unflattering portrait of Young, whom she portrays as 'nervous' and constantly intimidated by the overbearing Sir James Dewar. It is also presumably Young who should be identified as the 'white bearded old gentleman who sat day after day in the same chair reading and making notes in the margin of his book'.[86] Compared with his predecessor (Vincent) Yo' ng published little, although he did produce in his old age the first few printed sheets of what was presumably an extensive bibliographical project under the title *A list of bibliographies of authors (literary and scientific)* (London, 1932). By the time these appeared Young had already retired, having received a cheque and a silver rose bowl from the Institution's Treasurer, who praised his facility with 'facts, figures and names' and claimed that he 'had never found him at fault'. Young survived for another decade, his passing on 6 November 1939 being briefly noted at a General Meeting of the Institution.[87]

Just as Benjamin Vincent was probably responsible for Young's appointment, so Young may have assisted at least one member of a Sandemanian family gain employment at the Royal Institution. In 1890 the new assistant appointed in the Library under Young's aegis was Herbert C. Fyfe, but there is no evidence that he was connected with the Sandemanians, although there were a few members of the sect with that surname. However, Young's assistant in the Institution's accounts office during the years 1894–9 was G. A. Whitelaw, whose family were prominent members of the London meeting house.[88]

The above evidence shows that Sandemanians held important positions at the Royal Institution, and especially the key posts of Assistant Secretary and Keeper of the Library for a period of over a century, beginning with Faraday's induction into the Sandemanian church in 1821 and ending with Young's retirement from the Royal Institution in 1929. By holding such positions Sandemanians could influence appointments to the Institution's staff and other decisions formally made by the Managers. Yet the connection with the Royal Institution does not end there, since Sandemanians occasionally appear in the Institution's membership lists and on the lists of those receiving tickets for lectures (often provided by Faraday). For example, George Cargill Leighton was named as a Visitor in 1879 and Faraday's niece and companion, Jane Barnard, was presented with free admission tickets to lectures after Faraday's death. In 1879 the Managers considered a letter recommending that the Institution provide Jane Barnard with a pension. It is not clear whether this suggestion was acted on, but the letter came from William Douglas, MRI, of Glasgow – the name of a contemporary member of the sect.[89] The Library catalogue provides a few other clues. A substantial number of items were written or donated by members of the Vincent, Leighton and other Sandemanian families. Doubtless the long and close connection between the sect and the Royal Institution encouraged Sandemanian authors to donate copies of their works to the Institution's Library.[90]

Sandemanian names also frequently occur among the tradesmen hired by the Managers. Faraday's brother Robert was a gas engineer in Wardour Street who implemented some of Michael's discoveries at Buckingham Palace, the Royal Institution, the Athenaeum, and elsewhere. Robert, a Sandemanian, was first called to the Royal Institution in July 1838 to supply gas fittings in the lecture room.[91] Doubtless Faraday was to some extent responsible for his brother obtaining this and other commissions. Further work followed, and even after Robert's death in 1846 the firm of Faraday & Son, later Messrs Faraday, continued to supply and service the Royal Institution's gas appliances. In the period between 1846 and 1875 the firm was run by Robert's son James, who was also for many years in the sect.

Many Sandemanians worked in the book trade, and the Royal Institution frequently employed firms bearing Sandemanian names to bind and repair books. The firms Messrs H. & C. Barnard (1865–9), C. Baxter (1855), C.P. Baxter (later Baxter & Seaton and Baxter & Holcombe, 1882–93), Leighton & Son (1849–79), and Messrs J. & J. Leighton (1853–1914) feature in the Managers' minutes.[92] Thus over an extended period of time the Sandemanians in the book trade received regular, remunerative work from the Royal Institution, due presumably to the presence of Vincent and Young in the Institution's Library.

Faraday's own career at the Royal Institution began in 1813 when he was hired as Chemical Assistant. Through Faraday the connection became established between the Sandemanians and the Royal Institution. First Faraday, later Vincent and Young, were influential in bringing in Sandemanians to fill vacancies and in hiring their co-religionists. Thus the Sandemanians functioned not only as a spiritual fellowship and as a social network but also as a means of mutual help in obtaining employment and trade. While mutual assistance was certainly part of the sect's ethos, the moral code practised by its members ensured that one Sandemanian could unreservedly vouch for the honesty and reliability of another. We here find Faraday using his influence, although always careful to avoid appearing to act nepotistically, to benefit both the Royal Institution and the careers of fellow Sandemanians.

4.5 THE FINAL PARTING

One subject which features prominently in the extant letters between Faraday and other Sandemanians is death: indeed a recurrent theme for all Victorians. In a small, caring, closely interrelated group reports of death and approaching death and the comforting of the bereaved found frequent expression. By the time Faraday reached old age the significance of death was further emphasised by the declining size of the Sandemanian churches and the preponderence of elderly members, many of whom had grown up with Faraday. Shared illness and bereavement within this small community helped unite members in their grief. As one later Sandemanian wrote, 'When any of our number are very sick, they feel moved to call for our Elders, they go & pray with them & the mutual comfort of faith[,] hope and love is considered to be that "annointing with oil in the name of the Lord"'.[93] Moreover, they gained strength from understanding both illness and death from an identical biblical standpoint. Thus Faraday's father viewed his protracted illness as part of God's plan and he believed that his suffering should be borne with fortitude in the knowledge of God's mercy and grace.

The approach of death was viewed as the fulfilment of our role on earth. Writing to William Buchanan about a truly humble member of the London meeting house who was nearing death, Faraday described her as being 'very patient and comforted by the scriptures in the great & glorious hope of relief not merely from these things but from all sorrow & sighing [Isaiah 35:10] through Jesus Christ, and rejoices in her friends company'.[94] The death of his own mother in 1838 was endured with quiet fortitude: 'My mother died last Tuesday Morning', was the single sentence contained in one of Faraday's letters.[95] The sudden and unexpected death of his brother Robert in 1846 as the result of a carriage accident at the age of 57 deeply affected Faraday. Yet even this tragic event was conceived in religious terms: 'It was of the Lords will that he should be taken away'. Similarly, following the death of one of his nieces, he reminded her sister that we should 'look unto Him who rules all things according to the purpose of His own will'.[96] On the Sandemanian view we are merely players on a stage and we fit in with God's inscrutable plan. If most people understand the death of a near one only in mundane terms, nothing can comfort them for their loss; however, for a true Christian 'the very loss itself is but a step onward in that chain of events which he who rules all things according to the counsel of his own will, ordains to bring forth that state which we . . . desire and look forward to'.[97]

But what is this 'state' that the Christian should desire? To answer this question we need to probe further the Sandemanian view about how man can gain acceptance before God. Rejecting good works as providing the road to salvation, the Sandemanians instead insist that acceptance before God is to be attained only by living in accordance with the Bible and in imitation of Jesus Christ. '*Be perfect*; keep the commandments, and thou shalt live', urged Robert Sandeman.[98] But the life promised in the Bible to those who keep God's commandments is not limited to three-score years and ten or even four-score years; instead it is eternal life. In the Gospel according to St Matthew (19:16) Jesus promises eternal life to those who keep the commandments; again, it is the righteous who are promised eternal life (Matthew 25:46). Thus the desired 'state' in the above letter is eternal life, which is assured to the true Christian. Faraday entered more fully into this topic in a letter to one of his nieces, Caroline Deacon (née Reid). 'Death', claimed Faraday, 'has for the Christian everything hoped for, contained in the idea of reunion.' What Christianity offers is not the conventional view that death is the irrevocable parting but the certainly that the life of the soul will continue beyond the grave. From this perspective, 'What a wonderful transition it is!' This transition is exemplified by Christ 'who died, was judged, and who rose again for the justification of those who believe in Him'. What is then needed is total faith in God and the promise contained in the Bible, for although 'the fear of death be a very great thought, the hope of eternal life is a far greater' thought.[99]

This conviction sustains the mourner. After Robert's death members of the Faraday family immersed themselves in the 'glorious words of remembrance & comfort', especially in the epistles to the Corinthians and Thessalonians. Their religion, claimed Faraday, prevented them from wallowing in sorrow, 'as those who have no hope'; instead, they had the strength to console and comfort one another and to bear their grief with patience, gaining consolation from the Bible.[100] Similarly, in offering sympathy to the Genevan scientist Auguste de la Rive, who had become ill and melancholy after his wife's death, Faraday wrote: 'Surely the human being must suffer when the dearest ties are rent[,] but in the midst of the deepest affliction there is yet present consolation for the humble minded which (through the power that is over us) may grow up and give peace & quietness & rest'.[101]

Although mention of his illnesses frequently intruded upon earlier letters, concern with ageing and the approach of death only entered from the mid-1850s. One of the first occasions when he clearly manifested concern with these issues was in a letter written in the autumn of 1855 to Christian Schoenbein, who, in an earlier letter, had recounted a strenuous walk he had taken in the Swiss mountains. Faraday no longer felt capable of such exertion, adding, 'but my happiness is of a quieter kind, than it used to be and probably becomes more a man 64 years of age; and as we, i.e. my wife and I, go on our way together, our happiness arises from the same things and we enjoy it together, with, I hope, thankfulness to the Giver of every true and perfect gift'.[102] Although clearly drawing in on himself during the late 1850s, he continued to pursue research actively until the end of the decade. By 1861 his output had fallen almost to zero and the few publications of the 1860s were confined mainly to contibutions to reports drawn up by governmental bodies. A letter of September 1861 to his close and long-term friend Auguste de la Rive indicates his changed perception. Compared with their earlier correspondence, which was filled with scientific matters, Faraday now had nothing to report from the world of science. Yet the two men were bound by more than such transient interests, since Faraday felt that they were both now turning their thoughts to a far more important subject of mutual concern, 'the future [i.e. eternal] life which lies before us'.

> I am, I hope, very thankful that in the withdrawal of the powers & things of this life, – the good hope is left with me, which makes the contemplation of death a comfort – not a fear. Such peace is alone in the gift of God; and as it is he who gives it, why should we be afraid? His unspeakable gift in his beloved son is the ground of no doubtful hope; – and *there* is the rest for those who like you & me are drawing near the latter end of our term here below. ... I am happy & content.[103]

By 1865 his health was deteriorating rapidly. He still enjoyed the company of a few visitors but rarely left his rooms at the Royal Institution or his house at Hampton Court. Thus, typically, in

February 1865 he declined an invitation to Twickenham from the Count of Paris, adding: 'I bow before Him who is Lord of all, and hope to be kept waiting patiently for His time and mode of releasing me according to His Divine Word, and the great and precious promises whereby His people are made partakers of the Divine nature'.[104] With assured faith in God and God's promise of what lay in store for him, Faraday faced impending death placidly. By that time there was little he could do but wait, and wait patiently, since he was losing most of his physical faculties. His writing had deteriorated considerably and the few letters he did write during his closing years were short, disconnected and rambling. Often his niece Jane Barnard acted as his secretary, writing letters for him, negotiating with the Managers of the Royal Institution, and caring for him and Sarah. Thus in July 1866 she told William Barrett, then a young assistant at the Royal Institution and the descendant of a Sandemanian family, that Faraday 'continues quiet & calm though he gets very feeble as to bodily powers, the paralysis gaining ground; like all invalids he varies with the changes in the weather & other circumstances'.[105]

During those final years the Faradays spent much time at the house in Hampton Court made available to them by the Queen. He was frequently in pain and had to be administered morphine at night. Moreover he suffered various attacks. Nevertheless he was often reported as being contented and sometimes even in a positively cheerful mood. In the same letter to Barrett Jane explained that although Faraday's paralysis seemed to be gaining on him, 'between whiles he speaks most pleasant words shewing his comfort and trust in the finished work of our Lord' and that he had been repeating the twenty-third and fourty-sixth psalms. She had no doubt that ' "Thy will be done": – indeed the belief that all things work for good to them that believe, is the anchor of hope sure and steadfast, to the soul'.[106]

The end came on 25 August 1867. With a sigh of relief Jane wrote to Henry Bence Jones: 'Our cares are over: our beloved one is gone. He passed away from this life quietly and peacefully . . . He died in his chair, in his study; and we feel we could desire nothing better for him than what has occurred'.[107]

Sandemanians maintain that, since there is no biblical imperative, to hold a burial service is contrary to God's law. Moreover, they are opposed to all forms of spectacle and social frippery, such as very ornate hearses and tombstones, which were so prevalent among the early Victorians. The most celebrated of contemporary funerals was that of the Duke of Wellington in 1852, the elaborately decorated hearse being accompanied by a veritable battalion of mourners in their coaches and by a procession of military men. Many gigantic tombstones of the period are replete with carved angels and sentimental verse and would not look out of place in Forest Lawn. The Sandemanians, by contrast, turned their backs on contemporary modes of celebrating death.[108] As one descendant of the Whitelaw family noted, for Sandemanians there 'is no committal service, neither flowers nor music grace their passing. They die as they live, simply, reverently and unsung'.[109] Not surprisingly, therefore, Faraday specified in his will that his 'burial may be conducted in a moderate[,] sober and inexpensive way'. On Friday 30 August the cortege, consisting principally of family members and a few other mourners who joined it at the Royal Institution, made its way to Highgate cemetery where Faraday was buried in the unconsecrated area close to several other Sandemanians connected with the Barnard family. Faraday's tombstone manifests the same simplicity and unfaltering reticence, since it bears only his name – omitting all titles and honours – and the dates of his birth and death: 'Michael Faraday/Born 22 September/1791/Died 25 August/ 1867'.[110]

4.6 DEFENDING FARADAY

In the years following Faraday's death a cottage industry thrived on retailing information and anecdotes about him and his religious beliefs. His nephew Frank Barnard was a major source of information for his early biographers about the Sandemanians and their practices. The views of others who were either Sandemanians or had attended the meeting house were eagerly sought. Sarah supplied much information, especially to Bence Jones: for example, in a letter written some 3 months after Faraday's death she forwarded to Jones a copy of one of Faraday's exhortations and also sought to summarise his religious convictions in the following terms.

> but now he is gone, & I can only point to the New Testament as being his *guide & rule*; for he considered it as the Word of God (as you know) & equally binding on Christians at the present day as when written, so that, such scriptures as the following, were continually on his mind – 'If ye love me, keep my commandments [John 14:15] – Whosoever shall confess me before men, him will I also confess before my father who is in heaven [Matthew 10:32] – Do unto others as you would they should do unto you [Cf Matthew 7:12]'. And many others which I need not mention – for I feel myself quite unable to speak for him, but there is the *Word* to speak for itself ....[111]

Several people connected with the Royal Institution, including Bence Jones and Tyndall, continued to visit Sarah at her home in Barnsbury Villas during her declining years until her death in 1879.

As the ageing Sarah became less able to contribute to the demands of biographers, her unmarried niece Jane Barnard (1832–1911) accepted the role of protecting Faraday's image. Jane, the daughter of Faraday's sister Margaret and Sarah's brother John, was a regular member of Faraday's entourage from the mid-1850s, and she was admitted to the London meeting house in February 1857. During the last few years of Faraday's life she lived with him and performed the duties of nurse and secretary. Other unmarried nieces, especially Margery Ann Reid (1815–88), had performed similar functions at an earlier period. In Faraday's will 'my dear niece Jane Barnard who has now for many years been our affectionate companion and support' was left 'my love and remembrance' and also a proportion of his estate which was second only to the sum left to Sarah.[112]

Jane's role of mediating between Faraday and the outside world during the closing years of his life was continued after his death in 1867. Writing to Bence Jones shortly after the funeral, she perceived a great change in her situation: 'My occupation is gone'. However, a few months later she confided to a friend that she and Sarah 'seem to be still living in his atmosphere, all we do is connected with him'.[113] In one sense she continued with her earlier role after Faraday's death and for almost a further half century she policed the image of Faraday, portraying him in eulogistic terms to prospective biographers and other enquirers and helping to correct, if not suppress, views which did not cohere with this perspective. In a letter to Bence Jones shortly after Faraday's death she made an oblique reference to certain people who preferred a more public funeral, which would have bestowed more honour and praise on the great man.[114] It is clear that Jane was among that number. If he was buried in a quiet, simple manner, she devoted the rest of her life to the task of promoting the public image of Michael Faraday.

Although Faraday received a quiet and private burial, Jane was not alone in wishing otherwise. A public clamour soon arose and in June 1869 a meeting was held, chaired by the Prince of Wales, to raise subscriptions for a fitting memorial to the most eminent of British scientists. These subscriptions paid for a marble statue intended for Westminister Abbey, where a select handful of Britain's most outstanding scientists, such as Newton, are remembered. However, 'the religious scruples of his widow and of Faraday's family would not allow it to be placed either in the Abbey or in St Paul's, on the ground that he would not have entered in his lifetime a place of worship other than one belonging to his own denomination'.[115] Such was the strength of feeling among his close relatives that they would not even permit a statue of Faraday to adorn an Anglican church. The statue was then placed in the Royal Institution, where he lived and worked, yet the family's refusal is indicative of the way in which the memory of Faraday the Sandemanian had to be protected.

Ironically the public's insatiable appetite for formality and empty tradition eventually prevailed. During celebrations to mark the centenary of Faraday's discovery of electromagnetism in 1931 a sum of approximately £100 was somewhat grudgingly provided by the Royal Institution for a memorial tablet in the Abbey. On that occasion the remaining Sandemanians made no attempt to veto the proposal.[116] Likewise, a commemorative service is to be held at the Abbey to mark the bicentenary of Faraday's birth in 1991. In both these instances his memory has been honoured in public – but in a way he would not have approved.

5 *Faraday in Society*

The Sandemanian church offers its members both a supportive social environment and also the moral standards and practical norms by which to live. Owing to the sect's high social boundaries, its members tend to interpret events in very similar ways and to share almost identical responses to situations, including events in the world beyond the confines of the group. This perspective suggests that we can utilise the high degree of conformity among Sandemanians to understand better Faraday's attitude towards various issues, such as his response to politics. The example of politics is particularly instructive, since the subject rarely occurs in his writings, even in his extant correspondence. The secondary literature likewise contains few discussions of this topic, which has been addressed briefly by L. Pearce Williams and Morris Berman.[1] However, by setting Faraday's few explicit statements on political issues in the context of the attitudes shared by Sandemanians we are not only able to interpret his political position more fully and accurately but also show that his responses were not idiosyncratic but were closely related to his membership of this tightly knit sect with its biblically certified and historically conditioned norms.

While I wish to claim that we can usually understand Faraday's attitudes and actions against the background of his Sandemanian value system, this claim should not be taken to imply that he responded to these values in a simple, deterministic manner. He was not, as it were, preprogrammed. Frequently he found himself confronting situations that offered no simple or unambiguous solution in terms of his Christian values. At other times there were opposing, possibly egotistical, values operating. Moreover, spiritual growth is a slow process and learning from one's mistakes is an essential part of that process. As Faraday insisted, there are many situations in which it is better to make a decision, even a wrong one, than to prevaricate.[2] Faraday may not always have behaved in perfect accord with his Christian profession, yet his evident concern to live by Sandemanian values provides us with a key for understanding his actions in many different situations and his beliefs in a number of different areas, politics included.

The subject of politics may appear unimportant in the biography of a scientist who played no role in the organised politics of his day.

However, his reasons for dissenting from party politics shed considerable light on a number of related issues, especially on Faraday's attitude towards organised science and thus his involvement in such corporate, and politicised, bodies as the Royal Institution, the Royal Society, the Society of Arts and the British Association for the Advancement of Science. In this and the following chapter I shall argue that he was consistent in his responses to institutions, both scientific and non-scientific, and that the social philosophy of the Sandemanians offers a putative explanation of his views not only on politics but on a number of related topics. Most importantly, it enables us to comprehend Faraday's attitudes to science and the scientific community.

5.1 RELIGIOUS INSTITUTIONS

As noted in section 2.2, the formation of the Glasites was a highly charged political act in the context of early eighteenth-century Scottish ecclesiastical politics. Moreover, throughout their history the Sandemanians have endeavoured to keep themselves distinct from all other religious denominations, societies and groups in the belief that they alone are accurately following the directions given in the Bible. In asserting their exclusivity they draw a clear dividing line between the spiritual and the mundane, including in the latter category the religious institutions from which they differ.

Although extant documents contain little on Faraday's attitude to other religions and churches, he appears to have shared fully the views outlined above. While he numbered among his scientific colleagues many clergymen, such as the Rev John Barlow (Secretary to the Royal Institution and sometime Chaplain-in-ordinary at Kensington House) and the Rev William Whewell (Master of Trinity College, Cambridge), Faraday generally avoided religious issues in his correspondence with them or with others who were not Sandemanians. Indeed he was reticent about discussing his religious beliefs, although he would respond accurately and politely to any honest enquirer.[3] There were, however, a few non-Sandemanian scientists with whom he discussed religious matters, and principal among these were, significantly, three foreign scientists of long and close acquaintance – Christian Schoenbein, Jean-Baptiste-André Dumas and Auguste de la Rive.

Rarely did he comment on other denominations, but when he did, he spoke as a typical Sandemanian. For example, in a commonplace book, dating from shortly before he made his confession of faith in

1821, he quickly dismissed the pretensions of *soi-disant* ministers of religion. Under the heading 'Preaching Trade' he cited four examples taken from the *Evangelical Magazine* for August 1819. While one of these examples concerned the closure of a church for repairs, the other three notices heralded anniversary services. Faraday would have been disturbed by the thought that religious worship could simply be adjoured while bricks and mortar were restored, while the latter three notices were based on the assumption that the anniversary of a mere building possessed some religious significance. By contrast the only church that Faraday acknowledged was the invisible church described in the Bible and instantiated in the Sandemanian community. The most outrageous of these announcements in the *Evangelical Magazine* ended with the following note:

> NB. *A dry commodious and secure Burial ground* has just been *opened* contiguous to the above Chapel, where the religious public may find a *most interesting repository* for their *dear deceased relations* or *friends* by applying to the *Minister* or *Trustees* of the Chapel!!![4]

The three exclamation marks are Faraday's.

In a similar vein Faraday criticised the local practices of celebrating Christmas during his visit to Rome in 1814. Writing to his elder sister Elizabeth he contrasted the 'sincerity' of his own family with the 'hypocrisy' of the Romans, who devoted the whole of Carnival week to revelries which raise 'all my expectations, for the accounts I have heard of it make it a scene of *confusion* and folly. Professed fools (deserving of the title) parade the streets, and hold fearful combats with sugar plums. Religious clowns and every other kind of character fill the streets, and the whole world goes in masks'.[5] Having grown up in a strict, serious and very private Protestant sect, Faraday was appalled by this apparent mockery of religion in public. He was also shocked by the French, who insisted on opening their shops on the Sabbath, thus failing to distinguish it from the other days of the week.[6]

He was equally horrified by the religious revivals that started in America and swept through Britain in late 1859, and he responded to a request for his views in typically Sandemanian fashion. Placing his faith and trust firmly in the Bible, he dismissed revivalist preachers as lacking in any spiritual power to instruct or offer assurance. Moreover, revivals – with all their noise and commotion – are irrelevant to true religion since they appeal solely to the natural and social aspects of man and not to his spiritual nature, which can be touched only by God's revelation.[7]

Another pertinent indicator of Faraday's uncompromising attitude towards other churches, especially the established Church, is his opinion on funerals. In 1859 he wrote to the Swiss chemist Christian Schoenbein (a Catholic) concerning the funeral arrangements for Schoenbein's daughter, who had died while attending a school in Stamford Hill run by Jemima Hornblower (a Sandemanian). As he explained to his friend, his conscience would not allow him to attend the interment, since that would mean attending a Church of England service. Being a dissenter, he also declined the invitation to have his name inscribed on Miss Schoenbein's tombstone.[8] He was, moreover, inflexible in his refusal to attend religious worship at other churches, and thus was absent from services held to mark the passing of either friends or the famous. While the rest of the nation mourned and scrambled for seats, he declined the invitation to the Duke of Wellington's funeral at St Paul's in 1852.[9] Similarly, in 1865, he wrote to the son of the late Samuel Christie, who had been one of the Secretaries of the Royal Society, explaining that he could not attend the funeral since 'such things are to me only formal'.[10] Here Faraday implicitly contrasted the mere conventionality of the Anglican burial with the requirements laid down in the Bible. Since the New Testament makes no mention of funeral services and, moreover, Jesus was buried without ceremony, Sandemanians disavow the burial service (and also the marriage service) as contrary to the Bible's dictates. Such refusals placed Faraday outside the mainstream of the country's social and religious life.

5.2   RENDERING UNTO CAESAR

As already noted, the fundamental distinction for Sandemanians divided the spiritual realm from the mundane. In this context Glas, like other Sandemanians, cited John 18:36, where Jesus categorically states that 'My kingdom is not of this world'. This world, the mundane world, which most people accept as reality, is merely our temporary, transient dwelling, whereas true reality lies in the spiritual world, which is eternal. The Christian must not be seduced by the things of this world, otherwise (s)he will be drawn into the domain of pride, ambition and avarice. Since the Sandemanians try to lead their lives in imitation of Jesus's, they too refuse to commit themselves to the finality and values of a worldly existence. Instead, they believe the Christian must be dissociated from the world, its apparent pleasures and its vices, and recognise that salvation lies only by being committed to the code of behaviour laid down in the Bible and exemplified in Christ's own life. God's kingdom – the kingdom of heaven – is where we find all true pleasures and eternal life. Glas praised the excellency of God's kingdom 'for power and glory, number of subjects, and their quality, prosperity and peace; and for stability. Of this kingdom there shall be no end, for it hath the things that cannot be moved'.[11]

Throughout this chapter we shall frequently encounter this fundamental distinction, but for the present it should be noted that Glas conceived the kingdom of heaven in terms of an ideal state, a perfect political system. Standing aloof from the sordid wrangles of politicians and the greed of kings, Glas and his followers turned their eyes to the kingdom of heaven.

> Here are the best subjects . . . and an order and government infinitely excelling that of other kingdoms; absolute government without compulsion or oppression; perfect liberty, and a willing people, without any *confusion* or *disorder*; a government of rich grace, reigning through righteousness unto eternal life; unparalleled laws, written on the hearts of the subjects; and most righteous judgement, rendering unto every one according to his works.[12]

The utopian government of heaven, then, bears little relation to any kingdom in the mundane world, and for the Sandemanian worldly politics was a trivial, sordid game bereft of morality. The true Christian, then, does not meddle with politics.

In the Bible the absolute distinction between the worldly and the unworldly was also sanctioned by Matthew 22:21: 'Render unto Caesar the things which are Caesar's; and unto God the things that are God's'.[13] This passage was frequently on the Sandemanians' lips when they reflected on the social and political strife which appeared to be consuming so much of the world during the French Revolution or the insurrections of 1830 and 1848. For example, on surveying the plethora of riots at home and revolutions abroad, James Morison (1798–1878) of Perth wrote to a fellow Sandemanian in late 1830:

> It is almost impossible for one to think or write upon any other subject at present than the remarkable doing among the nations (including our own Country), which are daily passing before our eyes. Violence is indeed fast o'erspreading the earth [Genesis 6:11] – the waters roar and are troubled, and the mountains, more and more shaking by the swelling thereof [Psalms 46:3]. . . . In all that is going forward, I hope the Breathren both in Scotland and England will be found taking no part; but watching and keeping their garments [cf. Revela-

tions 16:15] – rendering unto Caesar the things that are Caesar's and to God the things that are God's.[14]

As this passage makes clear Sandemanians consider that they must stand aloof from events in the social and political realm and place their faith entirely in God. Convulsions in this world are none of their business but are, as the tone of Morison's letter suggests, the unfolding of God's plan as contained in the prophetic books. Prophecy and history again mingle in William Buchanan's letter of 1848 to one of the American brethren, in which he pointed to the accuracy of prophetic vision:

> what can more accurately describe the present state of the continent of Europe involved in convulsions from one end to the other than the Lords words[:] distress of nations with perplexity[;] the sea and the waves roaring[;] mens hearts failing them for fear and for looking for the things that are coming on the world[:] for the powers of the Heavens are shaken [cf. Luke 21:25–26] we not only see the powers of the Heavens shaken but the stars falling from Heavens and every where the restraints of law and authority weakened withdrawn or overturned and it would seem as if those who would not retain the blessings of Gods ordinance of Government in the way that He gave it but would exercise their own wisdom in that matter were to be given over to *confusion* & destruction so as the earth may be filled with violence as it was in the days of Noah . . . that time of great distress & tribulation may be brought on which is spoken of by Daniel and is referred to by the Lord himself when he says he has shortened those days [Matthew 24:15–22].[15]

In these letters by Morison and Buchanan we see the familiar intertwining of contemporary events with passages from scripture. Moreover, by describing political events in terms of confusion, disorder and destruction, Sandemanians were not only appealing to the apocalyptic visions contained in scripture but were implicitly drawing the contast between the necessary consummation of the mundane kingdom and the Bible's promise of salvation for the righteous. Recoiling in horror from the bloodshed around them, they found their faith and their apoliticality fully confirmed.

If social and political turmoil was to be kept at a safe distance and understood in prophetic terms, most forms of political life were likewise eschewed. Although often respected members of the community, Sandemanians rarely accepted civic positions, and in the few cases where they did, they avoided engaging in party politics. This point was emphasised in the obituary notices, dating from the early 1860s, of two highly respected members of the church who were also active and successful in the secular sphere. Dr John Crichton was an eminent surgeon who was particularly renowned for the impressive success rate of his lithotomy operations. Elected a magistrate in Dundee, he 'never troubled himself much about politics, least of all about burgh politics'.[16] An even more elevated civic post was held by William Buchanan who, early in his career, had been a supporter of Charles Fox and was a member of the Speculative Society, and was associated with the group of young Edinburgh Whigs who had founded the *Edinburgh Review*. He subsequently joined the Glasites, and although he rose to become the Solicitor of Tithes in Scotland and the Queen's Avocate and Solicitor in the Court of Teinds, he 'sought no personal advantages from political connections, and for many years he had ceased to take any interest in party struggles'.[17] The careers of both Crichton and Buchanan offer instructive examples of how Sandemanians could operate effectively in the life of the community and yet adhere firmly to the sect's rules of conduct.

Even though Sandemanians claim to be above party politics, their reading of the Bible contains explicit doctrines that need to be interpreted in political terms. In his first epistle Peter exhorted Christians to submit 'to every ordinance of man for the Lord's sake: whether it be to the king, as supreme; Or unto governors ...' (I Peter 2:13–14). This was one of the scriptural passages referred to by Samuel Pike in 1766 when reporting the practices of the London church:

> We think every Christian must be a *loyal subject*, submitting himself to civil concerns to every ordinance of man for the Lord's sake, punctually regarding the rules laid down (Rom. xiii. 1–7; 1 Peter ii. 13–17). This was required of the disciples and churches, when they were under a tyrannical and persecuting government; and it cannot be less a duty, under the present mild and peaceable one.[18]

Loyalty to the civic authorities was not only a requirement laid down in the Bible but, as Glas noted, the same doctrine was also exemplified by Jesus:

> He does not allow his subjects to disturb the kingdoms of this world, by taking the sword to advance or defend his interest and kingdom in the world, but calls them to be subject to the

powers that be, to *pay tribute* to them, *to pray for them, and to lead quiet and peaceable lives in all godliness and honesty.* ... And it has been observed, that the Christians, while they were sufferers, had no hand in the insurrections that were in the [Roman] Empire.[19]

Here Glas contrasted the New Testament with the Old, for while the ancient Israelites were frequently embroiled in battle, Jesus and his followers did not take up the sword but responded to conflict by bearing their sufferings with fortitude and patience. The Christian should not engage in bloodshed even to defend his religion, since matters of the spirit are totally divorced from the mundane realm in which battles are fought and blood is spilt. However, there is another form of warfare in which the Christian must engage, and this is the eradication of sin, starting with one's own. Thus it is in our power to alter our moral state through the exercise of will but we are not empowered to affect the mundane kingdom. Here we again encounter the distinction between the worldly and the spiritual and the Sandemanians' insistence that they must stand aloof from worldly affairs, which will be resolved through God's plan, and devote themselves to spiritual perfection.

While ignoring other social demands, the Sandemanian had thus to obey the laws of the state and the edicts of civic leaders. A summons to the Palace, for example, could not be refused or ignored, since the Bible required obedience to royal command, whether that command came from a just king or a despotic ruler. The question of loyalty was, of course, particularly convoluted in Scotland, where at least one mid-nineteenth century family of Glasites 'cherished Jacobite sympathies, and possessed several relics of Prince Charles Edward. Jacobite songs were first favourites, and no shortening of "Bonnie Dundee," or other lengthy songs in that vein, was allowed'.[20] In this example the paradox in the Sandemanian position is all too apparent.

In many political contexts the Sandemanians' loyalty to the Crown led others to characterise them as Tories. For example, during the American Revolution the small bands of Sandemanians in New England were generally recognised as loyalists and Tories, although, unlike other loyalists, they did not bear arms. Moreover, they were despised and somewhat feared since they belonged to a small exclusive sect with unusual religious practices. The suspicions against them led to their being interrogated, and in their deposition dated 14 September 1777 a group of Sandemanians at New Haven proclaimed that it was their duty 'to be faithful & loyal Subjects to our Sovereign King George the Third ... to whose Government we are heartily attached'. Although they noted that the Bible does allow the bearing of arms for defence, they expressed their desire 'to live peacefully with all men; . . . to love & pray for our Enemies; never to avenge ourselves, nor to bear ill-will to any men; to be no busybodies in other men's matters'.[21] Despite, if not because of, such assertions, they suffered contempt, were shot at, jailed and eventually exiled by the patriots.

Although the Tory party has not been static throughout British history, there are some recurrent features of conservatism that are traditionally (if sometimes incorrectly) associated with that party. Among these has been the opposition to all forms of dislocation, especially radical political and social changes, and the maintenance of the status quo. Moreover, the political conservative is a person of order reacting against uncongenial external events rather than actively trying to shape the world. Finally, since many conservatives believe in the imperfection of man, they place their trust in a transcendental moral order and a static view of the social order in which the monarchy stands supreme.[22] We can appreciate that in these important respects, although not in all others, Sandemanians were conservatives and thus, within the context of British (and Revolutionary American) politics, they have usually been branded as Tories.

The interpretation offered here is at odds with the connection which has sometimes been drawn between Sandemanians and radicalism. E.P. Thompson, for example, has argued that the intellectual roots of working-class radicalism in the late eighteenth and early nineteenth centuries are to be traced to such dissenting groups as the Unitarians, Swedenborgians, Inghamites and Sandemanians.[23] The case for the Sandemanian connection with radicalism rests principally on the examples of Thomas Spence and William Godwin. However, in these two instances the evidence is inadequate to support Thompson's thesis. According to Malcolm Chase, Spence was not a member of the Newcastle church which his brother Jeremiah only joined in the year that Thomas Spence departed Newcastle for London. Moreover, his name does not appear on any of the Sandemanian rolls that I have been able to locate.[24] Godwin's exposure to the Sandemanians may have been more extensive than Spence's, for he studied and lived in Norwich with Samuel Newton, who is described as a Sandemanian, for a number of years. Although Newton clearly was influenced considerably by Sandeman's writings, his church was not formally associated with the Sandemanians and did not come under their discipline. Moreover, there is no evidence that either Newton or Godwin made his confession of faith in a manner acceptable to the Sandemanians. Thus in the strict and

correct sense of the term Newton and Godwin were not Sandemanians. However, in his biography Peter Marshall detects an
ambivalence in Godwin, on the one hand reacting against the
restrictions imposed by his stern religious upbringing, and on the
other absorbing some of its theological doctrines. Most interestingly,
Marshall suggests that the Sandemanian emphasis on equality and
on mutual aid may have exerted a positive influence on Godwin's
radical thought.[25] Even if this were the case, the examples of Spence
and Godwin do not refute the view that the Sandemanians were
loyal to the Crown and that they practised a non-factional form of
conservatism.

Against this background we can interpret Faraday's few extant
references to political issues, since he, like other Sandemanians,
consistently refused to accept politics as part of his life. When the
subject did impinge on his consciousness, he, like James Morison and
William Buchanan, reacted with indifference, if not disdain. For
example, while touring Italy with Humphry Davy in 1815, he
received news of Napoleon's escape from Elba. While others
responded with alarm and foreboding, his diary entry is unchar-
acteristically phlegmatic: 'Being no politician, I did not trouble
myself much about it'.[26] Similarly, in a letter of July 1849 to his
close friend Auguste de la Rive Faraday claimed that he never
meddled in politics, which he dismissed as 'one of the games of
life'.[27]

Faraday evinced a strong moral distaste for all forms of politics.
For example, hearing that the French chemist Jean-Baptiste-André
Dumas had been appointed to political office, Faraday expressed in
June 1849 his concern that 'so much turns up near & about you that
seems to me incompatible with your habit of mind & [your
scientific] occupation that I mourn a little at times'. Implicit in
this letter is not only the contrast between science and politics but
also that between the obedience to divine law and man's arrogant
self-serving decisions, since the former, unlike the latter, lead to
moral improvement. In making this point he compared the recent
serious illness of Dumas' son with 'that which is evolved through the
tumults & passions of man'. While we can understand the former
event as caused by 'the hand of God' and can bear it with patience,
the latter 'does not bring with it that chastening & in some degree
alleviating thought'.[28] Engagement in politics therefore produces a
thoroughly negative moral effect.

Of the various forms of politics, Faraday found revolutionary
politics particularly threatening and disagreeable. Although the
social and political turmoil of Europe in the late 1840s barely
impinged on him, it was occasionally reflected in his correspondence

with an international coterie of talented scientists, some of whom were in close proximity to scenes of violence. Most interestingly in letters of 1848–9 to de la Rive, Schoenbein and Dumas, he strongly contrasted revolutionary events in politics with the positive values he associated with science. Thus in one remarkable letter he contrasted the quiet and philosophical pursuit of science and 'the pure and harmonious beauties of nature' with the sight of the degraded revolutionary mob. Viewing the situation in stark moral terms Faraday expressed his delight that the Swiss chemist Christian Schoenbein was not

> fighting amongst the crowd of black passions and motives that seem now a days [December 1848] to urge men every where into action. What incredible scenes everywhere, what unworthy motives ruled for the moment, under high sounding phrases, and at the last what disgusting revolutions.[29]

In the light of the Sandemanians' concern with morality and prophecy, it is hardly surprising that Faraday should interpret cataclysmic political events as evidence of the necessary moral degradation that occurs when man forsakes the kingdom of Christ. Faraday's reaction in 1848 also bears a close similarity to the response, cited above, by William Buchanan to contemporary events and that of James Morison to the social and political upheavals 18 years earlier. Like these two brethren, Faraday looked down on the worldly kingdom, judging human actions in moral terms while viewing the kingdom of Christ as offering the sure, long-term and stable future for mankind. However, Faraday did not employ the biblical language so evident in the letters of Morison and Buchanan. Yet Faraday thought in those terms and extensively utilised biblical quotations in exhortations and in letters to fellow Sandemanians. When writing to non-Sandemanians, such as Schoenbein, he could translate his beliefs into a secular and more widely acceptable idiom.

Faraday was concerned not only with the adverse moral effects produced by the political climate in the late 1840s but also with the physical dangers faced by some scientists in blood-stained Europe. Thus he expressed concern for Giovanni Majocchi, an Italian physicist who had been forced to leave Turin in 1849 in the wake of the invasion of Italy by France and internal strife. The life and welfare of individual scientists were being threatened, as was their opportunity to pursue research. However, his unease also stemmed from the essentially moral incompatibility between social and political disruption and the scientist's quiet search for truth.[30] This contrast underlines Faraday's belief in the truth and reality of scientific knowledge as compared with the ephemeral nature of politics and the passions on which it depends. Indeed, the events of 1848 were seen by him as inimical to the very values of science. Writing to Dumas in June 1849, Faraday contrasted the recent distasteful happenings with his friend's propensity for '*peace*[,] *order & science*'.[31] Here Faraday was projecting his own value system in which science was intimately associated with peace and order. Science also connoted progress and truth, in contrast to politics, which was morally regressive and ephemeral. As further discussion in section 6.1 will show, the association of these values with science (and their opposing values with politics) provides one crucial reason why Faraday found science so congenial and pursued it as his vocation.

Faraday's distaste for revolutionary politics appears also in a list of anagrams taken from his Commonplace Book: 'Revolution' becomes 'to love ruin' and 'Radical Reform' is transposed into 'Rare Mad Frolic'.[32] These and similar anagrams constitute some of the evidence Morris Berman has used to confirm his thesis that Faraday 'was not apolitical, but very political – a Tory'.[33] Berman's inference is false; Faraday shared with other Sandemanians an abhorrence of radical ferment, but this does not make him a Tory, although it is certainly true that he, like other Sandemanians, was often considered to be a Tory. Indeed, as discussed above, a distinction needs to be drawn between the kind of conservativism espoused by Sandemanians, Faraday included, and their rejection of party politics including the patronage system that went with it. It should now be clear that Faraday was in this sense a conservative by inclination, but not a Tory by persuasion.

One point needs to be clarified. The form of conservatism encompassed by Faraday and other Sandemanians does not imply that they simply accepted the status quo on all issues. While opposed to revolutionary change, they also were critical of many social conventions, particularly where these had an inhibiting effect on the individual or were not sanctioned by the Bible. For example, as has already been discussed, Sandemanians rejected national churches and they were highly critical of the pre-1837 Marriage Acts, which required marriages to be registered in an Anglican church. A further example emerges in Faraday's evidence to Her Majesty's Commission on Schools in 1862. Here he argued vehemently that all children possess the ability to learn science but that the men who had done most for science had not received a conventional education centred on classics. By contrast, those who received a conventional education not only failed to appreciate science but remained in later life antipathetic to it and unable to understand even the simplest physical principles.[34] Thus, far from endorsing contemporary educational practice, Faraday was highly critical of it and of the distorted values which the upper classes tended to assume.

The one political term that can be applied appropriately to Faraday is 'Loyalist', since Sandemanians accept as a requirement of Scripture that they must be loyal subjects. Many members of Sandemanian families (although not necessarily Sandemanians themselves) served in the armed forces, although not usually in combative roles. For example, General Robert Turnbull Sandeman, who entered the military service of the East India Company in 1824, was probably a member of the sect, but his son, Colonel Sir Robert Sandeman did not join, although he greatly respected its tenets. The latter served principally on the Indian frontier, where he achieved distinction as a humane but effective administrator.[35] Like these two contemporaries, Faraday served his country by teaching science to cadets at the Royal Military Academy, Woolwich, and also by pursuing a number of scientific investigations on behalf of the Admiralty (see section 6.5).

Faraday's loyalty cannot be doubted, particularly during the latter part of his life, when he enjoyed the company and respect of royalty. Prince Albert, the young princes and other members of the household attended Faraday's lectures on a number of occasions, and Albert was present in 1854 when Faraday made his famous 'Observations on mental education', and he subsequently sent the Prince a copy of the text. He gracefully accepted numerous calls to the Palace and to Windsor, usually at Albert's request, and he was invited to a ball at Buckingham Palace, although it is not known whether he attended.[36] Kings, queens and emperors, at home and abroad, bestowed honours on him: a knighthood from the King of Prussia, a Chevalier of the Légion d'Honneur from Napoleon III, and a house at Hampton Court placed at his disposal by Queen Victoria, who even undertook the necessary repairs.

5.3   ON PATRONAGE AND REWARDS

If Faraday withdrew from cataclysmic social and political events, he also stood aloof from the less violent and more commonplace politics of his day, refusing to align himself with any party or faction. To adopt this neutral position consistently was by no means easy, since his post at the Royal Institution, together with his reputation as a leading scientist, brought him into frequent contact with those who either possessed or were seeking political power. Had he sought power, influence or status, he could easily have attained them through such alliances, but, particularly after his apprenticeship under Davy, he sought to remain outside the circle of patronage which threatened to embroil him in the mundane world. Not only was the social and political world not the kingdom of the true Christian but its values were inimical to the kingdom of God. Men who allowed themselves to be seduced by worldly goals became prey to avarice and other vices which cut them off from brotherly love and from the perfect kingdom revealed in the Bible.

This concern with moral values found expression in a letter to the Swiss physicist Auguste de la Rive, one of the few scientific correspondents with whom Faraday shared his religious thoughts and feelings. In this letter of 1855 Faraday expressed his abhorrence of contemporary social and political events and the prevalent lack of morality. This was one of the rare occasions in writing to a non-Sandemanian when Faraday employed biblical language and he unselfconsciously expressed his understanding of world events through biblical prophecy:

> What a world this is! How the whole surface of the earth seems to be covered with the results of evil passions. – Ambition — contest – inhumanity – selfishness. – Thou shall love thy neighbour as thyself [Matthew 19:19], how the extreme reverse of this shapes its self into the forces of Honor, Patriotism, Glory, Loyalty, Romance &c. &c. Happy for us that there is a power who overrules all this to his own good ends, and who will one day make manifest the truth & cause the light to shine out of the darkness [2 Corinthians 4:6].[37]

In this letter Faraday looked down on the worldly kingdom, chastising human actions in moral terms while viewing the kingdom of God as offering the only sure, long-term future for mankind. Moral contrasts also found expression in an earlier letter to William Vernon Harcourt, in which he praised the late Henry Cavendish's uprightness and contrasted him with 'those whose craving is so great for fame that it leads them almost to the verge of dishonesty'.[38] As these examples indicate, Faraday repeatedly conceived moral issues in terms of sharply drawn dichotomies.

The first two evil passions listed in the letter to de la Rive – ambition and contest – were both manifest in the scientific community and in others who sometimes appealed to Faraday for patronage. For example, he was often asked to write letters in support of candidates for scientific and other posts and to assess various technical processes. He consistently refused to support applicants for salaried posts, although, if he was approached by an appointing committee, he was often prepared to offer his advice, though in fairly general terms. On only two occasions did he act as surety – for Benjamin Vincent and George Whitelaw, both respected members of the London Sandemanian community.[39] What he particularly disliked was being propositioned by those interested only in selfish gain. When, for example, he was requested in 1860 to report on the best means of preserving the stonework of the Houses of Parliament, he was horrified when one of the contenders for the contract visited him. '[I m]istook him for another person, and saw him for a few moments. Refused to discuss [the matter] with him. He threatened legal proceedings. Sent him out, and all his papers.' The issue troubled Faraday deeply, since he believed that the question of stonework preservation could be decided only by experiment and not by petition. As he subsequently wrote to Baron Mount-Temple, Melbourne's nephew and the Commissioner of Works, 'Whenever you give me the pleasure of being [in] any way useful to you again, I hope you will help me to keep clear of the parties, whose object is, of course, profit'.[40]

Although Faraday sought to maintain moral purity by avoiding entanglement in 'parties', he possessed an international reputation in science which few in his generation could equal, and he was therefore offered honours and rewards on numerous occasions. His response to these offers provides important clues to his general views on patronage, and hence we need to understand why certain honours were acceptable to him and others were not.

Faraday believed that those who devoted their lives to science should be eligible for honours, but he wished to see scientific honours dissociated from the political superstructure and freed from the base passions of ambition and competition. Honours, he argued, should be given to the successful cultivators of science as a just reward for their toil. These rewards (which bear no relation to God's assessment of an individual's actions) are purely the fruits of this world. By awarding civil honours to its scientists the leaders of a country show that they possess the intelligence to value the activities of its citizens. Moreover, they also thereby indicate that science is deemed worthy of pursuit. Thus to honour scientists reflects well on both the country's values and also on science itself. Moreover, those who receive the honour should be 'worthy of those whom the sovereign and the country should delight to honor',[41] and they will then stand out as shining examples and encourage others to enter wholeheartedly into the unselfish pursuit of science. Scientific honours should not be given to those who merely possess wealth or rank, but to those who deserve them by the fruit of their labour.

Indeed, Faraday opposed conflating scientific with existing civil honours, since the award of knighthoods and baronetcies to scientists degrades their work by setting them on the same level as men possessing hereditary power or gross worldly ambition. For this reason he consistently expressed his disinterest in a knighthood, since he believed that the British honours system was corrupt. Instead he wanted to see scientists recognised by a separate system of honours, or ranks, similar to those used in the army or the judicial system. Perhaps rather naively he considered that honours from abroad were untainted and given solely on merit; he was thus delighted to accept the knighthoods offered by the Prussians and the French and numerous other awards from foreign societies. While viewing British honours with contempt, he asserted that he wished to remain 'plain *Mister*' Faraday. Moreover, by usually refusing to use his titles, even the honorary doctorate he received at Oxford in 1832, Faraday affirmed his typically Sandemanian disinclination to over-value the rewards of this world.[42]

While generally in favour of rewards and honours, Faraday nevertheless considered that some forms of honour damage science and its practitioners. For example, he refused to sign a memorial testimonial to honour the famous German scientist Alexander von Humboldt because he considered that the system of writing testimonials had been so often abused as to dishonour the recipient's memory. More generally he accepted that honours should be given for proven success, such as the discovery of a law of nature. However, he warned that honours are likely to degrade the practice of science if they are offered for research to be carried out at some future time, since scientific talent then becomes 'a thing marketable & to be bought & sold', and when this state has been reached, 'down falls that high tone of mind' which provides the essential motivation for science and a greater spur 'than any common place reward'. Faraday, it should be noted, here expressed his concern that science might be degraded and become like a trade, whereas he considered that science should be associated with elevated moral values. The scientist must retain moral purity and not be contaminated by the base desire for worldly success. Worldly goals had, however, a deleterious effect on his mentor Humphry Davy who had used science to move into high society, obtain a knighthood, a baronetcy and the Presidency of the Royal Society, but had in the process debased himself and forsaken the laboratory for the salon. By contrast, the true scientist must shun society's conventional reward system; as he wrote to Thomas Andrews in 1843, 'such rewards will never move the men who are most worthy of reward'.[43] Science has to be its own reward.

We are investigating a rather grey area where it is difficult to disentangle inconsistent evidence. Despite Faraday's assertions about dissociating himself from mundane things, he clearly gained considerable pleasure from the tokens of esteem he received from kings, societies and individuals throughout the world. He also compiled lists of his awards: by 1844 he had been elected to some seventy scientific societies.[44] Moreover, he took pride in his impressive collection of approximately 250 portraits of famous contemporaries, many of which were accompanied by personal letters. This collection Faraday carefully arranged in two large volumes.[45] In pursuing such activities was he displaying his attachment to this world? Was he completely unaffected by the judgements of others? Could he believe that the honours he received were his just deserts and not the result of political intrigue among the Academicians in France or Sweden? Was Faraday totally consistent or did he, like most of us, sometimes fall short of his ideals? Merely to question his consistency may help prevent biography shading into hagiography. Yet my concern here is not to judge Faraday but to show that his views and actions were strongly affected by certain clusters of values derived from his religion.

In principle the line between correct and incorrect action was drawn in terms of his own values and not those of the world. In late 1837 he was approached by Thomas Spring-Rice, the Chancellor of the Exchequer, who enquired whether Faraday had found useful his Civil List pension, which he had received a couple of years earlier, and whether such pensions should continue. He responded positively on the assurance that the Prime Minister was offering the pension for merit and not from any political motive. Moreover, as if to substantiate this assessment, he submitted a list of the honours he had received as 'evidence in favour of my own character'. With one exception, all the honours listed were, he asserted, 'spontaneous offerings of kindness and goodwill from the bodies named'. The exception was his Fellowship of the Royal Society, which was 'sought and paid for'.[46] (As discussed in section 10.2, below, Davy attempted to veto Faraday's nomination because he claimed that Faraday had plagiarised the work of William Hyde Wollaston.) The implication here is that Faraday had not justly earned the letters FRS and that the searing confrontation with Davy was punishment for forsaking his high-minded principles.

One occasion when Faraday may have experienced particular difficulty in steering a clear course was his famous, or infamous, interview with the Whig Prime Minister Lord Melbourne in 1835. According to Bence Jones, Faraday was informed in the April of that year that, had Peel remained in office, he would have been awarded a pension. Faraday's first response was to make clear his refusal to accept any such a pension because it would not have been earned legitimately and because he was still able to obtain an adequate income from the Royal Institution. Although his reply had been uncompromising but polite, he evidently discussed the matter with his father-in-law, Edward Barnard, an elder in the Sandemanian church, who persuaded him to write another letter containing a less decisive refusal. As a result of this correspondence, he was summoned to meet Lord Melbourne on 26 October. What occurred in the Prime Minister's presence is less than clear, but the arch Tory *Fraser's Magazine* subsequently published what it claimed was a rough transcription of the dialogue. After having kept Faraday waiting for a long time, Melbourne opened the conversation by adverting to Faraday's religion as the reason that would prevent him from accepting a pension. Then Melbourne is said to have burst out:

> I look upon the whole system of giving pensions to literary and scientific people as a piece of gross humbug. It was not done for any good purpose; it never ought to have been done. It is gross humbug from beginning to end.[47]

Although Faraday might have agreed with some of these sentiments, he was placed in a compromising situation and therefore took his leave.

While much political capital was made from this incident in the Tory press, both Melbourne and Faraday were at pains to smooth matters. Initially Faraday declined the pension, owing to Melbourne's views about scientific and literary pensions, but he accepted it after the Prime Minister had expressed not only his complete confidence in Faraday's worth, but also his assurance that the honour had been earned and was not the result of political jobbery.[48] If there was some misunderstanding at the interview, it was soon overcome, with Faraday agreeing that the award was morally acceptable.

A few months later Faraday was the subject of a sketch in *Fraser's Magazine* in which the 'future Baronet' was described as 'a very good little fellow – a Christian, though, we regret to add, *a Sandemanian* (whatever that may signify) – a Tory (as might have been inferred from Rat Lamb's [Melbourne's] hostility)'.[49] The writer was off the mark in considering Faraday to be a likely 'future Baronet'; indeed, had he understood the meaning of Sandemanianism, he would neither have made that mistake nor called him a 'Tory', which, as we have already noted, is a problematic description of Faraday.

5.4   LAYING UP FORTUNES: FARADAY ON WEALTH

In his book *Social change and scientific organization* Morris Berman rightly argues that any adequate interpretation of Faraday must take account of his participation in a number of practical projects, ranging from the chemical analysis of substances brought to the Royal Institution, to research on disinfectants for use in prisons and the testing of lighthouse illuminants on behalf of Trinity House. Faraday thus emerges as a man of this world, a man deeply involved in numerous projects serving the capitalist economy. Moreover, in his interpretation of Faraday Berman appeals to *The protestant ethic and the spirit of capitalism* in which Max Weber portrays the capitalist as aiming primarily to acquire wealth. Concentrating on such groups as the Methodists, Weber conceived that there is no conflict between capitalism and Christianity, or between riches and righteousness. As Wesley had argued, far from being ashamed of wealth, the Christian should not only try to obtain salvation but should also amass a fortune, provided that some of this money is donated to charity. Salvation can be attained by good works and hard work. Here, then, was a brand of religion tailor-made for the capitalist, and the Methodists succeeded in attracting merchants, tradesmen and manufacturers by the thousand. Yet, as Berman recognises, the Sandemanians do not cohere with Weber's famous thesis, since they neither subscribe to this 'protestant ethic' nor do they encompass the 'spirit of capitalism' – the belief that the acquisition of wealth should be man's foremost aim.[50] By concentrating exclusively on this apparent opposition Berman offers not only a misrepresentation of the Sandemanians and their doctrines but also a problematic account of Faraday.

While Berman is certainly correct to emphasise the practical aspect of Faraday's work, he is less successful in relating Faraday's civic and industrial activities to what he sees as an antipathetic facet of Faraday's character. Since the Bible teaches 'My kingdom is not of this world' (John 18:36), he rightly seeks to portray Faraday as living an 'inner' life dictated by spiritual, not material, values. He is also perfectly right to stress that this 'inner' Faraday, the Sandemanian Faraday, rejected the 'spirit of capitalism'. At this point in the argument Berman generates an historiographical problem. Faraday's belief in God and his participation in a fundamentalist religious sect is sufficient evidence to convince Berman that Faraday was engaged in some 'mystical' activity irreconcilable with the exploitation of science within the capitalist economy. Berman thus offers two, apparently contradictory, interpretations of Faraday, and he is thus forced to confront the issue of how to interrelate the unworldly Faraday, who studied the Bible meticulously and prayed at the Sandemanian meeting house each Sabbath, with the worldly Faraday, who advised Trinity House and performed chemical analyses for industry. Berman resolves his problem by adopting the drastic strategy of positing not one but 'two Faradays' – one worldly, the other unworldly – who were 'antagonistic to each other'.[51]

Berman's portrayal of Faraday and other Sandemanians might lead us to suppose that members of the sect were prophets, poets and dreamers standing on the periphery of economic activity. However, while few Sandemanians figure among the captains of industry, they did not eschew capitalist enterprises but only the spirit of capitalism. Moreover, as discussed more fully in section 3.3 with respect to the London Sandemanian community, they are to be found occupying a wide range of economic functions. Turning to the Sandemanians throughout Britain, we find that many of them were skilled or semi-skilled workers; for example, among their number were many weavers, shoemakers and tailors, and Faraday's father was a black-smith. Moreover, a small but significant proportion of Sandemanians were shopkeepers or businessmen. A number were in the book trade, such as the Morisons of Perth, George Whitelaw of Fleet Street and the Boosey family, which became famous for selling sheet music. The Waterston family of Edinburgh were stationers and sealing-wax makers, a branch of the extended Sandeman family made its name as wine importers, while the name of Bell is associated with whisky. Moreover, a few of their number, Faraday included, were successful professional men. The fact that the Sandemanians encompassed labourers, successful manufacturers, professional men and businessmen points to a difficulty in Berman's thesis, which appears to imply that Sandemanians should not be engaged actively in the capitalist economy. To clarify this issue we must now turn to the bearing of the churches' doctrines on worldly wealth and success.

Glas and Sandeman argued at length that the true Christian must recognise that conventional morality is as naught before God. Sandemanians must instead turn to the Bible for their moral code and place the highest premium on the morals exemplified in the lives of Christ and the Apostles. Although in the eyes of many Christ appears 'humble, dispised and obscure', the Sandemanians argue that, if rightly understood, Christ's life contains a deep moral meaning and is exemplary of those eternal values by which the true Christian must live.[52]

The distinction between true Christian morality and the dubious moral values adopted by contemporary society forms part of a broader concern, which we have already encountered, to differenti-ate the religious and the secular in their multifarious forms. Extolling the virtues of the Christian way of life, Sandeman warned:

> But as to all earthly things [such as business ventures], it must be owned, that the best founded hopes are often frustrated by accidents, which no human foresight can prevent. And here the excellency of the gospel stands distinguished: For he who so knows the bare report thereof, as to love it, and run all risks upon it, shall in no wise lose his reward. 'Heaven and earth shall pass away,' but he shall not be disappointed.[53]

Here Sandeman does not forbid anyone engaging in worldly events, provided they are not prohibited by the Bible. What this and similar passages emphasise is the vast superiority of spiritual pleasures over mundane ones – thus inverting the priorities held by Weber's capitalist. Among the favourite subjects for Sandemanian exhortations were such scriptural passages as 'Love not the world, neither the things that are in the world' (1 John 2:15) and keeping oneself 'unspotted from the world' (James 1:27), which warn the Christian against becoming seduced by the attractions offered by the mundane world. Thus, by allowing oneself to become preoccupied in, say, business ventures, one would not only run the risk of neglecting religious duties, but also of engendering the sins of pride and ambition. Moreover, worldly (as opposed to spiritual) things are impermanent, and will, as prophesied, pass away.

In respect to the accumulation of wealth, the Bible warns: 'Lay not up for yourselves treasures upon earth' (Matthew 6:19). This injunction is not taken as requiring a vow of poverty, since such a vow would be a dereliction of faith and, moreover, self-imposed poverty would prevent one from fulfilling one's duty of charity. Instead Sandemanians interpret it to mean that one should neither love wealth nor pursue it as one's ambition. As Sandeman expressed the matter to John Barnard,

> it is easy to see how he who residing constantly in one place, possessing some means of living, be it a stock in trade, the utensils and materials of a mechanic, or whatever his fund be, makes use of it not to gratify worldly lusts, in pursuit of wealth, honour, pleasure, &c., but soberly, to support himself in fellowship with Christ's disciples, or so as he may best minister to the saints and do good to all men.[54]

Another writer stressed the unlawfulness of laying up treasures 'by setting them apart for any distant, future, and uncertain use'.[55] What the Sandemanians repudiate is the 'spirit of capitalism', but they could, and necessarily did, engage in business activities. They were, however, required to take cognisance of the proviso that they should not perceive such activities as their life's aim. They were free to pursue their trades and businesses but not to hoard their profit. Moreover, in their business dealings they were required to be detached and free from pride, avarice or worldly ambition.

The warning against covetousness was often linked with the giving of alms. Sandemanian meeting houses had three collecting boxes: one for the community, the second for the poor without and the third for the maintaining the building. The collection was not an optional extra but, in accordance with Biblical precept, an integral part of the Sabbath service. 'Every one, therefore, is to look upon all that he has in his possession as open to the calls of the poor and the church, to contribute according to his ability, as everyone has need.'[56] The sharing of goods was deemed essential in maintaining unity and fellowship within the church.

That is the ideal. However, not surprisingly, difficulties sometimes occurred in practice. One Sandemanian succinctly expressed the matter through the antithesis, *'ye cannot serve [both] God and Mammon'*.[57] In 1770 John Barnard noted with regret that elders in both Scotland and America had been 'put away', and he expressed his concern that 'the spirit of trade, without which they [the elders] can do no good as masters, should too much entangle them with affairs of this life'. The elder's office, he contended, requires 'such a very different (though not inconsistent) current of mind from that of conducting trade to advantage, that I have not seen both done well by the same man'.[58] A cautionary tale is provided by the small church at Trowbridge in Wiltshire, whose members were poor labourers employed in the clothing industry. In the late 1770s the London congregation, which was very rich, 'liberally supplied their wants'. The effect was dramatic: the Trowbridge community rapidly lost sight of Christian values and had to be separated from the London meeting house until two elders from London brought the situation under control.[59]

In Faraday's day Thomas Boosey vacated the elder's office owing to some problem relating to his music-publishing business.[60] At Aberdeen in 1834 one of the elders was separated, together with several members of his family, because he received income from a friendly society; the other members of the meeting house considered this to be 'an improper way for Christ's disciples to be supported, and especially an Elder'.[61] Other Sandemanians are known to have refused to take out life assurance policies and to have declined legacies, since both were considered incompatible with true faith in God's providence. The latter custom does not appear to have operated in London, for in his will Faraday left sums of money to a number of close relatives and other Sandemanians.

The Sandemanian ideal on the subject of worldly wealth provides the key to understanding Faraday's views on this and related subjects. Thus when a publisher approached him in 1859 requesting him to publish some juvenile lectures, he insisted that 'money is no temptation to me. In fact, I have always loved science more than money; and because my occupation is almost entirely personal, I cannot afford to get rich'.[62] Unlike the true jewels to be found in science and Christianity, worldly riches offered no particular attraction. Nevertheless, as discussed above, Sandemanians did not renounce money but recognised that they must engage in financial transactions, while at the same time adhering to the values and discipline of the church.

Although Faraday often offered his services gratis, he expected fair and modest payment from his employers. Thus, when negotiating with Trinity House in 1836, he recognised that, although the appointment could command a much higher salary, the sum of £200 was 'sufficient for necessary purposes'.[63] However, the remuneration he received from the Royal Institution was exceedingly low: 21 shillings per week in 1813, rising to £100 per annum in 1825, when he became Director of the Laboratory, and to £300 in 1853. Other related income was provided by the Fullerian endowment, which was worth £100 per annum after 1833, and the courses of lectures, most particularly the juvenile lectures, for which he was paid a further £100. His income was further boosted after 1835 by £300 per annum from his Civil List pension.[64] He rarely offered lectures outside the Institution, but in 1829 he was approached by Lt-Col. C.W. Pasley of Chatham and Col. Percy Drummond of the Royal Military Academy at Woolwich, who invited him to deliver an annual course on chemistry. Faraday capably negotiated the terms, expressing his unwillingness to accept payment lower than that offered by the Royal Institution. There is evidence to suggest that he felt undervalued at the Royal Institution and uncertain about his position following Davy's death. Although the sum agreed was £200 per annum, he stressed in negotiations with the Academy that good laboratory facilities were of crucial importance to him.[65] One other relevant piece of information is that, according to Tyndall, Faraday's early income from business sources was not insubstantial, but by the early 1830s (when he accepted the deacon's office) it rapidly dropped to zero.[66]

At the peak of his career Faraday's income would have amounted to at least £1,000 per annum. He had no children to support and his living accommodation at the Royal Institution was provided gratis, together with coal and candles. About his expenditure we know very little. He supported Sarah, his mother and one or two nieces. He also spent a great deal of money on science, particularly on apparatus, and in the late 1820s and early 1830s he paid for his assistant, Charles Anderson, out of his own pocket. He certainly bought books, went to the theatre and travelled away from London a few times each year. Yet he lived simply and unostentatiously. In a print of his tidy living room the only hint of affluence was a richly embroidered tablecloth. Faraday's income was, nevertheless, more than sufficient to cover the relatively modest needs and the few sober pleasures that he and Sarah enjoyed. In terms of income he must be counted among the financially secure early Victorians and, as Frank Barnard later pointed out, he was 'for some time the wealthiest man' among the London Sandemanians.[67] How great his wealth was is difficult to estimate but after his death his estate was valued at about £6,000. This was certainly not a princely sum, but it does indicate that Faraday had accumulated moderate wealth. There is clearly a problem here that I have not been able to resolve satisfactorily: how to reconcile his income of several hundred pounds a year, low expenses and his membership of a sect that denigrated avarice and pre .ched against laying up treasures on earth.

The problem might be easily resolved if Faraday's account books could be found. He donated significant sums to the deserving, upright poor in the Sandemanian community and to the charitable work carried out by the Sandemanians in the wider community. Moreover, he would have contributed each week to the upkeep of the meeting house. The size of his weekly donations is impossible to determine, but they may have represented a considerable proportion of his income. How large was his surplus income after deducting domestic essentials, science related expenses and contributions to the Sandemanians? While this is impossible to determine with any exactitude, his letter of resignation from the Athenaeum in 1851 casts some light on the subject. He could not, he claimed, continue his membership owing to 'the diminution of income and therefore necessarily the diminution of pleasures depending upon it'.[68] If he could not afford the fee of 6 guineas, his surplus income may have been quite small, and then the problem of laying up treasures would not have arisen. However, since his estate was worth £6,000 at his death, his response to the Athenaeum may not have expressed his real reason for resigning. Whatever his disposable income, the sums he earned were small compared with what he, as a leading scientific lecturer and researcher, might have obtained. He could have used his position at the Royal Institution and his scientific reputation to amass a considerable fortune, as did many other scientists, such as William Thomson (Lord Kelvin), who reaped the commercial benefits of their inventions.[69] That he did not tells us a great deal about Faraday's views about wealth.

A letter of 1822 provides further insight into Faraday's attitude towards money. Writing to Sarah, he recounted his meeting with a visitor to the Royal Institution who

> insisted on my accepting two ten-pound bank-notes for the information he professed to have obtained from me. . . . Is not this handsome? The money, as you know, could not have been at any time more acceptable; and I cannot see any reason, my dear love, why you and I should not regard it as another proof, among many, that our trust should without a moment's reserve be freely reposed on Him who provideth all things for his people. Have we not many times been reproached, by such mercies as these, for our caring after food and raiment and the things of this world?[70]

The moral is clear. In time of need God will provide, but we should not actively pursue riches in this world. When others are in need, we should dispense charity freely, but only to those who deserve and not to those who beg for alms. As part of their Christian duty Sandemanians extended charity to the poor without, one collecting box being reserved for that purpose. Faraday would have contributed to that collection and also to other charities. Thus he donated £5 to the London Female Dormitory, although insisting that in accord with his principles he wished to remain anonymous.[71] However, like other Sandemanians, he was horrified by the practice of money-lending, and looked upon beggars as morally degenerate.[72]

Apropros Berman's argument, it should now be clear that there exists no fundamental incompatibility between Faraday the Sandemanian and Faraday the chemical consultant. Thus the 'two Faradays' thesis falls to the ground. But the foregoing discussion raises another problem for Berman. Although we do not have access to the full details of Faraday's financial position, Berman's assertion that Faraday was imbued with the 'spirit of capitalism' needs to be qualified by the *caveat* that, although he did earn a moderately sized income, he, like other Sandemanians, did not view the aquisition of wealth as a primary goal and did not strive to procure a fortune or luxuries. In this sense Faraday was no capitalist.

5.5   FARADAY IN CONTEMPORARY SOCIETY

Having discussed his attitudes to such issues as politics and wealth, we are now in a position to enquire how Faraday interacted with contemporary society beyond the clearly defined boundary of the

Sandemanian church. The crucial point here is that he engaged in such interaction principally on his own terms, or, more precisely, on the strict moral terms expected of a disciple of Christ. From this perspective, he was not part of this world but had to inhabit it and interact with it. However, he created a space in which he could safely interact with the world and thus keep threatening situations at bay. Within this protected space he led a highly ordered and well-regulated life and one that allowed him only limited social contact, even with other members of the scientific community. Especially when he was absorbed in scientific research, he became 'an anchorite', as he once described himself.[73] Even when research was not so pressing, he consistently avoided social gatherings, such as dinner parties, except when directed to attend by a member of the royal family or by the Presidents of the Royal Society and the Royal Institution, whose invitations he likewise considered as a command.[74] He made his attitude clear on one occasion when, declining an invitation to stay with friends in the country, he stated that he and Sarah 'eschew[ed] all the ordinary temptations of society'.[75] As he stressed to Tyndall, he did not avoid such events from 'religious motives as some imagine'[76] Instead he stated that he considered social and fashionable frippery to be a waste of time, and his time could be more profitably devoted to his scientific work. However, behind this antipathy towards social events lurks a motive with religious connotations.

The deep concern with deciding what is important can be seen in the numerous references to time in his correspondence. For example, in 1832 he offered 'willingly [to] give help to the Government Gratis . . . but *my time is my only estate*'.[77] Lacking a monied or titled background, Faraday possessed no estate in this world, but as a Christian he felt that no God-given moment should be wasted. His time had to be strictly controlled. He pursued both his science and his religion with total dedication. However, as a prominent member of the scientific community and an employee of the Royal Institution, whose doors were open to its members and even non-members, he was frequently being propositioned. Even as early as 1828, long before he became a celebrity, he wrote to Richard Phillips: 'I began this letter directly after breakfast and it is now three o clk. All the rest of the time has been wasted in nearly useless conversation with callers[;] there is no end to them in this house[.] They leave me no time to write my letters'.[78] At that time he was also hard-pressed in making and chemically analysing different types of glass for a committee of the Royal Society, under Davy's stern direction.

Valuing his own time, Faraday censured timewasting in himself and others. One of his early lectures at the City Philosophical

Society was entitled 'Observations on the inertia of the mind', and in this he warned his audience against idleness as a great moral failing which prevented many people from achieving their goals. Although he stated that he wished to avoid entering into the theological issues raised by this topic, it is nevertheless clear that his diatribe against mental idleness was founded on religious considerations. For example, early in the lecture he affirmed that 'The goal before us is perfection'.[79] Faraday's appeal to perfection needs to be read in conjunction with the *Letters on Theron and Aspasio*, in which Robert Sandeman asserted that the only path to spiritual salvation is to '*Be perfect*; keep the commandments, and thou shalt live'.[80] This concern with moral perfection found expression in Faraday's discussion of the progress of the human mind. For Faraday idleness – whether in scientific or religious matters – was an imperfection which has to be rooted out by the will. We are only granted a limited amount of time on this earth and that time must be used most effectively. The Bible offers a dire warning against those who waste time: in Proverbs 6:6 the sluggard slips into want and poverty.

In the same lecture Faraday defined morality as 'the natural impression of the Deity within us'.[81] In his intercourse with others Faraday sought to practise this morality and to live according to biblical precepts. One such precept is found in Titus 2:6: 'young men, likewise, exhort to be sober minded'. Starting from this text, Robert Sandeman composed a discourse extolling the moral precepts, especially sobriety, which should be inculcated into the young Christian. But sobriety should not lead to 'that solemnity, demureness, fantastic grimace, and affected gravity' adopted by many religious people, since such forced dispositions invariably lead to hypocrisy and conceit. Instead the sobriety called for by the apostle meant that the Christian should 'live and walk as becomes the gospel' and in following God's law and the example of Christ. In particular, the Christian should not be swayed by covetousness, pride or ambition which lead to 'the lusts of the flesh, the lust of the eye, the pride of life, or love of wealth, and honour, and pleasure'.[82] In short, the Christian must not be seduced by worldly desires. Although this discourse was not published until 1857, it probably encapsulates the attitudes that young Faraday would have imbibed early in his life and thereafter practised.

The Sandemanians were not ascetics and, although Sandeman had emphasised sobriety, they did not forsake worldly enjoyments, excepting those proscribed by the Bible. Thus they saw no need to abstain from such social pleasures as the theatre or alcohol, provided that these were undertaken in moderation. In a pamphlet outlining their practices it is stated that since 'amusements, public or private, are not forbidden [by the Bible, excepting games of chance], any such may be engaged in that are not connected with circumstances really sinful'.[83] If, for example, one of the members drank excessively or committed adultery, (s)he would be put away – drunkedness and adultery were the reasons given for several exclusions – but most Sandemanians enjoyed their pleasures, although in a fairly restrained manner. They were certainly not the cold, joyless sect – 'the most dismal people on earth' – that some of their critics have portrayed.[84] Although not often boisterous, they surrounded themselves with joy and music. For example, the opening of a new meeting house at Newcastle in 1861 provided the occasion for 'a very merry party' which Faraday attended, and another source records the regular parties held by the younger members of the Edinburgh meeting house.[85]

Faraday was no exception to these generalisations. For relaxation he sang, visited the theatre, concert hall or the opera. Together with these secular enjoyments, Faraday participated, though somewhat grudgingly, in 'a singing school attached to our meeting house'.[86] Popular novels constituted another source of enjoyment. On her visit to his private rooms at the Royal Institution in 1850, Cornelia Crosse was surprised to find Faraday 'half reclining on a sofa – with a heap of circulating library novels round him'. She was later told by Sarah that he 'reads a great many novels, and it is very good for him to divert his mind'. To another visitor he explained that he particularly enjoyed 'stirring' novels, 'with plenty of life, plenty of action, and very little philosophy'.[87] Such pleasures were to be enjoyed. As Tyndall later claimed, 'There was no trace of asceticism in his nature. He preferred the meat and wine of life to its locusts and wild honey'.[88]

While Faraday enjoyed reading popular novels and theatregoing, these private activities did not threaten the barrier between him and the mundane realm. Yet, as will be discussed further below, the public domain in the Royal Institution, and indeed any form of contact with the non-Sandemanian public, were potentially threatening, and Faraday was usually able to deal with such situations on his own terms. To take one example, by the early 1830s he had become a celebrity and yet, as many contemporaries noted, he was not spoilt by his success. To understand how this relates to his religion we should note that Sandemanians reject the view that man can gain salvation through good works. As Robert Sandeman stridently stated, in the eyes of God there is 'no difference betwixt the best accomplished gentleman, and the most infamous scoundrel: – no difference betwixt the most virtuous lady and the vilest prostitute . . .'.[89] In holding that good deeds and worldly success are as naught before God, the Sandemanians were frequently charged with the antinomian heresy: the heresy that Christians are exempt from the obligations of moral law. But the charge has little substance, since Sandemanians do not seek to dispense with morality but rather they insist on following the demanding moral code of the Bible in preference to the infirm moral judgments of mankind. Acceptance with God depends on living a morally pure life as exemplified by Christ's thoughts and actions, and by placing one's faith in the Word. Thus Faraday could ignore worldly success as of no ultimate significance, and remain aloof from both the plaudits and the demands of fashionable society.

This discussion leads to an apparently paradoxical conclusion. Although Faraday considered himself to be 'an anchorite' and was a sectarian standing apart from the mainstream of Victorian society, he was a celebrity. He drew crowds, including the titled, rich and fashionable, to the Royal Institution. He was invited to Buckingham Palace and knew and corresponded with the aristocracy and the famous. To resolve this apparent paradox we need to realise that Faraday the Sandemanian bridged the divide in two ways. Firstly, as will be discussed further in section 6.4, science provided a non-denominational form of discourse which allowed him to communicate with non-Sandemanians. Secondly, as discussed in section 10.1, he possessed and projected a number of qualities that were generally recognised as highly admirable. Even a cursory examination of the contemporary literature reveals that Faraday was almost invariably portrayed as a man of integrity who was unaffected by fame and who was charming, congenial and helpful to others. Eulogies abound. For example, one of Faraday's auditors later remarked 'how rare it is to find so much talent, with so much humility, so much appreciation of all there is to charm in this world, without being *spoilt* by it!'[90] While we must recognise that Faraday's public persona masked other aspects of his character (cf. chapter 10), we should also appreciate that Faraday effectively projected positive personal qualities that were readily recognised both by his friends and by those who only saw him from afar.

Faraday's public persona was shaped by his religion. Indeed, many of the moral virtues which others saw in Faraday were considered essential Christian qualities by the Sandemanians. As one of the sect's main documents emphasises, the 'grand central duty of Christians and the great evidence of true Christianity is *brotherly love* for the truth's sake'.[91] Although a highly exclusive sect, the Sandemanians are required to treat all people with complete love, and to avoid ambition, pride and covetousness. Certainly Faraday could be proud and occasionally irritable, but he was also exceptionally gentle and kind. Biographers are liable to trade in hyperbole, and yet Bence Jones was not indulging in mere hagiography when he claimed that Faraday's 'second great quality was his kindness *(agape)*. It was born in him, and by his careful culture it grew up to be the rule of his life; kindness to every one, always – in thought, in word, in deed'.[92] As his passing reference to the Love Feast (*agape*) makes clear, Bence Jones considered Faraday's kindness intimately connected with his religion. In this instance, as in many others, the strong impress of the sect can be seen.

To the public Faraday was known principally through his lecture courses and Friday evening discourses at the Royal Institution. As many testified, he was an outstanding lecturer. His audience commented not only on his style of lecturing – such as the clarity of exposition – but also on his public persona, which appeared an essential part of his performance. For example, one lady drew attention to

> his great talent, great goodness, and the wonderful simplicity of his nature. All this was evident, directly he spoke, in his bright, clear, truthful voice. His lectures were '*mind addressing mind*,' and you felt he was full of sympathy with his audience; and with his fellow creatures, though so far above them![93]

This auditor, like many others, was immediately attracted by Faraday's charismatic personality and by the lack of barriers between the lecturer and his audience. We also find Faraday identifying with his audience; for example, in 1859–60 he opened the first of his juvenile lectures by commenting that 'unfitted as it may seem for an elderly infirm man to do so, I will return to second childhood and become, as it were, young again amongst the young'. Elsewhere he emphasised the importance of maintaining distortion-free communication between science teachers and their students, and the danger of the teacher taking refuge in incomprehensible terminology or mathematical symbols.[94] In teaching science Faraday possessed the remarkable ability of communing with his audience and making them feel that no barriers stood between the lecturer and themselves – '*mind addressing mind*'.

In respect to the lack of barriers, the situation is strikingly similar to that at the Sandemanian meeting house, where all members were united in Christian profession and were equal. Thus, for example, all members were encouraged to read from Scripture and contribute their thoughts at the services and all major decisions were made after extensive and democratic discussion. However there are also important differences between the two cases. At the Royal Institution he was in a position of authority (although he was careful not to misuse the power that went with it) and his communication of science took place in only one direction; in respect to scientific knowledge he was 'so far above them'. But perhaps the most important dissimilarity between the two locations concerns his relation to others in the room. At the meeting house he was surrounded by his spiritual brothers and sisters. In their profession of faith there *could be* no barriers. However, standing behind the lecture table at the Royal Institution he only *appeared* to his audience to be in complete control of the situation and in direct, intimate communication with them.

This appearance is deceptive, and I want to explore the situation more fully, since it contains a deep, internal tension, arising from Faraday's public activities at the Royal Institution. I shall claim that in order to operate effectively in the public arena Faraday had to deploy a line of demarcation and adopt rules of conduct which enabled him to cope with potentially threatening public situations. That it was a threatening situation is attested by Sarah in a letter of 1864 in which she admitted that earlier in his career 'the thoughts of giving a Friday evening' discourse at the Royal Institution 'quite affected his health'.[95] Lecturing was a source of considerable anxiety, although his audience perceived him as a highly polished performer.

Before arguing this thesis in more detail, I shall propose it in metaphorical form, drawing on the interesting work of Sophie Forgan, who has argued that Faraday's activities at the Royal Institution should be understood in terms of two distinct spatial regions within the building. One space was *private*. As Superintendent of the House Faraday lived in the upstairs rooms, from which the public was barred, and he allowed entry to a few close friends – his family, co-religionists, and select members of the scientific community. Another private space was constituted by his basement laboratory, where he and his assistant Charles Anderson performed their honest toil. By contrast the middle levels of the building were *public*. The Library was open to members from 10am to 10pm; crowds flocked to public performances in the lecture theatre; and all sorts of callers filled the corridors for purposes of business or pleasure, for reasons social or personal. As Forgan has stressed, Faraday had to exert himself 'managing and controlling this . . . division between the two domains'.[96]

Different rules applied across this divide. In the downstairs laboratory Faraday pursued his scientific research and in the private rooms upstairs he leaned on the sofa reading popular novels, playing games with his young nieces or chatting to Tyndall and other friends. There were no social divisions in this private space. It allowed a simple, ordered existence with the door firmly closed on the world outside. Now let us watch Faraday walk down the connecting stairs into the public domain. As his foot reaches the final step, a mask slips over his face. He is in that other world, one in which he, a sectary, is not at home. Yet he can handle himself admirably in company – almost too admirably. The ladies and gentlemen seated stiffly in the lecture theatre comment on the excitement of his lecture, his faultless delivery and his impeccable manners. What most of them fail to notice is that he is not one of them and that it is more than the demonstration table that divides him from them. Although Faraday was not 'at home' in the public space at the Royal Institution, he knew how to deport himself and what he could, and could not, do. Not only could he not lounge about in the lecture theatre reading popular novels or carry out his research before the throng, he could not join in many activities that were considered normal in polite Victorian society.

These two types of space are metaphors for Faraday's realms of experience. The private rooms at the Royal Institution stand for his private sectarian existence (and also his scientific research). This is the kingdom of heaven, Christian brotherhood and the domain of peace and order. By contrast, the public rooms at the Royal Institution represent the mundane realm of social and political turmoil, of confusion, avarice, pride and ambition. Faraday the sectary needed to distance himself from that world and its values; so as he moves down the stairs from the private to the public domain, he has to manage and control that space and its periphery.

In the public domain at the Royal Institution Faraday avoided those aspects of the mundane world which were inimical to his Sandemanianism. He avoided involvement in politics, religious controversy, commerce and the fripperies of fashionable society. Instead science was the common coin. He appeared in public as the expert who possesses esoteric knowledge of nature's laws and moreover as an expert who is willing to impart that knowledge freely to any attentive listener. Moreover, his lectures were prepared in minute detail so that he was in control of himself and of the situation. In the public's eyes science was not sectarian; it was not relevant to that situation that Faraday was a Sandemanian. Instead, it was through his expertise in science that he made contact with his audience. On this one subject no barriers existed, whereas almost all other forms of communication were closed off.

Faraday was admirably suited to the Royal Institution, and the Royal Institution admirably suited Faraday; indeed there was probably no other place in British science where Faraday could have flourished. In the same building he could play out both his private and his public roles. He could preserve the periphery of his private, sectarian existence. He could appear in public on his own terms and not have to stray into those areas which severely challenged his Christian values. He was provided with an excellent research laboratory and with the freedom to pursue his research. He received an adequate income from performing his duties. For their part the Managers of the Royal Institution found in Faraday a charismatic, admired lecturer who drew crowds of respectable Londoners to his lectures. Moreover, he did not demand a large fee and the Managers initially paid him a small, barely adequate wage. The Institution gained an international reputation for research. The status of the Institution increased, as did the contents of its coffers. The Managers found in Faraday the perfect *loyal servant* who made minimal demands and positively sought to benefit the Institution.

Sophie Forgan has written that the 'keynote of Faraday's attitude towards the [Royal] Institution throughout his life was *loyalty*', a term I have already emphasised on several occasions.[97] When, for example, he was invited in 1827 to join the staff of the newly formed London University, he declined, because 'I think it is a matter of duty and gratitude on my part to do all I can for the good of the Royal Institution'. Moreover, he added, 'I possess the kind feelings and good-will of its authorities and members, and all the privileges it can grant or I require.'[98] Even in 1829 when he considered taking up a better-paid position at the Royal Military Academy and simultaneously freeing himself from the onerous experiments on glass under the absentee Davy's direction, he still insisted that out of gratitude he had to continue for some time lecturing at the Royal Institution.[99] Like the archetypal Jeeves, he put the welfare of his master (the Institution) before his own. He never criticised the Managers in public. Recognising that the Institution was frequently beset by financial difficulties, he did not make excessive demands and was satisfied with a salary far below what an eminent science lecturer could have expected. Apart from the laboratory placed at his disposal, he did not use the Institution to promote his career but saw his success as a lecturer as benefiting the Institution. Nor did he encroach on the Institution's politics, which would have brought him into conflict both with the Managers and with his own principles.

In short Faraday and the Royal Institution were very well suited to one another. The Managers accepted the clearly circumscribed role which Faraday was prepared to play and they respected his integrity. Except when he was offered the Institution's Presidency (see section 10.3) this arrangement worked very well for both parties for a period of more than half a century.

6 *Scientific Institutions*

6.1 THE REPUBLIC OF SCIENCE

L. Pearce Williams has suggested that Faraday's attitude towards the politics of science should be understood as an extension of his attitude towards politics in general.[1] This observation underpins the present chapter, which concerns Faraday's attitudes towards science and the part he played in such scientific organisations as the Royal Institution, the British Association and the Royal Society. First, however, we shall examine Faraday's conception of the role of the scientist, and here it will be argued that he considered that the scientist should hold values very similar to those advocated by the Sandemanians in prescribing Christian virtue. Thus, I want to delineate several specific ways in which Faraday the scientist reflected the values of Faraday the Sandemanian.

Faraday's choice of career offers a suitable point of departure for this discussion. When he was asked by Humphry Davy's biographer, John Paris, to describe his earliest connection with Davy, he related an interview in 1813 at which he had explained his reasons for seeking a career in science. 'My desire', wrote Faraday, was 'to escape from trade, which I thought vicious and selfish'. He then proceeded to contrast these negative associations with his positive assessment of science, 'which I imagined made its pursuers amiable and liberal'.[2] (The two activities were also unfavourably compared in a letter, dating from shortly after Faraday's return from his continental tour with Davy: 'Trade which I hated' was contrasted with 'science which I loved'.)[3] Doubtless thinking the young man somewhat naive, Davy 'smiled at my notion of the superior moral feelings of philosophic [i.e. scientific] men, and said he would leave me to the experience of a few years to set me right on that matter'.[4] Experience did not, however, diminish Faraday's high ideals; indeed, his close relationship with Davy could only have confirmed him in his determination to maintain scientific purity and not to be seduced by those social aspirations that had seized his (one-time) mentor.

It is significant that Faraday never applied the word trade to science, but in several letters written mostly while under Davy's tutelage he referred to their labours in the laboratory as a business. Thus, writing to Benjamin Abbott in 1816, he explained that their mutual correspondence offered him the best 'relaxation from business – business I say, and I believe it is the first time for many years that I have applied it to my own occupations'.[5] While trade was associated with lowly activities and lucre, business instead referred to the diligent, although sometimes dreary, laboratory work that Faraday performed. Moreover he contrasted business with relaxation. It may also be significant that on one occasion he referred approvingly to his nephew's occupation, as a gas fitter, as a business.[6] A further term that became more prominent in Faraday's later correspondence is profession. For Faraday professions were superior to trades, but they too implied a financial, and thus ultimately flawed, relationship with other people. Thus in a letter to Admiral Sir Thomas Byam Martin, Faraday declared 'There is no charge for I am not Professional'. In similar vein he wrote to a friend, 'I am not a Professional and do not undertake [chemical] analyses for any one – not even for the Government' – in order to earn a fee. He thereby dissociated himself from professional chemists who would earn their livings by undertaking chemical analyses.[7] Insufficient evidence remains to enable us to analyse Faraday's consultancy work in detail, but it appears that much of his work in this area was performed as part of his duties at the Royal Institution and, particularly after 1830, he usually refused a fee or chanelled it into the coffers of his employer, the Royal Institution.

What Faraday assiduously avoided was any situation in which he might appear to be soliciting consultancy work for personal gain. If a governmental body or a member of the Royal Institution approached him with a request to undertake such work, he would try to oblige out of a sense of loyalty. But these demands pressed heavily on the time needed for his research, and to scientific research he accorded far higher priority. True science, like true religion, was not to be tainted by money. Although, as stressed above, there is no necessary conflict between Sandemanianism and trade, Faraday clearly felt that any financial arrangement in which he would be exposed to base feelings was contrary to this ideal. Science was so inextricably linked with positive moral values that it beckoned him as the highest calling: a vocation, not a trade or a profession.

Not only did Faraday adopt the Baconian distinction between 'luciferous' and 'lucriferous' experiments, he also deployed the classic contrast between science and politics. Politics was, in his opinion, marked by moral depravity, and any scientist who entered that arena was likely to become tainted. As he wrote to Charles Babbage, who had actively and vociferously participated in the argument over governmental support for science with the publication of his *The exposition of 1851* (1851),

I grieve that your powerful mind ever had cause to turn itself in such a direction and away as it were from its high vocation & fitting occupation [i.e. science.] Still I know that we cannot avoid the checks & jars of a naughty world and that at times we are driven from our most direct courses by very unworthy objects under our feet.[8]

The moral status of the scientist was thus lowered by contact with politics, even the politics of science. The contrast was likewise apparent in Faraday's portrayal of the violence of the revolutionary mob as opposed to the peaceful scientist meditating on the order and harmony of nature. At work here was a version of the fundamental distinction: politics exemplifies man's fallen state – 'I think but poorly of human nature'[9] – and it leads to depravity whereas science (like true religion) leads to morality. In a strong sense, then, the pursuit of science is an expression of man's divine nature.

As will be discussed in section 10.4, Faraday sought to impose order on all aspects of his life. Yet the order he imposed on his reading, writing and laboratory work was not merely an aspect of his personality but also an expression of his belief in the ultimate order of the God-made universe and in God's general providence. The scientist, likewise, expresses the divine side of his/her nature when order is imposed over confusion, truth over error. Science was, then, mankind's discovery of this order, which was exemplified by the physical laws God had imposed on matter. In the case of Faraday the dual significance of law is very apparent, for he lived by God's (moral) law and discovered God's (physical) laws. Science could satisfy the soul and be a profound religious experience.

In contrast to the order evident in nature and in true Christianity, confusion was seen to reign in the mundane realm. Faraday experienced considerable difficulty in coping with all forms of confusion. He retreated from the world of politics, with its clash of opinion among different factions, which he viewed as exemplifying hopeless confusion from which nothing constructive could emerge. A similar scene was witnessed in religion, and Sandemanians were horrified by the way in which self-professed Christians create 'worldly factions', aggrandise their leaders and engage in power politics, even bloodshed.[10] The image of the confusion of tongues was almost everywhere apparent. The exceptions were true Christianity and science, in both of which truth and order could be discovered.

Faraday held that truth is singular and that we discover truth by a convergent process; that 'we are all advancing more or less towards truth though it may be by different roads'.[11] Scientific progress does not, however, occur when there is confusion, either in terms of confused ideas or if there is discord among scientists. (The latter interpretation gives a fresh, and I believe correct, gloss to Faraday's use of Bacon's famous aphorism, 'truth can more easily emerge from error than from confusion'.[12]) Faraday's rejection of discord as improper to science appears particularly in his comments on scientific controversy. If two scientists obtain different results, then their misunderstandings could, he believed, be ironed out easily and thus controversy could be avoided. Although he found controversy repugnant and sought to avoid it, Faraday recognised that any form of eminence, even scientific eminence, leads to 'the cavils and rude encounters of envious men'.[13] He refused to be drawn into controversies, except when he felt that his honour had been severely compromised. Writing to Carlo Matteucci, who was engaged in a dispute with Emil du Bois-Reymond, Faraday explained that 'it is only in the cases where moral turpitude has been implied, that I have felt called on to enter upon the subject in reply'.[14] Such a case had occurred early in Faraday's career when he had been charged with plagiarising the discovery of electromagnetic rotations from William Hyde Wollaston. Shocked by the charge, Faraday had fought to demonstrate his honesty and to preserve his honour.

In the same letter to Matteucci Faraday sought to explain why science was littered with controversies.

> These polemics of the scientific world are very unfortunate things; they form the great stain to which the beautiful edifice of scientific truth is subject. *Are they inevitable?* They surely cannot belong to science itself, but to something in our fallen natures. How earnestly I wish, in all such cases, that the two champions were friends.[15]

Here Faraday drew a sharp distinction between science, as the disinterested search for truth about nature, and the worldly reign of covetousness, pride and ambition. Science, true science (like true religion) puts us in touch with the kingdom of heaven, whereas factional science (like pharisaical religion) is the result of our fallen state. As he explained in another letter, controversies are 'exceedingly unwholesome in a moral point of view, for they generally lead each one in private to justify themselves, and so foster a pharasaical [sic] condition of mind, and a growing tendency to judge others rather than ourselves'.[16] In politics, where there is no truth but where prejudice and opinion abound, controversy and its attendant evils must reign. However in both science and religion, where the most important truths are to be found, he maintained that there is no legitimate place for controversy and should it arise then it is a sign that brotherhood has given way to degraded feelings between scientists. Tyndall related several occasions when Faraday disagreed with him on scientific matters but still extended the hand of friendship. On one such occasion Faraday is said to have stated that 'you [Tyndall] differ, not as a partisan, but because your conviction impels you' while in respect to a later disagreement Tyndall claimed that Faraday's 'soul was above all littleness and proof to all egotism'.[17]

For Faraday another striking similarity between true Christianity and idealised science is that each group of practitioners constitutes a brotherhood. Sandemanians consider themselves bound to practise brotherly love: 'A new commandment I give unto you, That ye shall love one another' (John 13:34). The church is the true spiritual brotherhood and unanimity must be maintained for the church to function. As Faraday explained to Tyndall, who was a frequent controversialist, the situation in science is very similar, since scientists should work together adding facts and inferences to the grand total edifice. 'When *science is a republic*, then it gains; and though I am no republican in other matters [being a loyal subject of the Queen] I am in that'.[18] The political disanalogy is important, since it underlines once again the dissimilarity between science and Faraday's conception of politics, which exemplified so much in the mundane realm.

A brief discussion of one of the Tyndall's many controversies offers further insight into Faraday's views. Writing to the German physicist Julius Plücker in 1856, Faraday explained his belief that scientists holding opposing views can, and should, be on good personal terms; as he noted, although he and Tyndall had many scientific disagreements, they remained the closest of friends. Friendship provided the essential basis for the resolution of such disagreements and the universal recognition of truth. Faraday concluded the letter by expressing his pleasure at seeing his work now carried on by others, whom he described as '*that band of brothers*'.[19] The phrase strongly suggests that Faraday considered that scientists, like Sandemanians, form a fraternity upholding the values of truth and morality. While the sect's discipline was aimed at preserving unity within the church, there was no such mechanism in science, which depended on the self-discipline of members of the scientific community. Yet Faraday considered that self-discipline was essential for the correct functioning of science.

I have argued that Faraday was attracted to a career in science because he considered it opposed to the 'spirit of capitalism', politics and social fragmentation. Science provided Faraday not only with an occupation but also with a social network that bore close similarity to that of the Sandemanian community. In particular it provided him with a number of friends who were bound together by common values, a common purpose and friendship. These friendships were a source of great pleasure and he greeted his closest scientific comrades with considerable warmth. His friendships with such men as de la Rive, Dumas, Schoenbein, Tyndall and Whewell were almost as intimate as those in the Sandemanian brotherhood. At its best the brotherhood among scientists was like the Sandemanian church – a united congregation bound together by mutual recognition of higher, God-given truths. Writing to Auguste de la Rive in 1855, he regretted that his scientific powers were failing, but added, 'when philosophy has faded away, the friend remains'. This emphasis on friendship was an echo of the claim he made four decades earlier when he told Benjamin Abbott that 'True friendship I consider one of the sublimest feelings'.[20] Significantly the scientific community provided Faraday with his sublimest friendships outside the Sandemanian community.

Faraday not only held that the progress of science was most rapid when there were no disruptions in society or in the scientific community, he also denied that science develops through conflict or revolution. Instead he believed that the history of science shows a slow and continuous progress towards truth. Although there may initially be some differences of opinion, he told Oersted, 'Time will gradually sift & shape them & I believe that we have little idea at present of the importance they may have in 10 or 20 years hence'.[21] Such differences will be ironed out; minor discoveries may prove of major importance; the circle of truth will be expanded. Just as in the social sphere, he opted for continuity, not revolution, in his account of the historical development of science. This connection between scientific and social change also emerged clearly at the conclusion of a series of lectures in 1837, when he argued that 'Progress of knowledge [does] not [occur] in floods – dangerous as floods of water, but [by] a calm and dignified process'.[22] Floods have, of course, strong biblical connotations and they signified to Faraday the dissolution of this fallen world, while in political terms they symbolised revolutions. By contrast, science was the peaceful, ordered study of the works of God and the repository of higher values. Society is always changing; some of its members experience decline, while others improve their mental and spiritual powers. Unlike politicians, scientists are able to make progress, and progress is, as he had told his audience at the City Philosophical Society some 20 years earlier, the natural state of man.[23]

In section 5.3 I developed the opposition between religion (the true brand of Christianity advocated by the Sandemanians) and the mundane realm, as exemplified in politics, the 'spirit of capitalism', and such passions as envy, pride, covetousness and ambition. It should now be clear on which side of this dichotomy we should place science. Although Faraday recognised that not all scientists are free from pride, ambition or other vices, he prescribed, and largely practised, a form of science which was intimately associated with religious values and opposed to the values of this world. 'Things run irregularly in the great world', he wrote to his mother from Geneva in July 1814 at the close of the Napoleonic wars.[24] Faraday's conception of science was the very antithesis of what he perceived occurring in that 'great world'. He created in the science he practised an oasis of regularity and order, just as religion offered him security from doubt and a system of absolute, unshakeable values. Both activities were also ideally pursued by communities dedicated to the highest moral ideals and not motivated by self-interest. Thus there is a strong sense in which Faraday's career as a scientist and his conception of science reflected the Sandemanian social philosophy.

While science was thus closely allied with Sandemanian values, it should also be noted that Faraday's conception of science was vested with psychological significance. Since Faraday appears to have felt threatened by chaos, revolution and controversy, he retreated to the safety of science and also erected a conception of science in which order, truth and friendship reigned supreme. With this ideal Faraday could pursue his science and keep at bay all uncongenial aspects of contemporary scientific practice.

6.2 SCIENTIFIC SOCIETIES

The period in which Faraday worked saw important extensions and changes to the institutional basis of British science. At the turn of the century London science was dominated by the Royal Society under the firm but uninnovative Presidency of Sir Joseph Banks. Meetings were usually lifeless and not calculated to stir the imagination or to encourage the rapid advance of natural knowledge. Banks was acutely conscious of the Society's role as a club that would bring together the more gentlemanly of the scientific community and potential patrons. The Royal Society Club prospered. If this role appealed to many Fellows, there was, during the opening decades of the century, an increasing concern that the Society failed in other, and more pressing, ways. One criticism arose from its failure to cater adequately for specialist interest groups, which grew impatient with its broad scope. The first major challenge to Banks and the primacy of the Royal Society came in 1807 with the formation of the Geological Society of London, which noisily proclaimed its independence and its intention to benefit geology actively by amassing the facts of British geology. Although the Geological Society of London later spawned a dining club, it had by the 1820s become an effective learned society, with its *Transactions*, a formal organisational structure and regular research-orientated meetings.[25]

Other specialist societies soon proliferated, most notably the Astronomical (1820), Zoological (1826) and Chemical (1841). At the turn of the century there were but four generalist and two specialist societies in the metropolis, as against five generalist and fifty specialist societies half a century later.[26] Not only did scientific societies mushroom in London but also in provincial settings. Major cities and many minor towns spawned their Literary and Philosophical Societies and Naturalist Clubs. In 1831 the British Association for the Advancement of Science was founded to encompass both the broad audience for science and its many subject-orientated interest groups. The growth of scientific societies not only indicates the increasing interest in science and the changing fortunes of both specialists and generalists but also provides a new social map of science. In London, in particular, one could by the 1830s fill every weekday evening during the season with a lecture, a dinner or some other meeting of a scientific society. The peripatetic British Association could occupy several days each summer. Scientists, particularly of the more gentlemanly variety, moved in a small but gradually increasing circle of acquaintances, and they discussed their experiments, fieldwork, the new theories in geology or optics and, of course, the politics of organised science.

The vibrant scientific life of the metropolis was constituted not only by the practitioners of science but also by the large lay audience that filled the lecture theatres of the Royal Institution, the Russell Institution, the Surrey Institution, the London Institution, the Literary and Philosophical societies and Mechanics Institutes.[27] To service this large, diverse and lucrative audience, a small band of lecturers offered short courses on any popular branch of science, often illustrated by experiments and exhibits.

Faraday lived and worked in this environment. The Royal Institution was the fashionable centre for London science and Faraday was in close contact with many of his scientific contemporaries in London, the provinces and abroad. Moreover, his relationship with most contemporary scientists seems to substantiate his ideal – that the scientific community should be a fraternity. However, a series of problems arise when we compare this ideal with Faraday's rather slight involvement in organised science. Why did he play a very limited role in scientific societies and other institutions? Why did such an ardent advocate of the republic of science rarely attend the Royal Society or the British Association? Why did Faraday refuse the Presidency of the Royal Society? In this and the following section I will try to answer these and similar questions by examining his involvement in organised science, particularly his role in the British Association for the Advancement of Science and the Royal Society of London.

Faraday was both part of the scientific community and also tangential to it. Evidence for this claim is provided by his irregular attendance at scientific meetings. In the 36 years between the foundation of the British Association in 1831 and 1866, Faraday was present at only fifteen of its annual meetings, his most regular attendance being in the 1840s and the late 1850s.[28] He even absented himself from the first meeting, held at York in 1831, claiming that he was busy with teaching commitments at the Royal Military Academy, 'pressure of business, and again I am not a social man'.[29] Moreover, particularly during his later years, Faraday also experienced physical and psychological difficulties in coping with large, crowded meetings. For example, he claimed that at the 1844 York meeting (while he was recovering from his protracted illness) his 'mind & memory became quite bewildered amongst the men & things & I sadly mistook one for another', and he was pleased to leave in the middle of the 1859 Aberdeen meeting owing to 'excitement leading to weariness'.[30] When he did attend, he often stayed for only part of the meeting, excusing himself by claiming ill-health, pressure of time or the need to return home in order to attend the Sabbath service at the Sandemanian meeting house. It was this incursion of the British Association on his orderly religious practices that led him to complain to Richard Phillips that the 1848 Swansea meeting was scheduled to run over 2 weeks and that 'I never travel on the Sabbath'.[31] On at least one occasion he felt impelled to attend a meeting out of a sense of duty: in August 1842 (which falls in the middle of his extended illness) he was appointed to represent the Society of Sciences at Modena at the Association's General Meeting, but 'having done that I left Manchester early in the morning in which the great body met and so escaped [to] London'.[32] One can detect a deep sigh of relief.

Although he never attended the main organising committee of the British Association, Faraday sometimes served on the sub-committees overseeing Sections A (mathematical and physical science) and B (chemistry and mineralogy), held the post of Vice-President of Section B on several occasions and was thrice its President. These were not onerous tasks and the committees were simply constituted by the participating scientists for the duration of the annual meeting. Faraday particularly enjoyed the 1837 Liverpool meeting, the only meeting at which he played a formal role. In a letter to Sarah he explained that he had been elected President of Section B and that he had thus become a 'most responsible' person. Almost apologetically he sought to downplay the honour that had been bestowed on him and he informed Sarah that had he and John Daniell 'refused altogether to join in the common feeling, we should have looked like churls indeed'.[33]

One London-based society with which Faraday was associated over a long period was the Society of Arts, whose main function was to assist art and technology through the award of prizes. Faraday had joined the Society in 1818, and although he did not serve on its main organising committee, he was a chairman of the committee on chemistry for a number of years during the 1820s and 1830s. This position required him to assess the chemical innovations submitted to the Society and to judge which deserved a prize or premium. The idea of justly rewarding merit would have appealed to Faraday, who appears to have enjoyed his long association with the Society and donated many gifts to its library. For its part the Society held Faraday in high esteem and, in 1866, he was awarded the Albert Gold Medal for his discoveries in electricity, magnetism and chemistry, which had been widely and successfully applied in industry.[34]

Faraday likewise played a role, though a much smaller one, in several specialist metropolitan societies. For example, he joined the Geological Society in 1824 and served on its Council from 1828 to 1830.[35] While geology was never one of his main interests, chemistry constituted the very centre of his research. As one of the foremost chemists of the day, we would expect him to have been among the main movers of the Chemical Society, which held its first formal meeting at the end of March 1841. Surprisingly his name does not appear on the early membership lists. His absence may in part be due to his extended illness at that time, and he was forced to spend the summer of 1841 recuperating in Switzerland. He did, however, join the Society in January 1842 and was later honoured by the positions of Vice-President and member of Council. However, according to the Society's historians, 'it does not appear from the minutes that he attended meetings in either capacity'. In 1851 he was invited to become the Society's President, 'but he declined the honour for reasons of health'.[36] Faraday did not honour the Society by publishing in its journals, but remained throughout his life loyal to the journals of the Royal Institution and Royal Society, and also to the *Philosophical Magazine* conducted by Richard Taylor. Despite

Faraday's negligible involvement in the Chemical Society, that body posthumously honoured him with the Faraday Lectures, the first of which was delivered in 1869 by his close friend Dumas, who dwelt at length on Faraday's impressive discoveries and personal qualities.

Faraday's connections with the Royal Society were more extensive, but also more fraught. As will be discussed in section 10.2, his initial attempt to enter the Society was darkened by the charge that he had plagiarised William Hyde Wollaston's work on electromagnetic rotations and that he had not adequately acknowledged Davy's contributions to that area. Although these charges had initially been laid in 1821, the issue smouldered for a further 2 years and resulted in the souring of relations between Faraday and his patron Humphry Davy. In 1823 Davy opposed his election to the Royal Society and demanded that his certificate be withdrawn.[37] However, he was elected to the Society on 8 January 1824 with his certificate supported by twenty-nine signatories – an extraordinarily large number, for most applicants were supported by about ten names – including Wollaston, Davies Gilbert and John Herschel. Yet Davy's attack on his honour continued to rankle for the remainder of his life; as he remarked a decade and a half later, his FRS was the only title he 'sought and paid for'.[38]

The incident permanently coloured his relationship with the Society, for although he belonged to it for more than four decades, he only participated actively in its affairs during the late 1820s and early 1830s. In November 1828, a year after Davy resigned the Presidency, he was elected to the Society's Council, but served for only 5 years: 1828–31 and 1833–5.[39] During that period he attended Council meetings fairly regularly and contributed to its proceedings. Yet his participation seems to have been confined to two main areas. One concerned scientific work initiated by Council. Thus at his first Council meeting he was appointed to a committee charged with producing glass for optical purposes; much of this investigation fell on his shoulders and it was not until July 1831 that he submitted the report. Again, in March 1834, he presented to Council a report on the specimens of platina and palladium donated by Wollaston. At the end of that year he was elected to the Excise Committee concerned with constructing instruments and tables to determine the strength of alcohol.

While Faraday served the Society loyally in pursuing such scientific investigations, his other main activity on Council was to press for the more just distribution of honours (cf. section 5.3). Thus in December 1830 he was appointed to a committee examining the methods of making awards. Possibly as a consequence of this committee's deliberations, Faraday moved a resolution in January

1835 which was intended to 'effect a proportionate and impartial distribution of the honour of the Royal Medals' by ensuring their rotation between six areas of science and preventing any particular paper from being accepted for the competition in more than one area. At the same meeting he gave notice of his intention to introduce a resolution which would allow only papers communicated to the Royal Society to be eligible for the award.[40] This proposal was in keeping with his feelings of loyalty towards the Society and his insistence on submitting his entire series of electrical researches for publication in the *Philosophical Transactions*. While he was active in trying to improve the system of awards, he also proposed a number of individuals for medals or membership. In 1834 he announced his intention to propose Macedoine Melloni for the Rumford Medal and William Snow Harris for the Copley Medal. Likewise he supported the applications for Fellowship of a number of scientifically able men; for example, in the early 1850s he signed the membership certificates of such young physical scientists as James Prescott Joule, William Thomson (later Lord Kelvin) and John Tyndall.[41]

Significantly Faraday's first period on Council coincided with the concerted efforts of a vocal lobby to reform the Royal Society. As Roy MacLeod has argued, the question at issue was not whether science should be amateur or professional but whether science should be bound by aristocratic patronage or should be directed by the scientists themselves. At stake, then, was the question of which political model was suitable for the organisation of British science.[42]

This question found its most incisive articulation in Davies Gilbert's statement that the 'Great Contest . . . is the Conflict of Aristocratic and Democratic Power'. Gilbert was in no doubt where he stood since he fervently 'wished the Royal Society rescued from the latter'.[43] With uprisings at home and revolutions abroad, 1830 was a highly charged year in which to announce his intention to resign the Presidency. Not surprisingly he began to manoeuvre for the Duke of Sussex to become his successor. When this plot was discovered, a number of Fellows demanded that Gilbert make public his correspondence with the Duke. Although he related these events at the Society's meeting on 11 November 1830, the assembled Fellows proceeded to pass two rather mild resolutions. The first required that future office-holders were *au fait* with science, while the second enabled the Fellows to exercise some power in Council elections by their being presented with a list of fifty contenders, comprising both existing Council members and other Fellows (and not merely accepting the slate that Gilbert and his supporters pressed on the membership). The latter motion was proposed by

Herschel and seconded by Faraday. As an expression of their dissatisfaction with Gilbert's underhand way of proceeding, the reformers also pressed for a presidential candidate who would place the interests of science above those of patronage. Their choice was the accomplished astronomer and natural philosopher John Herschel. However Herschel was not keen to stand, since he, like Faraday, was primarily concerned with research and did not relish the prospect of reducing his research effort in favour of the administration and politics of science.[44]

Faraday adopted an unusually high profile in supporting Herschel's candidature. Although Herschel proved unsuccessful when the votes were counted on 30 November 1830, in a letter written to Carlo Matteucci in 1843 Faraday indicated his reasons for that keen but atypical participation in the reformers' lobby:

> I think you are aware that I have not attended at the Royal Society, either meetings or council, for some years. Ill health is one reason, and another that I do not like the present constitution of it, and want to restrict it to scientific men. As these my opinions are not acceptable, I have withdrawn from any management in it (still sending scientific communications if I discover anything that I think worthy). This of course deprives me of power there.[45]

It is clear from this letter that Faraday objected so strongly to the Society's domination by 'Aristocratic Opinion' and patronage that he almost severed his connection with it, although continuing to direct many of his papers to the *Philosophical Transactions*. He held that the Royal Society should be for scientists, for the benefit of their republic and unsullied by external interests. In adopting this position Faraday certainly cannot be understood in terms of incipient 'professionalisation'; indeed, as discussed above, he was thoroughly opposed to science being a profession. If some of Herschel's supporters conceived of science moving closer to commercial and industrial interests and away from the aristocracy, this was not Faraday's position – he wanted science freed from all forms of corruption. Indeed, as in the Sandemanian community, he wanted the scientific church bound together by the bonds of truth and brotherhood.

This view is strikingly similar to the Sandemanians' insistence that religious fellowship can only occur among those who have made their confession of faith and that communion must be refused to members of other churches.[46] While such exclusiveness might severely limit membership, the Sandemanians maintain their separation from other churches as a Biblical requirement: 'Be ye not unequally yoked together with unbelievers: for what fellowship hath righteousness with unrighteousness: and what communion hath light with darkness?'[47] (1 Corinthians 6:14). Just as the fraternity of true believers is exclusive in its religious practices, so scientists reading the book of nature must keep their own house in order and free from the pull of patronage or economic or political interest. Yet Faraday's position was paradoxical, since on this one occasion he entered the politics of science in order to try to free science from politics and patronage. We do not know whether he recognised this paradox.

In the 1830s the Society became less insular, drawing closer to the country's commercial and political concerns and away from aristocratic privilege.[48] If this change received little encouragement from Faraday, he explicitly condemned the Society's failure to reform its inefficient and ineffective administration. Having been away from Council for a year, he declined re-election in 1832 on the grounds that he needed to devote his time to original research, 'not that I undervalue the character of that high and responsible office'.[49] A year later he was back on Council, largely, it appears, at the insistence of John Lubbock, one of the Secretaries. Writing to Lubbock the day following the first Council meeting of the session, Faraday expressed his ambivalence towards serving on Council and his antipathy towards its procedures. While he wished to assist Lubbock's attempts to improve the Society, especially through a reduction in fees, which would have enabled poorer men of scientific ability to join, he admitted that he was 'unfit for a Council or a committee man'. He continued:

> I think business is always better done by few than by many – I think that the working few ought not to be embarrassed by the idle many. and further I think that the idle many ought not to be honored by association with the working few. – I do not think that my patience has ever come nearer to an end than when compelled to hear an examination of witnesses &c &c in committee. the long rambling mal-apropos enquiries of members who still have nothing in consequence to propose that shall advance the business. But in all this I will promise to behave as well as I can.[50]

With the failure of the reformers and the inability of Council to advance science effectively, Faraday finally withdrew from Council at the end of the 1835 session and thereafter rarely attended the Society's meetings, although channelling many of his papers in its direction. In other words, he related to those functions of the Society that were commensurate with his vision of science but withdrew from those which were associated with the vices of the mundane realm.

Closely associated with the Royal Society was the Athenaeum, founded in 1824 with the aim of providing a meeting place for those eminent in literature, science and the arts. The main mover was John Wilson Croker, who complained that while London was well-endowed with clubs for the fashionable and the military, these other groups were neglected. Croker soon enlisted the support of Davy, who, as President of the Royal Society, doubtless encouraged many other Fellows, Faraday included, to join. Faraday served as the club's first Secretary, initially offering his services gratis, though he was soon awarded a small salary. However, he found the duties too demanding and resigned a few months later in favour of his close friend Edward Magrath, who held the post until 1855. Although Faraday resigned his membership in 1851, he, together with his brother Robert, was responsible in the early 1840s for introducing a new gas lighting system that not only improved illumination but also reduced unpleasant emissions that affected both humans and book-bindings.[51] It is not clear how frequently Faraday visited the Athenaeum or who he met there, but it may have provided him with many contacts and correspondents both among scientists and those in other fields. More generally, the role of the Athenaeum in British science deserves further investigation.

Whatever Faraday's early association with the Athenaeum, he generally stood aloof from the social side of the Royal Society; he rarely graced the parties, soirées and dining clubs that were such a prominent feature of early Victorian science, especially metropolitan science. For example, in 1847 his friends William Grove, John Herschel and John Gassiot founded the Philosophical Club, which aimed to promote 'the scientific interests of the RS and to facilitate intercourse between fellows cultivating different branches of natural science'. Although the Club's purpose was to advance science and its membership was confined to those who pursued this aim, it convened over dinner in a London hotel before the Royal Society's monthly meeting. On being invited to join the Club, Faraday only accepted 'under the express understanding that he is not expected to *dine*', thus undermining one of the Club's main aims, if not the main one.[52]

On two occasions Faraday was considered for high office in the Royal Society. In 1827, when Davy relinquished the Presidency, he was proposed as one of the Society's Secretaries, but was unsuccessful in gaining that position at a time when Gilbert was seeking to mould the Society's administration.[53] A very different situation occurred in February 1848 when Faraday, now an eminent scientist, declined Herschel's invitation to be considered for the Presidency.[54] However, in May 1858, when the reigning President, Lord Wrottesley, indicated his intention to resign, a serious search began for his successor. A deputation, consisting of Wrottesley, Gassiot and Grove, sought to persuade him to undertake the Presidency. This approach to Faraday – perhaps the most respected scientist of his day – was considered of such significance that a group portrait was painted. It now hangs in the Royal Society's rooms (Figure 6.1). According to Walter White, Faraday 'requested a day to consider, but the impression is that he will refuse'.[55] No contemporary evidence indicates his reasons for refusing the Presidency, only John Tyndall's account in *Faraday as a discoverer* (1868) provides some clues. According to Tyndall, despite his remonstrations, Faraday finally refused the honour after taking Sarah's advice. Perhaps Sarah reminded him of his religious obligations and prompted his often-cited reply to Tyndall, 'If I accepted the honor which the Royal Society desires to confer on me, I would not answer for the integrity of my intellect for a single year'. This aspect of Faraday's response exhibits his fear of being swallowed up by the politics of science, and it is therefore of a piece with his terrified reaction to the offer of the Presidency of the Royal Institution 6 years later. (See section 10.3.) But Tyndall's account alludes to a further issue that Faraday raised, and this was whether he possessed the time, energy and ability to change the Society.[56] Although many reforms had been instigated a decade earlier, he evidently felt that he could not immerse himself in the Society in 1858.

According to the material presented in this section, it is clear that during the latter part of his life Faraday generally refused to take part in the management of British scientific institutions and played a very limited role in the Royal Society, the British Association and the Society of Arts. However, this peripheral role needs to be set against his much more extensive involvement in scientific organisations during the 1820s and early 1830s, when he even served on several councils and committees. Why, then, did this marked change in attitude occur in the early to mid 1830s? There are several factors that appear to have contributed to this change. Since Davy was intimately involved in London science, his death in 1829 lessened Faraday's connection with several scientific institutions. Faraday's experience in supporting Herschel's attempt to gain the Presidency of the Royal Society in 1830 must have made him doubt the value of such unreformed organisations. Faraday worked under considerable pressure, particularly after his research career had taken off in the early 1830s, and he was very reluctant to waste time on what he

Figure 6.1 Deputation, consisting of Lord Wrottesley, J. P. Gassiot and William Robert Grove, inviting Faraday to become President of the Royal Society in 1858. Portrait by E. Armitage.

Source: Royal Society of London.

considered to be unproductive activities. Finally his appointment as deacon in the London Sandemanian meeting house in 1832 not only placed a further constraint on his time but also made him particularly wary of involvement with worldly matters and demanded the highest moral standards.

6.3   IN THE SCIENTIFIC COMMUNITY

In their cogent analysis of the early years of the British Association for the Advancement of Science (founded 1831), Jack Morrell and Arnold Thackray characterise the men who formed the nucleus of the Association as 'Gentlemen of Science'. While the authors rightly emphasise that other, even opposing, interest groups were active in the science of the period, they portray these Gentlemen as the cutting edge of science and as exemplifying its new social relationships. Drawing the term from Coleridge, the authors conceive their Gentlemen as constituting a new scientific 'clericy' that would spearhead the improvement of science and thus of civilisation. It will therefore be helpful to outline the Morrell–Thackray thesis and to enquire whether Faraday can be included among the Gentlemen of Science.

One factor which immediately disqualifies Faraday is that he was not a member of the gentlemanly stratum of society. The Gentlemen, who were mostly minor gentry or from the highest echelon of the middle classes, were men of leisure who were able to travel and pursue science. To do this demanded a secure, although sometimes relatively modest, income and a freedom and independence which were denied even to the middling ranks of the middle class. Of the twenty three men who formed the Association's core all but two were either university professors, unencumbered clerics, gentlemen of leisure or aristocrats. Moreover, the vast majority had attended university, Trinity College, Cambridge, being the most prominent on the list, with Trinity College, Dublin, in second place. By contrast Faraday was an artisan's son who in later life might rather anachronistically be called middle class. Even then, what security he possessed arose from his unique position at the Royal Institution and from the smile of princes rather than from a solid bank balance. Unlike the Gentlemen, he was dependent on science for both his income and his social position: had he remained in the book trade he would probably have been as unknown to most of us as Thomas Boosey, John Leighton or David Morison of Perth (all Sandemanian printers and booksellers). Again, while he was granted the time and facilities to pursue his research, he did not bask in the kind of leisure that is usually associated with luxury and the availability of free time after completing one's exertions. The time and energy that Faraday was able to direct towards science was all the more limited by illness, particularly after about 1840. Moreover, unlike most Gentlemen of Science, he was not university-educated.

The areas of religion and politics provide further crucial contrasts between Faraday and these Gentlemen. The latter, or at least the vast majority of them, were Liberal Anglicans, committed to a rational, anti-evangelical theology which emphasised the argument from design as illuminating God's role in the physical universe. Moreover, they were opposed to Biblical literalists and especially such writers as Scriptural Geologists, who founded their views about natural phenomena on detailed interpretations of the Bible. Another useful contrast is with the Oxford Movement, which attacked the British Association and especially its leading Gentlemen on the grounds that natural theology encouraged pantheism and that devotion to science led scientists to ignore the high moral message of the Bible. Yet, as Morrell and Thackray stress, the Gentlemen repeatedly asserted the independence of science from theology while at the same time arguing that science works in harmony with religion. Thus their conception of natural theology helped create a space in which scientists could pursue their research without interference from the Church: indeed, for Liberal Anglicans the scope of intellectual curiosity and the authority of science depended on this freedom. While natural theology was attacked by Evangelicals, Tractarians and Scriptural Geologists, it also provided a common ground on which Liberal Anglicans could meet men of other religious persuasions, especially nonconformists. The early meetings of the British Association attracted large numbers of Quakers and Unitarians, one from each of these denominations being included among the twenty-three Gentlemen of Science.

Critics of the Association sometimes perceived a necessary connection between its proclaimed latitudinarian theology and its support for political reform. While the Association often explicitly proclaimed itself to be non-partisan, it, and its leaders, leaned clearly towards a moderate, middle ground in British politics, thus aligning itself with the respectable classes in society. Many of the Gentlemen of Science were active Whigs and a few were Peelite Tories, a profile which encouraged cautious, conservative reform in both science and politics, but without challenging existing institutions. Not surprisingly, the Association found little support from either ultra-Tories or Radicals, and even Henry Brougham was viewed as being too progressive to hold high office in the Association.[57]

Faraday was not a Liberal Anglican, moderate Whig or Peelite conservative. As a Sandemanian he was ambivalent towards natural theology and he was closer (theologically speaking) to many of the Gentlemen's biblical literalist opponents than to the Gentlemen themselves. Yet it is also important to note that there was much common ground between them and Faraday. While Faraday could not identify with the British Association as an institution, he nevertheless attended its meetings, used its platform and discussed scientific matters with both its leaders and its lesser luminaries. He would have found particularly attractive the spirit of toleration that prevailed and also the Association's stated commitment to pursue science untrammelled by political or religious dogma. The Gentlemen of Science actively sought his support and welcomed him to the Association's meetings. For his part Faraday particularly looked forward to meeting scientific friends at those gatherings. In this sense he was as much a member of the scientific clericy as were any of the Gentlemen. Yet, coming from a sect that eschewed the trained ministry, the very notion of a clericy would have worried him, since he would have denied that scientists should either replace or stand alongside the church. The republic of science is not like the established Church, since it possesses no political or ecclesiastical authority but is an association of like-minded people who are committed to discover the laws of nature. Moreover, science could not, in Faraday's opinion, replace true religion or perform its spiritual functions.

The period during which Faraday flourished was one in which the process of professionalisation was very marked.[58] This is not to say that science became a profession, in the sense of a recognised, paid body of practitioners, but rather that there emerged a number of scientific specialisms, each with its attendant community of committed activists who interacted and recognised each other's work. There was little room for the generalist who took all of nature as his canvas (although a few like Whewell and Humboldt remained); instead most scientists worked in one or two clearly defined areas, and these areas were likewise demarcated in the organisation of science. Thus, for example, the gentlemanly geologists of the period formed the backbone of the Geological Society, where they fought over the divergent interpretations of phenomena and published their papers in the Society's *Proceedings* and *Transactions*.[59] Faraday was likewise a specialist, although his areas of specialism were far less institutionalised: his work does not correspond to the domain of any single scientific society. Moreover, few gentlemen were apparent in the specialist communities in which he moved: if geology bears a relationship to land ownership and the countryside, electrochemical experiments were performed in the laboratory and often were seen as having practical, industrial applications. We must now examine more closely the scientific communities in which Faraday was active.

As we have seen, Faraday played a highly circumscribed role in the scientific institutions of the day, even those based in London. Just as he rejected the established Church, he also turned his back on the scientific establishment, except in so far as it accorded with his own conception of science. Yet Faraday was part of a diverse group of people who were institutionally located at the Royal Institution. This was not a specialist group but consisted of scientists and also the porters, the housekeeper (Miss Savage), the redoubtable Charles Anderson (who served as Faraday's assistant from 1832 to his death in 1866), the Librarian (Benjamin Vincent) and the several others who maintained the Institution's services and enabled Faraday to pursue his research and lecturing. Among the scientists in this group Humphry Davy played a crucial role for Faraday, initially as mentor, although their relationship subsequently became strained. Davy's successor to the chair of Chemistry was William Brande, who served from 1812 to 1852. Faraday's relation to Brande is far from clear but their published letters suggest that they were on close personal and professional terms. For example, Brande generously acknowledged the popularity and scientific reputation of his younger colleague.[60] The other main scientific figure at the Royal Institution was John Tyndall, who was appointed to the chair of Natural Philosophy in 1853 and became an admirer and later a biographer of Faraday. One may add to this group a number of scientists, such as William Grove, John Barlow and Henry Bence Jones, who served the Institution in a variety of ways.

It would be incorrect to portray Faraday's role in the scientific community solely in terms of its institutions, since his most frequent and sustained intercourse with other scientists operated through informal, rather than formal, channels. As his extant correspondence attests, his postbag was filled with the exchange of scientific information and he freely, and often at length, responded to correspondents who shared his interests, while others – such as table turners and petitioners – received short shrift. Since it is impossible to reconstruct his complete correspondence, let alone his conversations, any analysis of his relation to other scientists must be incomplete. Nevertheless, with the aid of his extant correspondence we can obtain a chart of the scientific network round Faraday. Moreover, it offers some insight into the informal scientific groupings to which he belonged.

One further difficulty in characterising the divisions within the scientific community during Faraday's day is that the different areas of study (or emergent disciplines) have changed considerably during the intervening century and a half. Thus the map of the different branches of science in, say, 1841 looks very different from today's list of scientific disciplines and, moreover, the map changed significantly between the 1810s and 1860s. Only towards the end of that period did physics emerge; indeed some would claim that William Thomson and Peter Guthrie Tait's *Treatise on natural philosophy* (1867 – the year of Faraday's death) was the first physics textbook.[61] It would be anachronistic, then, to call Faraday a physicist. However, there are three terms which appropriately characterise him: chemist (or, perhaps, chemical philosopher), electrician and natural philosopher. The first two terms are relatively unproblematic and they correspond to selfconscious, if largely overlapping, groups of practitioners.

While Faraday openly repudiated the role of a professional chemist who earned his living by performing analyses, he clearly aligned himself with philosophical chemists who, according to a contemporary definition, 'investigate the nature and properties of the elements of matter, and their mutual actions and combinations

... [and who] enquire into the *laws and powers* which preside over and effect these agencies'.[62] In acknowledging this long-established view of chemistry Faraday was in the company of a small but thriving international community. Many of Faraday's correspondents, such as Thomas Andrews, Auguste de la Rive, Jean-Baptiste Dumas, Justus von Liebig and William Hyde Wollaston, can be considered chemists in this general sense. Yet the conception of chemistry underwent an important but subtle change around the beginning of the century, when it was found that the arrangement of certain substances in, say, the form of a voltaic pile, gave rise to an electric current which, in turn, possessed the power to decompose chemical compounds. As one commentator wrote in 1800, the study of galvanic electricity was 'a fresh hare just started; the Royal Society hounds are now in full cry'.[63] Humphry Davy was off and running with the pack, as were several of Faraday's other correspondents, such as J. J. Berzelius.

In 1820 Hans Christian Oersted opened a further major area of research on electricity. When he passed an electric current through a wire aligned in a north-south direction, a neighbouring suspended magnet turned so as to align itself perpendicular to the wire.[64] This phenomenon, which demonstrating a link between the forces of electricity and magnetism, rapidly attracted the attention of a number of scientists, including both Davy and Faraday working at the Royal Institution. In the early 1820s Faraday included among his correspondents three of the major contributors to this thriving area of research, André-Marie Ampère, J.J. Berzelius and Peter Barlow, and later he exchanged letters on electromagnetism with George Biddell Airy, John Herschel, James Clerk Maxwell, Joseph Plateau, Julius Plücker, William Thomson and numerous others. David Gooding, who has analysed the London-based scientists concerned with electromagnetism in the 1820s, has noted that no formal network or 'invisible college' existed but that several of these practitioners met and corresponded on an informal basis. Faraday stands out from this group of approximately two dozen scientists in two interesting respects. Firstly, only he and William Sturgeon pursued a long-term research interest in the subject, most of the others engaging the subject for just a few years. Secondly, while most of these electricians concentrated on one or two specific aspects of electromagnetism, Faraday and Peter Barlow pursued the full range of electromagnetic phenomena.[65] Thus Faraday stands out as idiosyncratic in researching all of electromagnetism in the period from the early 1820s until the twilight of his career.

The trail after these two hares – electrochemistry and electromagnetism – was still fresh in 1837 when William Sturgeon edited the first volume of his *Annals of Electricity, Magnetism & Chemistry; and Guardian of Experimental Science*. Indeed, Sturgeon's journal, which continued for eleven volumes over 7 years, exemplifies in its title most of Faraday's scientific interests. Not surprisingly, the first volume contained two extensive extracts from Faraday's 'Experimental researches in electricity'. However, Sturgeon appended to these papers two short but sharp critiques of Faraday, questioning his competence as an experimenter.[66] These criticisms illustrate the distance that separated Faraday from Sturgeon and his London-based group of electricians who were practical men concerned with exploiting science for economic advantage. To these electricians Faraday may have appeared a dreamer cocooned in the safety of the Royal Institution and out of touch with the realities of life.[67] Sturgeon was also aggressive towards Faraday and probably resented his success both as a lecturer and researcher. Despite this distance separating Faraday from Sturgeon, his journal is helpful in identifying many of the chemists and electricians of that period. Although the contributors to that journal included some of Faraday's correspondents, many had little or no contact with him. This indicates the diversity of concerns among those who pursued research in the areas of chemistry, electricity and magnetism. It is, moreover, clear that Faraday was very selective in his correspondents, avoiding contact with those who were aggressive towards him or were using science purely to line their own pockets. His network of correspondents was confined principally to those who he considered constituted the elite – the 'band of brothers'[68] – in the scientific community.

Despite Faraday's proclaimed aversion to trade, we must be careful not to portray him and his network of correspondents as 'pure scientists' totally disinterested in practical problems. He certainly placed the discovery of the divinely ordained laws of nature above the application of science and, unlike some of his correspondents, he repudiated the laying up of fortunes from scientific endeavour. Nevertheless he considered that within the scheme of divine providence we must use scientific innovations for the benefit of mankind (but not the selfish entrepreneur). As he told the Royal Commission on public schools in 1862, the physical sciences are 'most valuable in their application. They make up life. They make up our artificial state'.[69] Evidence for Faraday's long-term concern with the responsible application of science includes his cooperation with a number of industrialists and engineers. Thus among his correspondents were the Birmingham glass manufacturer James Chance, who discussed with Faraday the design and manufacture of optical systems for lighthouses, Marc and

Isambard Brunel and James Nasmyth. Under this heading should also be included his extensive correspondence concerning Trinity House, the Royal Society's committee on optical glass, and various civic agencies which will be discussed in the next section.

The majority of Faraday's scientific associates worked in the areas of electricity or chemistry; moreover, a significant number were, like Faraday, particularly concerned with electrochemistry and electromagnetism. Yet there were several correspondents, such as William Whewell and Charles Babbage, whose scientific interests are not adequately specified by the scope of chemistry plus electricity. While these writers cannot be called physicists, the term 'natural philosopher' is appropriate. An eighteenth-century dictionary defines natural philosophy as 'that Science which contemplates the Power of Nature, the Properties of Bodies, and their mutual Actions upon one another'.[70] This definition accurately describes Faraday's conception of science, since he was concerned not only to identify the properties and mutual actions of bodies but also the different manifestations of power in nature. While for Faraday the first aim of empirical research was to discover the laws regulating particular classes of phenomena, these broader natural philosophical interests were always present and were often reflected in his correspondence.

I have used the terms (philosophical) chemistry, electricity and natural philosophy fairly loosely to indicate clusters of scientific interests, as indicated by the extant letters exchanged with Faraday. It is important to notice that the chart of Faraday's associates does not correspond to any particular scientific institution (although the Royal Institution is obviously well represented); rather, it is defined in terms of scientific interests, thus cutting across institutional, national and other boundaries. In this sense he was part of an invisible college comprising the members of an intellectual elite. The vast majority of his main correspondents were Fellows of the Royal Society and many were based in London, although there were a number of provincials, such as David Brewster and William Scoresby, and several Cambridge men. Indicative of Faraday's international standing is the large proportion of extant letters (approximately one third) with foreign scientists.

By concentrating on his correspondents we are liable to underestimate Faraday's highly ambiguous role in British science. Although one of the most celebrated scientists of his day he was, paradoxically, a rather isolated figure. Yet from that position of isolation he established close friendships with a number of other scientists and actively aided their research. Moreover, several young men owed considerable debts to Faraday for the stimulus they received during their early careers. William Barrett, for example, testified 'to the enduring impression made by his words of kind encouragement and acts of thoughtful friendship'. A second example is provided by Henry Deacon, scion of a Sandemanian family and later a successful chemical manufacturer in Widens, who in his mid-teens was taught by Faraday and allowed access to his laboratory. Likewise the marine engineer and shipbuilder Alfred Yarrow related how Faraday would often spend half an hour after his juvenile lectures talking to the boys who attended and making them repeat some of the experiments by themselves.[71] Nevertheless it is important to note that Faraday worked alone, with Charles Anderson as his sole assistant, and that no other scientist shared in his research at the Institution once he had served his apprenticeship under Davy. Indeed, of the nearly 500 printed items listed in Jeffreys' bibliography, only three early entries refer to jointly written papers.[72] Faraday's independence is also illustrated by his refusal to accept Ada, Countess of Lovelace, as his student. Although he excused himself on the grounds of ill-health, it seems likely that had he been in the best of health, she would still have received a negative response, since he pursued his research in a personal and independent manner.[73]

Although Faraday had many scientific acquaintances, they occupied very different positions on his scale of friendship. At the very top was the Genevan chemist Auguste de la Rive (1801–73). Faraday had met his father, Gaspard, while travelling with Davy in 1814 and had received from him considerable encouragement. Thereafter Faraday remained in close contact with the family and met Auguste both on the continent and in London, and their correspondence continued for more than three decades. What is unique about this correspondence is that it went far beyond the men's mutual interest in science. De la Rive addressed Faraday as 'Monsieur & tres cher ami', even 'Mon tres cher & excellent ami' and signed himself 'Votre tout dévoué', while Faraday addressed the other as 'My very dear Friend'.[74] They took great interest in each other's family lives and also on many occasions they discussed their religious feelings in their letters,[75] a subject which Faraday usually kept from his scientific acquaintances. These two men were in complete personal harmony and no subject was taboo; it was as if de la Rive had been an honorary Sandemanian.

Of Faraday's other close scientific acquaintances several were foreign: Jean-Baptiste Dumas, Christian Schoenbein and Julius Plücker. Indeed, he may have experienced less difficulty in coping with these men living abroad than with those he met regularly in London. Judging from the extant correspondence, he was slightly more guarded with Dumas, Schoenbein and Plücker than with

Auguste de la Rive and religious sentiments intruded less frequently. Moreover, there were other barriers; for example, Schoenbein did not discuss with Faraday the concern he shared with Berzelius over exploiting the benefits and rewards for developing gun-cotton, since the subject would have met with Faraday's disapproval.[76] With his British correspondents Faraday shared his views on science and expressed friendship, but this was often tempered with a degree of formality even in his letters to such close friends as Whewell and Tyndall. Formality even entered into the mode of address: 'My dear Whewell', 'My dear Tyndall' or 'My dear sir'. Even the content of letters was clearly circumscribed; Faraday could only meet these friends on the field of science. Not only was the subject of religion avoided but all forms of personal exposure were kept to a minimum. Always friendly and correct, Faraday knew how to circumscribe his friendships.

This line of demarcation was easily recognised and generally respected by others. They knew that Faraday held peculiar views on religion, that the social and political issues of the day were not on his agenda and that he was opposed to the exploitation of science for personal gain. Those who strayed into areas that he was not prepared to discuss received short shrift. When, for example, he was badgered by table-turners or by someone seeking his patronage, he could be rude and dismissive. For example, as mentioned above, Faraday was asked in 1860 to prepare a report on three methods that had been proposed for preserving the stonework at the Palace of Westminster. Mr Daines, one of the contenders for the contract, managed to confront Faraday who '[r]efused to discuss [the matter] with him. He threatened legal proceedings. Sent him out, and all his papers'.[77] Unwittingly the forceful Mr Daines had placed Faraday in an uncomfortable position and Faraday responded by abruptly terminating the interview in order to preserve himself. A similar case was reported by Walter White:

> Dr. Stenhouse says Faraday was selfish and narrow-minded. That a man once went to him, as he himself had gone to Davy, and that F. sent the young man to [the chemist Thomas] Graham, of which incident Graham made a standing joke.[78]

There were probably other such incidents which the Victorian hagiographers managed largely to suppress. Despite his obvious charm and congenial nature, few scientists could approach Faraday at close quarters and it is noticeable that some contemporaries with similar scientific interests, such as William Sturgeon, never penetrated into Faraday's closed world. Indeed, as Joseph Henry remarked after his visit to Britain in 1837, Sturgeon and Faraday were 'not on good terms'.[79]

Faraday's most effective weapon was silence. Personal attacks were common in early nineteenth-century science and anyone publishing a paper might expect to have to defend not only his claims but also himself from severe criticism. William Sturgeon was no gentleman but a self-made man who was carving out a career in electrical science. He attacked Faraday viciously in the first volume of the *Annals of Electricity* (1836–7), claiming that Faraday was incompetent at making measurements with a galvanometer and that in his electrochemical researches he had overlooked the significance of both the size of the electrodes and the variation in the cell's operation over time.[80] Such an attack on the reputation of a leading scientist might ordinarily have brought a rapid and closely argued response. Not so with Faraday, who, in public at least, failed even to acknowledge this assault on his competence. It may have been Christian polity to turn the other cheek, but it was not how scientists were expected to behave. They found it very difficult to deal with this man who set perfection as his ideal in life and who generally stood aloof from scientific controversies and from the rough and tumble of the scientific community.

While Faraday's scientific contemporaries engaged in the cut-and-thrust of science they also often held strong political views, which, particularly in the 1830s, were evident in their correspondence. Since politics (including ecclesiastical politics) was a no-go area for Faraday, these acquaintances must have experienced great difficulty in communicating with him. But there were also certain *scientific* subjects which were not on his agenda. Faraday's career spanned the most profound changes in theorising about the earth and biological species. He was 8 years older than the geologist Charles Lyell, and 18 older than Darwin, and the furore over the *Origin of species* (1859) broke while Faraday was still able to take an interest in scientific matters. Although he attended Brande's lectures on geology, joined the Geological Society and even served on its Council during an earlier era, his correspondence is singularly lacking in references to these rapidly changing sciences.[81]

Faraday's silence on these issues is easily understood. As a Sandemanian who accepted the biblical account of how the world and mankind had come into being, he had no need for recourse to the bold hypotheses of his contemporaries. He may have shared the views expressed in *The philosophy of the Creation*, the work attributed to Robert Sandeman but only published in 1835, when, according to its editor, a number of 'men of eminence in the scientific world, have appealed successfully to the Scriptures for the truth of geological theories'.[82] What followed under Sandeman's name was a fairly standard example of Scriptural Geology based almost entirely on the opening verses of Genesis. Whatever sympathy Faraday felt towards this approach to geology he would have repressed it and remained silent in the presence of Lyell, Darwin and their followers. Silence was a strategy to cope in public with these problematic theories held by his scientific contemporaries. How he coped internally is difficult to assess, but unlike some later biblical literalists, he may have ignored or compartmentalised these theories so that they would not clash with Genesis.[83]

While turning his back on such theories, he was an amateur naturalist, frequently visiting zoological and botanical gardens and taking a keen, excitable interest in the flora and fauna encountered during country walks. One of his nieces recounted that he took her and a copy of John Galpine's *A synoptical compend of British botany . . . after the Linnean system* (1806) on such walks and taught her not only to name the plants but also to give them their Latin, i.e. scientific, names.[84] Galpine's book is noticeably lacking in any discussion of botanical theories, consisting instead of long tables listing each species according to its class and order (in both Latin and English) together with such data as the type of soil and the colour and duration of the flowers. The simple method of ordering employed by Linnaeus and his concise (and apparently theoryless) use of language must have appealed greatly to Faraday. For one who saw 'Sermons in stones, and good in everything', the flora that Faraday studied must have provided him with further evidences of God and His providence.[85] To see these he did not require uniformitarian theories in geology or Darwin's theory of evolution. Instead he may have viewed these theories as distractions from the truth and as indications of human pride.

Faraday's anomalous position in the scientific community can, then, be understood in terms of the Christian values he held. In general he met his fellow scientists on his own terms, constructing barriers to prevent uncongenial subjects (even people) invading his privacy. Within these self-prescribed limits communication was unencumbered, but beyond them it was impossible. He was therefore generally perceived as a highly individualistic member of the scientific community. In this sense his sectarianism becomes apparent.

6.4   PUBLICIST FOR SCIENCE

There was considerable public interest in science in Faraday's day – witness the large number of lectures and books aimed at that audience and the prominence of scientific articles in journals ranging from the *Quarterly* and *Edinburgh* reviews to the *Penny Mechanic*. Thus most scientists not only pursued research but also appeared before the wider public in a number of different guises. In his generation of outward-looking scientists Faraday became the foremost science lecturer. His celebrated lectures at the Royal Institution, particularly his Friday evening discourses and his courses for juveniles, have firmly connected Faraday's name with the successful public exposition of science. (The Royal Society has recently instituted the Faraday Award for those who have made major contributions to the public understanding of science!) Hence Faraday was not only a member of the invisible college of chemists and electricians concerned with developing new and esoteric knowledge claims but he was also both skilled at and committed to the exposition of that knowledge before a non-expert audience. There is an analogy here with Faraday's religious experience, since he not only participated in prayer and scriptural reading with the other members of the Sandemanian community but he also sought to convey Christian values to others through his daily actions and social intercourse.

Faraday held a firm conviction that the truths of science were not just for the *cognoscenti* but had to be shared with others, one reason being that public lectures 'facilitate our object of attracting the world, and making ourselves with science attractive to it'.[86] In being able to pursue their research, scientists were dependent on the wider public, and were therefore obliged to reciprocate by placing their wares before that public. Moreover, the truths of science had to be shared truths and not the prerogative of any particular social, political, religious or scientific group. This was a theme on which Faraday elaborated in a Royal Institution discourse on mental education attended by Prince Albert in 1854, and also in his evidence before the Public Schools' Commissioners 8 years later. On the first of these occasions he stressed the importance of self-education and particularly the need to subject oneself to the most stringent intellectual honesty and discipline. With his own self-education in mind he was implicitly arguing that the basis for a person's potential to comprehend science depended on that individual's moral qualities, and not on his or her social background or formal education. However, the right educational experience, be it formal or informal, was important for actualising that potential.[87]

Standing before the Earl of Clarendon's Commission, he expressed astonishment that children were largely ignorant of the major scientific developments of the previous half century and he asserted the need for adequate science education. Although the Commission was concerned primarily with public schools, Faraday clearly did not consider that science should be taught only to the upper ranks of society. Indeed, while bemoaning the ignorance of the general public, he found both the educated and uneducated wanting. A more specific example was provided by British lighthouse-keepers, who were, he claimed, far less knowledgable about science than their French counterparts. The army officers he taught at Woolwich were equally deficient, and he complained that they 'were very often altogether, or nearly, unable to give an answer to the questions which I asked them'.[88] Moreover, he asserted that the juvenile audiences that flocked to his lectures at the Royal Institution lacked an adequate scientific background. His remedy was science for all, and he annoyed the Commissioners by failing to limit his comments to the public schools (the Commissioners' only brief), instead discussing the need to educate everybody in science.

Faraday also criticised the domination of the school curriculum by classics and the misplaced emphasis on pure mathematics. Society devalued the men who invented the electric telegraph, the steam engine and the railway system, and it 'neglected and pushed down below' the knowledge that is necessary for developing and maintaining these socially important technologies. Most people who had been educated in the classics and in pure mathematics could not comprehend the laws of nature or understand how a simple machine works. By contrast, the sciences should be taught, which are the 'most valuable in their application. They make up life. They make up our artificial state. They make up the body of physical science, and are to us most important in life'.[89] Notice that Faraday was here stressing the importance of applied science in maintaining the fabric of modern society and that the technologically sophisticated society in which he lived constituted man's 'artificial', not his natural (fallen) state. In other words, technology – not classics – is the pillar supporting modern culture.

Another of Faraday's main arguments for teaching science to all was the necessity to guard against such popular prejudices as mesmerism and table-turning which afflicted those untutored in the physical sciences (including some mathematicians, such as Augustus de Morgan). The first rumblings of modern spiritualism were felt in 1848 in up-state New York, where two young sisters received communication from beyond in the form of rappings. Not only was America both socially and spiritually ready to latch on to reports of the Rochester rappings and turn them into a craze comparable in size with health foods or jogging in our day, but Britain was equally fertile ground. From America travelled Mrs Hayden in 1852, David Richmond and Mrs Roberts in 1853 and D.D. Home – a veritable saint among mediums – in 1855, bringing with them an array of raps, spirit voices and gyrating tables. Soon indigenous mediums arose and spiritualism became not only the talk of the town but also a preoccupation stretching from high society to the suburban sitting-room and to the Workmen's Hall at Keighley. Britain, like America, was in the grip of the spiritualists, if not the spirits.[90]

Faraday entered the fray with a letter published in *The Times* of 30 June 1853 and a somewhat longer piece in the *Athenaeum* of 2 July. This intervention was necessitated by the many requests he had received asking him to pronounce on the subject of table-turning and by the circulation of rumours according to which he endorsed the topic and sided with the spiritualists. He therefore inaugurated a series of experiments to settle the matter and these showed, to his total satisfaction, that the movement of the table was 'a quasi involuntary muscular action (for the effect is with many subject to the wish or will)', and not the presence of a natural force, such as gravitation or electricity, or even a supernatural one. While he believed that the question of cause could be readily resolved by experiment, what particularly disturbed Faraday was the public's gullibility and its unwillingness to subject the issue to critical evaluation. He concluded his letter by pointing an accusing finger at the existing education system, which had so manifestly failed to provide the public with a grounding in science. Without an adequate education in science the public was prey to every charlatan.[91]

Not only was spiritualism not a science but those who sought supernatural explanations of so-called spiritualistic phenomena were courting thoroughly anti-Christian sentiments. Shortly after penning his letter to *The Times* on table-turning he complained to Schoenbein that the human mind is 'weak, credulous, incredulous, unbelieving, superstitious, bold, [and] frightened', and he compared the average mind unfavourably with that of a dog. Yet there is 'One above who worketh in all things, and who governs even in the midst of misrule to which the tendencies of men are so easily perverted'. That misrule included the belief in spirits moving tables, but, he insisted, the only unclean spirits were those 'working in the hearts of men, and not, as they credulously suppose, in natural things'. In the realm of natural things God ruled by constant, incorrigible laws which were not subject to the human whim. Moreover, Faraday considered that the Bible proscribed spiritualism, which evoked the 'unclean spirits' of Acts 8:7: When Philip preached in Samaria 'unclean spirits, crying with loud voice, came out of many that were possessed with them'. Spiritualism thus evoked for Faraday the demonic influences that were totally opposed to Christianity. To rout the heretic he turned to the powerful swords of science and science education.[92]

Since these popular superstitions did not simply fade following these public pronouncements by the leading scientist of the day – indeed, the fashion seems to have increased throughout the 1850s – Faraday felt impelled to make table-turning the butt of his lecture on mental education delivered before Prince Albert. As we have seen, Faraday possessed an immense psychic drive to differentiate between truth and error, good and evil. Sandemanians utilised this distinction in moral and theological matters and they frequently asserted the need to apply accurate judgement to all aspects of our lives.[93] Moreover, the Bible emphasises that judgement enables us to distinguish right from wrong, and it also associates correctness of judgement with righteousness, for example, in Psalm 72. It is therefore highly significant that Faraday asserted that *'deficiency of judgment'* was the major fault in contemporary mental education, producing such manifest errors as the belief in table-turning. He continued,

> I know that in physical matters multitudes are ready to draw conclusions who have little or no *power of judgment* in the case; that the same is true in other departments of knowledge; and that, generally, mankind is willing to leave the faculties which relate to judgment almost entirely uneducated, and their decisions at the mercy of *ignorance, prepossessions, the passions, or even accident.*[94]

Here again we find Faraday discriminating not merely between truth and error but also between informed judgement and prejudice.

In his evidence before the Royal Commission Faraday asserted that he did 'not see clearly how classical studies educate the mind' in the appropriate exercise of judgement, whereas studying the 'laws of matter . . . and the true logic of facts' provide a rounded training for life and an appreciation of science both 'as regards its own applications, and as regards the general judgment of man'.[95] The same theme had been stated at greater length in the earlier lecture on mental education, in which he also emphasised the need for self-discipline in the development of our judgement. With this instrument and a firm grasp of the laws and facts of science we can cut away error and improve ourselves. Most importantly, he emphasised, the exercise of the judgement will enable us to destroy our desires, inclinations and temptations so that we do not submit to false doctrines and hypotheses. In short, a well-formed power of judgement is both essential for moral virtue and the best antidote to the serpent of prejudice, against which Faraday continually fought.[96]

Faraday's crusade against prejudice (in all its forms, not just table-turning) therefore made education in the physical sciences a necessity, not a luxury, for all people, not just the upper classes. Yet the practical problems of implementing a system of science education were immense. As he told the Commissioners, such a system would have to be managed effectively. In the classroom children should be taught how phenomena are brought about by laws. The teaching of facts by themselves is of little value, whereas laws provide faithful guides to understanding natural phenomena. He also warned that science should not be taught by pedants (=hypocrites?) who are enamoured by abstract terms, technical terminology or mathematics, which would be incomprehensible to children beginning their science studies. Most importantly, lectures 'depend entirely for their value on the manner in which they are given. It is not the matter, it is not the subject, so much as the man'.[97] The onus, then, fell firmly on the character of the lecturer and, in line with this prescription, Faraday strove to develop (and largely attained) the highest standard in his own science lectures.

In a letter written soon after he joined the staff of the Royal Institution, Faraday claimed that he had been studying the styles of the lecturers he heard, noting their 'various habits[,] peculiarities[,] excellences and defects' with a view to ascertaining the qualities that are required for a pleasing lecture. Even at that time (1813) Faraday held firm views about how a lecture should be conducted and over the years he tried to implement these ideals. There is a wealth of common sense in Faraday's thoughts on lecturing – the lecturer should speak slowly and clearly, he should not read from lecture notes or turn his back on his auditors, the audience's interest is likely to be lost if there are long digressions or if the lecture lasts more than one hour, etc. Yet for Faraday the aim of lecturing was not simply to deploy the best mode of exposition; instead it was 'to gain completely the mind and attention of his audience and irresistibly to make them join in his [the lecturer's] ideas'.[98]

How successful he was in achieving this is difficult to gauge, but one of his auditors, who has already been quoted, commented on his 'clear, bright, truthful voice', and claimed that in his lectures 'mind [was] addressing mind'. This listener also felt that 'he was full of sympathy with his audience; and with his fellow creatures, though so far above them'.[99] One facet of his success as a lecturer was his ability to gain a deep and instant rapport with his audience. Although vastly more knowledgeable of scientific matters than most of his auditors, he strove to appear humble and avoided giving himself airs. This view corresponds to one of Faraday's major aims, which was to 'evince a respect for his audience'.[100] The lecturer should be dignified and should not 'angle for claps & ask for commendation' but at the same time he should know his audience and accommodate his delivery accordingly. Most importantly, the audience must like and trust the lecturer and build a mutual bond.[101]

In providing his audience with access to nature and nature's laws Faraday played the role of a sage. A man of immense insight and wisdom he mediated between his audience and nature (and thereby also God). Not only did he portray in clear terms how nature behaved but he sought to evoke a sense of wonder in his audience and also a feeling for God the Creator. He constantly pointed to the wonders of nature and the beauty of phenomena.[102] Some found that his lectures came close to religious experiences, for here was a man who 'had the *deepest* sense of religion, and . . . was one of those happy mortals "who could read" sermons in stones, and good in everything'.[103] However, the religious message was rarely stated explicitly, and Faraday usually confined himself to demonstrating and expounding the laws of nature. Yet even this activity had strong ideological connotations, since he sought through this subject to link all people, irrespective of religion, sex, age, rank or nationality. Scientific truth transcended all barriers. Thus in a lecture of 1860 he began by expressing his pleasure at having served Trinity House, for in the improvements of lighthouse illuminants many nations had worked in harmony for the good of the mariner.[104] In His providence God had created the laws of nature for the good of all.

In his lectures Faraday communicated not only through words but by his deft manipulation of nature. As a virtuoso he could make the operations of nature manifest before the audience's eyes. His audience thus both heard about the laws of nature and also saw them in action. It is significant that in an early letter Faraday had placed the power of the eye far above that of the ear in conveying ideas clearly to the mind.[105] Thus observable experiments played a central role in Faraday's didactic repertoire. For the audience the immediacy of this experience proved one of the most potent aspects of Faraday's lectures, as William Crookes testified:

The pleasure which all derive from the expositions of Faraday is of a somewhat different kind to that produced by any other philosopher [i.e. scientist] whose lectures we have ever attended. It is partially derived from his extreme dexterity as an operator, – with him we have no chance of apologies for an unsuccessful experiment, no hanging fire in the midst of a series of brilliant demonstrations, producing that depressing tendency akin to the pain felt by an audience at a false note from a vocalist. All is a sparking stream of eloquence and experimental illustration.[106]

Likewise, when Joseph Henry, the American scientist, attended some of Faraday's lectures in 1837 he was particularly attracted by the lecturer's 'inimitable tact of experimenting'.[107] What emerges strongly from these and similar accounts of Faraday's discourse and experimental demonstrations is his apparent total command of himself and therefore of the proceedings. This is not to say that experiments did not sometimes fail to function as expected, but on such occasions he could turn the apparent failure to advantage and not lose control of the situation.[108] As in so many other aspects of Faraday's life, he was able to assert order over confusion and thereby maintain his self-integrity. This order was not tense and fragile, since he brought deep understanding and inner strength to his lectures and managed to communicate these qualities to his audience.

The zeal and commitment that Faraday brought to his science teaching may have had its roots in his religion; 1 Timothy 3:1-7 specifies the elder's personal qualities, including both his high moral calibre and his being 'apt to teach'. For a community that depended on an intimate knowledge of the Bible the elder's teaching function was essential. While there is no necessity to extend this aptitude to teach to secular subjects – only a few Sandemanians were teachers or lecturers – Faraday clearly conceived his scientific role as manifesting those personal qualities which were specified in the Bible. In accepting teaching as part of this role Faraday extended the audience for this branch of secular knowledge beyond the bounds of the scientific community so as to include the public at large. Faraday also brought to his teaching an inspiration and commitment that has rarely been surpassed by other scientists. Moreover, he offered science to the public not solely to benefit science by making it more acceptable in society, but also because he believed firmly in the social value of science and in the central role of science education in the mental education, and thus fulfilment, of men and women of all classes.

However, were Faraday's ideals compromised by his position at the Royal Institution? At a time when popular science lecturing was rapidly speading to the working classes through such organisations as Mechanics' Institutes Faraday did not contribute directly to that movement but remained loyal to the Royal Institution, with its clientele drawn almost exclusively from the upper and middle classes. Mechanics were unlikely to have seen him lecturing, although they would have been familiar with his name. Despite his reputed congeniality towards newspaper boys and his ministrations to the poor in the Sandemanian community, he must have appeared a distant figure. This apparent distance did not spring from any disdain for the working man, since he also distanced himself from his middle- and upper-class auditors at his lectures at the Royal Institution. Although he displayed consummate skill in projecting science from behind the U-shaped demonstration table in the lecture theatre of the Royal Institution, he combined intimacy in delivery with social distance. As a sectary, he stood apart from all ranks of society. However, his loyalty to the Royal Institution ensured that he directed his energies to supporting it rather than the many other organisations that catered for the wider public. Thus while a major advocate for universal education in science, his own efforts were directed to a select audience.

6.5 CIVIL SCIENTIST

Of the many significant changes that occurred in science during the early Victorian period, one of the most important was the creation of a new role, that of the civil scientist. As part of the greatly increased power of central government, scientists' expertise was, for the first time, extensively used for the benefit of the state. Indicative of this trend were the numerous Royal Commissions that examined and pronounced on so many aspects of Victorian life, and whenever an issue relating to science was engaged, a bevy of FRSs would be called to give evidence. Moreover, the period marks the beginning of the fledgeling scientific civil service, with the appointment in 1864 of Robert Angus Smith as the first alkali inspector. Smith's job was to police the alkali works in order to prevent emissions exceeding those specified in the Act. Yet his role was a complex one, since he had to mediate between the government, the alkali manufacturers and the public. Moreover, he was a professional chemist who was concerned not only with ensuring that the Act was implemented but also with researching several chemical problems relating to the manufacture of alkali and devising and installing technological improvements that would reduce emissions. This was the paradigm case of a governmental policy utilising science for regulatory purposes.[109]

More generally the alkali inspectorate reflected the Victorian belief in the efficacy of science for solving pressing social and technical problems. Science was powerful precisely because it was seen to provide the government and its agencies with readily applicable solutions. It was also widely accepted that science transcended matters of class, religion or party. Thus the advice proffered by the expert scientist was non-partisan and carried with it the imprimatur of true, objective knowledge. If we no longer share this naive faith in science, we should still recognise the historical importance of this message and the new-found civic responsibility which many early Victorian scientists readily accepted.

Faraday willingly adopted this civic role. His attitude towards government was but an extension of the Sandemanian views about supporting the monarchy and upholding the rule of law. As a British citizen, Faraday thus appears to have felt obliged to accept unquestioningly the law of the land and the demands made on him by the civil authorities (provided that these demands did not conflict with God's word). Writing to Lord Auckland, the First Lord of the Admiralty, in 1847 Faraday explained that although he wished to avoid all other calls on his time so as to concentrate on research, 'I have always, as a good subject, held myself ready to assist the Government'. Moreover, as he explained in the same letter, he was usually prepared to offer his services gratis to governmental agencies. An indication of his demand as a civil scientist is that he had recently been approached not only by the Admiralty but also by 'the Ordnance, the Home Office, the Woods and Forests, and other departments'.[110] Moreover, particularly during the latter part of his life, he sat on the Royal Commissions charged with considering gas lighting in art galleries, the site of the National Gallery and the state of the Gallery's pictures, and he also gave evidence before Royal Commissions dealing with harbours, the management of harbour lights and public schools.

Far more demanding on Faraday's time were other long-term connections with governmental agencies, such as his appointment in 1829 as Scientific Adviser to the Admiralty with a fee of £100. However, during the previous 4 or 5 years he had performed an extensive series of experiments aimed at improving optical glass for navigational purposes, especially telescopes. These experiments began under Davy's direction as a fairly routine investigation in response to a joint committee (chaired by Davy) of the Royal Society and the Board of Longitude (which was abolished in 1828). In May 1825 Faraday had joined this committee and had been named as one of the three members of an experimental sub-committee responsible for supervising the work of the glass man-ufacturer, Messrs Green and Pellatt, and testing samples. The other two members were the astronomer John Herschel and the optician George Dolland. The majority of the work fell to Faraday, and since it proved very onerous, he was forced to divide his time between the Royal Institution and the glassworks, some 3 miles distant. Faraday therefore requested that a furnace be constructed at the Royal Institution so as to enable him to perform all his work in one place. With the agreement of both the Royal Society and the Royal Institution a room was set aside, the furnace constructed in September 1827 and Charles Anderson hired as Faraday's assistant. Yet the project progressed slowly and many problems were encountered. For example, the glass samples were often inhomogeneous, since denser chemicals, such as lead monoxide, sank to the bottom of the trough. Again, samples often contained bubbles or striae. There were also major structural problems with both the troughs and the furnace.[111]

Following Davy's death in May 1829, the intensity of Faraday's optical glass research waned, and in November he presented the fruits of his onerous labour to the Royal Society.[112] Although he continued work in this area for a further year and a half, he was clearly relieved to send his experimental notebooks to the Royal Society in July 1831, together with a letter expressing his willingness to continue, should the Royal Society wish. However, he was far from sanguine whether the manufacture of optical glass could be improved significantly. Moreover, he expressed to Peter Roget, one of the Society's Secretaries, his wish to 'lay the glass aside for a while, that I may enjoy the pleasure of working out my thoughts on other subjects'. While he had devoted 'the whole of' his 'spare time' to optical glass research, he had been forced 'to resign the pursuit of such philosophical enquiries as [had] suggested themselves to my own mind'.[113] Eight weeks later, while pursuing one such philosophical enquiry, he carried out his famous experiment on electromagnetic induction (see section 9.1).

Another civic position that Faraday held was that of Scientific Adviser to Trinity House, from 1836 until 1865. He was required to examine the various proposals for lighthouse illuminants and to decide which were safe and practicable. Given the significance of shipping for the Victorian economy, the provision of an adequate lighthouse system was of crucial importance, and numerous proposals for improving the visibility and identification of lighthouses were considered.[114] Faraday was also concerned with the more mundane questions of the ventilation of lighthouses, 'their lightning conductors and arrangements, the impurity and cure of water, the provision of domestic water, the examination of optical apparatus, etc'.[115] Since the archives at Trinity House have been destroyed, it is not possible to trace in detail Faraday's activity in this area. However, from extant letters it appears that Faraday was particularly active during the late 1850s and early 1860s. We find him spending several days at Whitby in October 1860 working on the light and waiting for the wind to change in order to observe the light from a boat. Explaining the situation to Benjamin Vincent, Faraday expressed his concern that he would be working 'up *to* the Sabbath or within an hour of it'.[116] Earlier that year the train he travelled in was caught in a massive snowstorm near Dover, and although he could not reach the South Foreland lighthouse by road he eventually arrived 'by climbing over hedges, walls, and fields' in order to make 'the necessary inquiries and observations'.[117] At that time he was in his 69th year and his health was in marked decline.

While his role as Scientific Adviser to Trinity House made considerable physical and temporal demands on Faraday, it also brought him into close contact with industry. He made frequent visits to James Chance's glassworks in Birmingham in order to experiment on and improve the lens system designed especially for the Whitby lighthouse. The correspondence with Chance offers an all-too-rare glimpse of Faraday working in close collaboration with an industrialist on a series of problems that involved both science and technology.[118]

The final example concerns Faraday's participation in the coroner's inquest into the explosion that occurred at Haswell Colliery, in the centre of the Durham coalfield, on 28 September 1844, killing ninety-five men and boys. This disaster occurred shortly after a strike, so that there was considerable enmity between the miners and mineowners at the inquest. During the first 3 days of the proceedings several witnesses – miners, inspectors, viewers and underviewers – concurred that the mine had been one of the safest and best ventilated in the area. It was also agreed that the mine should be examined and a report laid before the coroner and jury. However, Mr Roberts, who was identified as the 'pitman's attorney-general', insisted on the first day of the inquest that the examination be carried out by miners. This proposal was rejected by the mineowners' representative, who claimed that the miners were biased against the owners and that the owners objected to some of those named by Roberts. On the third day Roberts pressed that Matthias Dunn be one of those sent to examine the state of the mine. The owners once more objected since they believed that Dunn held a grudge against them. The coroner, faced with this impasse, adjourned the court and applied to the government for impartial inspectors to be appointed.

The government acted quickly. Sir Robert Peel, the Prime Minister, together with Sir James Graham, the Secretary of State for the Home Department, scouted for scientists to offer expert evidence and cross-examine witnesses at the Inquiry. As the geologist Charles Lyell was told, 'they want scientific men known in your and

Mr. Faraday's line to be there' – Faraday's 'line' being chemistry, and he knew little about geology or coal mines. Despite the government's clear preference for Faraday among the chemists, he was reluctant to accept the commission.[119] However, a week later, on Wednesday, 9 October, Faraday and Lyell were present in the courtroom listening to the witnesses. On the Thursday Faraday, Lyell and Samuel Stutchbury (a colliery viewer from Bristol) inspected the mine for several hours. The inquest continued on the Friday, with Stutchbury attributing the explosion to a build-up of 'fire-damp' – a hydro-carbon that is highly explosive when mixed with the appropriate quantity of air – in the 'goaf' – the underground chamber created by the extraction of coal and roof-falls. This accumulation of gas had been sparked accidentally, possibly by a damaged Davy-lamp. The jury of twelve 'respectable' men from neighbouring villages now claimed that the experts had provided sufficient evidence to enable them to reach their decision on the cause of the accident. Roberts, however, wanted to call further witnesses from among the miners but the coroner refused and cleared the court, presumably owing to protests from the public. The jury then returned its verdict that the explosion had occurred purely by accident and that no one was to blame. It is important to notice that the task of a coroner's court is to establish the circumstances in which death occurred and to apportion responsibility. Its aegis does not include decisions about guilt or punishment; that lies in the domain of a criminal court.[120]

Faraday played only a minor role at the inquest, being present during the final 2 days and, according to the report in *The Times*, making a small contribution to the last day's proceedings. (He was also at that time afflicted considerably by loss of memory.) However, the presence of Lyell and Faraday at the Haswell Inquiry possessed symbolic significance, since, as Morris Berman claims, they conferred on the decision 'the status of scientific respectability'.[121] This should not be taken to mean that the evidence of Lyell and Faraday was lacking in substance or that their analysis was necessarily correct; rather it indicates the social significance of the testimony of scientists such as Lyell and Faraday. As Faraday later claimed, he was not 'a professional man as regards coal mines; I am, however, a man who has looked at the laws of nature, and as far as observation and practice have enabled me, I have applied them in working out their results'.[122] To know nature's laws and to apply them in the solution of problems carried great kudos in early Victorian society.

The report that Faraday and Lyell wrote for Sir James Graham, the Secretary of State for the Home Department, was published in the *Philosophical Magazine*. The authors outlined their analysis of the cause of the explosion and recommended ways of ventilating the goaf so as to draw off fire-damp and thereby prevent similar occurrences. They also concurred with the decision of the jury to exonerate the mineowners, although they were critical of the 'evils' that can result from concern with profits by mineowners in general; these evils were, firstly, not taking sufficient precautions to prevent the communication of fire-damp between neighbouring workings and, secondly, the wasteful policy of plundering only the richest seams.[123] Faraday also delivered a Friday evening lecture on the subject before a packed audience at the Royal Institution in January 1845. Here he made only passing reference to the deaths at Haswell and the role of the miners and owners. Instead he engaged in a form of discourse that did not engage the social context of the Haswell accident but instead concentrated on general questions of fire-damp explosions and forms of mine ventilation. These were strictly scientific problems which could be dealt with in terms of the laws of nature and illustrated by profuse experiments; for example, he exploded samples of coal-gas mixed with air. The accident that killed ninety-five men and boys at Haswell and many similar accidents were rarified into the realm of scientific discourse and a technological fix offered. For Faraday science held the key to understanding such situations and remedying them.[124]

While the Royal Institution may often have performed an utilitarian function, Faraday was no Benthamite. Instead, as suggested above, his concern that science serves the general good sprang from a religious sense of service and, in particular, the Sandemanian insistence on duty towards the civic authorities. (As we shall see in section 7.5, Faraday considered that science fulfilled a much higher moral role than its technical and social utility.) Yet Faraday's attendance at the Haswell Inquiry and also his research into lighthouses and optical glass, to take but three examples, illustrate not only his public role as an expert, at a time when scientists were beginning to play a major role in public administration, but also his appropriateness for that role. Fulfilling the role of expert provided Faraday with a way of participating in the world, while, as a Sandemanian, remaining detached from it. He could avoid having to align himself with any class, party or faction – Whig or Tory, miner or mineowner – and instead bring to bear on the situation the laws by which God had chosen to govern the physical world. Speaking in the Royal Institution lecture theatre before a fashionable audience, Faraday could distance himself from the social clamour and explain the physics and chemistry of fire-damp, and how explosions occur and could be avoided. The 'objective' discourse of science may have served various social purposes but it also enabled Faraday to play a public role that was thoroughly consistent with his Sandemanian beliefs.

7 A Theology of Nature

Although Faraday's early biographers emphasised his empiricism, historians of science have more recently sought to understand Faraday in terms of his theoretical commitments and have argued that his success resulted from his rich and insightful use of theoretical constructs such as the lines of force which he conceived permeating space. If this latter approach undervalues his commitment to detailed empirical research, and especially his sophisticated use of experiment, it also often minimises the importance of metaphysical presuppositions in Faraday's thought. However, Faraday's presuppositions have recently begun to receive the attention they deserve, particularly in the writings of David Gooding, who has argued that Faraday's science was crucially dependent on a number of metaphysical assumptions which both directed and constrained his research. Furthermore, Gooding claims that these assumptions (a number of which will be discussed below) were theological in origin.[1]

The aim of the present chapter is to extend this line of argument so as to locate Faraday's metaphysics in his religion and, in particular, in his views about the structure of the divinely created physical world. These views, it will be claimed, coloured Faraday's highly idiosyncratic theories about matter and force.[2] Indeed, his theories on these topics differed significantly from those of most of his contemporaries, who conceived the world as composed of small, hard, impenetrable particles with forces of attraction or repulsion acting between their centres. While the mature Faraday rejected this model, it would be incorrect to portray him as isolated from the theories of his contemporaries. L. Pearce Williams, in particular, has argued that Faraday's views on matter and force were learnt from Davy, who, in turn, was the heir to Kant and Coleridge on questions of force and a disciple of Roger Boscovich, the eighteenth-century Jesuit, who had propounded the theory of point atoms.[3] Doubtless Faraday learnt a great deal from Davy and others and many useful insights can be gained from locating him with reference to the various traditions in natural philosophy. However, such an approach portrays the scientist as principally the recipient of views already in circulation. By so doing, it underplays the integrity of the individual and the ways in which a scientist can develop his or her own views about both nature and science. Since Faraday was not institutionally constrained to follow any existing programme, especially after

Davy's death, we should be careful not to tie him too closely to any particular natural philosophical tradition. Moreover, his own views about matter and force developed in a highly creative but individualistic manner throughout his long research career. While not ignoring Faraday's intellectual debts to existing natural philosophies, this chapter draws attention to the ways in which Faraday conceived the physical world as God's creation. He was, of course, not alone in holding this belief, but it will be argued that his conception of the divine origin of the world placed certain constraints on how he conceived nature and natural phenomena.

In section 8.1 it will be claimed that while Faraday utilised the argument from design as indicating God's wisdom and power, he did not make exclusive use of reason, as would a natural theologian. Like other Sandemanians, he dismissed unmitigated natural theology as unscriptural and indicative of man's pride, if not prejudice. Instead, while attributing a limited role to reason, he based his religion firmly on revelation. Of particular concern to him were questions concerning the interrelation between God and the world – between Creator and created. Such questions constitute the theology of nature, a subject which should not be confused with natural theology.[4] Thus Newton's arguments about the immanence of God in the creation or the disputed interpretations of the opening verses of Genesis would be included within this topic. Steeped in the Sandemanian interpretation of the Bible, Faraday imbibed a number of metaphysical conceptions about the physical world which he took to be necessary truths about God's universe and how it operates. These assumptions, which helped direct his scientific research, were usually implicit (even subconscious) in Faraday's work, although he occasionally articulated them in public, particularly in his later years. It would, however, be incorrect to think of these metaphysical assumptions as rigidly held propositions which formed the bedrock on which the edifice of Faraday's science was constructed. Rather they were intuitions which guided and helped shape his science. Yet it is important to recognise that his religiously based metaphysics informed his science via a number of what I shall call 'metascientific principles', such as his belief that physical force is conserved. I use this phrase to distinguish these principles from metaphysical beliefs, such as nature being God's creation, and also to emphasise that they could readily be expressed in empirically meaningful terms.

7.1   SCRIPTURAL PHYSICS

In chapters 5 and 6 Faraday's social views were interpreted as instances of the Sandemanian social philosophy, which was, in turn, predicated on their reading of certain biblical passages. In this and the following two chapters we shall be analysing Faraday's science. Here the situation is significantly different, since there was no strong Sandemanian tradition in science and, moreover, a specifically Sandemanian exegesis of the relevant biblical passages cannot be identified. If other Sandemanian writers offer us little assistance in illuminating Faraday's science, we must look outside the Sandemanian sect and briefly examine a leading tradition among theologians of revelation to see whether Faraday can be located among their number. As we shall see, this tradition will prove of rather limited use in understanding Faraday.

The term 'scriptural geologist' is often used to refer to those who founded their understanding of geology on the biblical narrative, particularly the first chapter of Genesis, which they interpreted in strict historical detail. From the mid-seventeeth to the mid-nineteenth century numerous theories of the earth in which evidence from Genesis was accepted as central were published, although they were often supplemented by various secular physical theories, such as Newton's or Descartes' mechanics. Other geologists insisted that geology must not be founded on a highly literal interpretation of the Bible and this secular approach became increasingly dominant during the nineteenth century, although scriptural geology continued to attract a large public following even after its credibility had declined among *cognoscenti*.

While Faraday accepted the truth of Scripture, he left no written account of how he interpreted the opening verses of Genesis. In the absence of such evidence we might try to reconstruct Faraday's views from the numerous exegeses offered by other eighteenth- and early nineteenth-century authors, particularly by biblical scholars. However, the situation is complicated by the great diversity of interpretations and by the manifest points of contention. Within this considerable literature there are two works by Sandemanians which offer interpretations of Genesis. One is James Morison's *An introductory key to the first four books of Moses* (1810), which has little to say about the physical processes which occurred at the Creation. The other is a small book issued by one of the Perth Sandemanians in 1835 with the title *The philosophy of the Creation, as narrated in Moses' principia, Gen. Chap.I. v.1 to 18* and attributed to Robert Sandeman. As its title indicates, the author drew considerably on the interpretation of Genesis offered by the early eighteenth-century writer John Hutchinson, whose best-known work was *Moses's principia* (1724).[5]

While Faraday would have been familiar with *The philosophy of the Creation*, this was an atypical work in the Sandemanian corpus, since it dealt in some detail with the process God had deployed in forming the earth. Moreover, unlike the sect's explicit social philosophy, Sandemanians had no tradition of enforcing consensus on specific aspects of geology. Faraday would certainly have accepted the truth of Genesis but that alone would not have made him sympathetic to any specific theory of the scriptural geologists. Moreover, by the 1830s scriptural geologists were increasingly excluded from the community of geologists. As an established, if idiosyncratic, scientist Faraday would not have aligned himself with the scriptural geologists.

There was another scriptural tradition which requires discussion, since it pertains more directly to the sciences on which Faraday worked. This concerned natural philosophy – roughly what we would now call physics – and the tradition can legitimately be termed 'scriptural physics'. Again there were many incompatible strands within this tradition, but one writer stands out above all others because the scriptural physics he enunciated gained an extensive following in the eighteenth and early nineteenth centuries. This was John Hutchinson – the numerous writers who adopted his form of natural philosophy are usually referred to as the Hutchinsonians. In order to determine whether Faraday can be subsumed within this tradition we need to outline Hutchinson's scriptural physics.

Hutchinson considered that the Bible, in its original Hebrew text, contained the true, pure word of God. However, during its subsequent history the text had become corrupted by the addition, for example, of the Masoretic signs which represent vowel sounds. By removing these intrusions, Hutchinson considered that he had returned the text to its original form from which he could decode the hidden meanings of the Hebrew words with the aid of some apparently rather arbitrary rules to interrelate words with similar Hebrew forms. Moreover, he claimed that the text contained two parallel meanings, one referring to the physical, the other to the spiritual, realm. *Moses's principia* contains Hutchinson's commentary on the opening verses of Genesis, according to which the physical system was completed by the end of the sixth day. The finished system is a perfectly operating machine consisting of particles of matter in motion but no void space. As well as the particles comprising gross bodies, the remainder of space is filled by a fine, subtle matter which can take on three forms – fire, light and air. In the form of light, the particles are in fast translational motion, whereas in the form of air, groups of these particles congeal into 'grains'. The sun at the centre impels particles in the form of light radially towards the firmament, from whence they return towards the sun in the form of air (i.e. as grains). The sun then melts the grains and the individual particles are again projected towards the firmament as light. It should be noted that for Hutchinson both matter and motion are conserved and thus God's creation is a perfect machine. Moreover, Hutchinson conceived the three forms of ether to correspond to the three persons in the Trinity: fire, light and air (or spirit) corresponding, respectively, to the Father, Son and Holy Ghost.

Hutchinson's scriptural physics proved very attractive in the eighteenth and early nineteenth centuries to those who associated Newtonian science with natural theology, latitudinarianism, the Low Church, and even deism and atheism. It was particularly popular among the advocates of revealed theology, including many High Churchmen. However, the attitude of the Sandemanians is less clear-cut. One of the first Independent ministers to join the London Sandemanians in the early 1760s was Samuel Pike, who a few years earlier had written a clear exposition of the Hutchinsonian system.[6] The author of the anonymously-published *Philosophy of the Creation*, possibly Sandeman, also drew on the Hutchinsonians in trying 'to trace in the works of the Creation the impress of the Almighty hand'.[7] Like Hutchinson, this author filled space with an ethereal fluid which is in motion; darkness is not an absence of this ether but is merely the ether at rest, whereas light is a specific motion of the ether. The author interpreted the sixth verse of Genesis in terms of this ether, which is subject to two conflicting powers comprising expansion and contraction. The whole system is a conservative one in that both matter and motion cannot be created or annihilated by any natural process. Finally the author viewed the system as 'shewing forth the wisdom and power of Elohim' and that it is 'upheld continually by the same powers that created it'.[8]

Sandeman, to whom this book is attributed, was elsewhere highly critical of Hutchinson. In his *Letters on Theron and Aspasio* he chastised Hutchinson for opposing 'the apostolic doctrine concerning the divine sovereignty, the person of Christ, and [the doctrine of] acceptance with God'. In particular he was repelled by Hutchinson's portrayal of Christ as a mere human but possessing divine qualities to an inordinate degree. Sandeman also found Hutchinson morally objectionable, since he was totally lacking in Christian love. Finally, Hutchinson was so preoccupied with 'the amusement of accommodating Hebrew words to his scheme of philosophy [science]', that he both underrated the importance of the New Testament and also failed to acknowledge that the true moral message of the Bible was perfectly comprehensible in translation.[9] John Glas had likewise attacked Hutchinson's hermeneutics as unscriptural and naive in the *Edinburgh Magazine* of 1759.[10]

In the light of these strident criticisms by fellow Sandemanians it seems unlikely that Faraday had much sympathy for Hutchinson's method of framing a theology of nature. He must have known of the Hutchinsonians not only because they were referred to by both Glas and Sandeman but also because there were so many works in this tradition published in the eighteenth and early nineteenth centuries. Although he made no specific comments about Hutchinson or the Hutchinsonians, his oft-quoted phrase – 'There is no philosophy in my religion'[11] – can be interpreted as rejecting this type of enterprise, since Hutchinson's natural philosophy (or science) was, literally, in his religion. By contrast, Faraday did not derive his science from the Bible. Instead, he would have agreed with the more moderate tone adopted by Sharon Turner, who had written *The sacred history of the world* (1832–9), that the brief opening verses of Genesis do not contain a detailed account of the process of Creation and, more generally, the Bible does not provide a complete specification of the physical world, although it does contain various physical truths which cannot be eroded by modern science.[12] Moreover, many of the Hutchinsonians' leading physicalist doctrines are not apparent in Faraday's science, such as the particulate ethereal fluid capable of three modifications. Faraday *cannot* then be located in this scriptural physics tradition.

Although Faraday was *no* Hutchinsonian he, like Hutchinson, Turner and many other writers, accepted the Bible as indubitably true and placed revelation far above reason. This commitment to revelation identifies one important issue on which these writers were in agreement – they did not doubt that God had created the universe and that Genesis contains the true account of how the physical world was brought into being.

This postulate provides a point of entry into Faraday's theology of nature, since, at a very fundamental level, he conceived the physical universe from a theological perspective. This is apparent in several of Faraday's unpublished writings where he took cognisance of the Creation. For example, in his 1844 memorandum on 'Matter' he speculated that 'by his word' God might have spoken 'power into existence' round points in space, which constitute atoms.[13] That he should speculate in terms of what God might have done at the Creation shows how Faraday conceptualised the history of the universe: he considered that the present state of the physical universe is the direct result of the creative act. Yet this was not a topic confined to Faraday's speculative moments. In a lecture series of 1846 he stated that the 'properties which belong to matter depend upon the power with which the Creator has gifted such matter'. Moreover, the laws of nature 'were established from the beginning' and are thus 'as *old* as creation'.[14] These phrases were not merely literary ornaments and it is clear that Faraday considered the laws of nature to have been constant since the Creation. More importantly, what he and other scientists were examining in their laboratories were not just nature's laws and powers but the final state of the creative act. In other words, he was experimenting on the resultant condition produced during the six Days of Creation as recounted in Genesis. Furthermore, he was immensely aware that he was contemplating God's creation and thus coming close to God himself. As a Sandemanian committed to live by the Bible, he could not ignore those opening verses, which gave a true, but brief, account of how the physical world, which he examined in the laboratory, came into existence.

This provides us with one sense in which Faraday's understanding of the physical world was strongly theistic. We would also expect that, since Faraday had committed himself to live by the Bible, other scriptural passages would similarly have influenced his understanding of nature and his practice of science. As we shall see, there were a few occasions when Faraday cited Scripture in relation to scientific issues. However, such occasions were limited to his private correspondence and his public lectures, particularly his more speculative ones dating from the 1840s and 1850s. Although his research papers read before the Royal Society sometimes contained general meta-scientific principles, which we can interpret theologically, he avoided drawing on Scripture, which would have been inappropriate.

However, there were a number of other places in his writings where the Bible was not explicitly cited but where resonances with appropriate biblical passages can be detected. The warrant for searching for such resonances is provided by Faraday's firm commitment to the Bible and his intimate knowledge of its text. However, to select any particular biblical passage as relevant and to suggest that Faraday had that passage in mind necessarily calls for speculation. Yet such speculation is not out of place, provided that it is appropriately signalled. Therefore when pointing out such biblical resonances I shall adopt an explicitly speculative mode of expression. It should also be made clear that the majority of the ensuing analysis does not depend on linking Faraday's science with particular scriptural passages, since the main line of argument is that his science was permeated with religious (although not explicitly biblical) meaning.

There is one further aid which can be used, but should be used judiciously, when trying to ascertain where Scripture is relevant to Faraday's science. This is the way other writers, such as the Hutchinsonians, have construed biblical passages when framing their natural philosophies. While the Bible is open to diverse interpretations, sometimes emphasising one passage at the expense of another, there are a few very general points of agreement between Faraday and the Hutchinsonians which may help us understand better how biblical texts could have informed Faraday's science. This is *not* to make Faraday a Hutchinsonian or a scriptural physicist. The aim of comparing Faraday with the Hutchinsonians is to help identify certain specific similarities in outlook which were shared by these theologians of revelation who allowed only a limited role to natural theology.

One common feature has already been identified – the awareness that the present-day physical world was created by God according to the Genesis narrative. To this can be added other concerns that he shared with the Hutchinsonians. Firstly, the belief that God did not act arbitrarily but with purpose so that the physical world is a designed system. Thus Faraday conceived the laws of physics and chemistry as willed by God at the Creation. Moreover, the world manifests the aim of its designer. Secondly, since God created a perfect system both matter and 'force' are conserved and the system is self-sustaining. As we shall see in the next section, Faraday was firmly committed to the view that nature is economical. Thirdly, Faraday shared with the Hutchinsonians an antipathy towards the notion of void space and also emphasised the plenitude of the physical world. Finally, both Faraday and the Hutchinsonians did not consider mathematics to be the language of nature. These aspects of Faraday's science will be discussed further in this and the following chapter and, where appropriate, reference will be made to the Hutchinsonians as a comparison group.

7.2   THE ECONOMY OF NATURE

For scientifically orientated advocates of revealed theology, Faraday included, the Bible contains not only the Creation but also the account of the divine plan as it is manifested in history, particularly human history. The Old Testament relates how that plan had unfolded from the time of Adam, while the New Testament is centrally concerned with the history of Jesus. The prophetic books, particularly the books of Daniel and of Revelation, relate how the divine plan will unfold in the future. While the Creation of the physical world is given in Genesis, the other books of the Bible make but minimal claims about its subsequent history. Yet it was clear to Faraday and others of a similar persuasion that God had created a perfectly designed system and one which therefore manifests His perfection. It is crucial to understand that for Faraday the natural world constituted an economy in the sense that all events are tightly ordered by divine providence. This 'economy of nature' – an apposite term widely used in the eighteenth and early nineteenth centuries – thus manifests God's plan.

Throughout his writings Faraday utilised a number of terms in describing nature which indicate how he conceptualised the operation of this divine economy. Below are listed some of these terms which, taken together, indicate how he conceived natural processes. These terms are closely related and are sometimes difficult to distinguish but they were all attempts to express in a scientific vocabulary Faraday's religious conception of the physical world which God had created according to His plan. Yet these terms constituted what I have called 'metascientific principles', since they underpinned much of Faraday's scientific research. They were also his attempts to articulate what he believed to be an incorrigible truth about nature and the essence of God's relation to the physical world, as implied in the Bible. These metaphysical principles sometimes functioned as necessary truths and sometimes as guidelines in his scientific theorising and, ultimately, in his research.

(The reader who does not wish pursue this matter in detail is advised to scan the italicised terms at the beginning of each paragraph and to proceed to the final paragraph of this section.)

A. *Equality of cause and effect.* This principle is illustrated in Faraday's earliest letter to Benjamin Abbott in which he reported a series of experiments with a battery comprising zinc and copper discs separated by flannel discs soaked in sodium chloride solution. At the end of the experiment he found that the metal discs were coated; 'I think this circumstance well worth notice, for remember no effect takes place without a cause'.[15] Thus a scientist who observes an effect can be sure of the operation of an appropriate cause, and not merely a cause, for the economy of nature *guarantees* that the effect perfectly matches the cause. They are balanced and equal, although it is important to realise that Faraday rarely conceptualised equality in strictly numerical terms. An alternative formulation of this principle is the equality of action and reaction. No cause (or force) could operate without necessarily invoking its effect (or an equal opposing force).[16]

B. *Direct proportions.* On those occasions when Faraday used numerical data he invariably expressed them in terms of simple ratios or proportions, which played a significant role in Faraday's chemical work. Thus if 65 grams of liquified chlorine contains 18 grams of chlorine and 47 grams of water, then 100 grams of the liquid would contain 27.7 grams of chlorine and 72.3 grams of water. Similarly Faraday's first law of electrochemistry asserts a '*direct proportion[ality]*' between '*the chemical power of a current*' and '*the absolute quantity of electricity which passes*'. Both these and similar proportionalities illustrate that nature is an economical, conservative system.[17]

C. *Lawlikeness*. The role of law will be discussed in section 8.2, where it is emphasised that Faraday conceived the laws of nature to have been framed by God as integral parts of His plan. To this discussion must be added the requirement, when considering the natural economy, that in its operation any law must embody a constancy – a constant relation between cause and effect.

D. *Invariability*. Nature is thoroughly lawlike and only God possesses the supernatural power to change or suspend the laws of nature. The forces in nature and the laws which natural things obey are constant through space and time; they are, in other words, universal and unchangeable.[18] The constancy of nature's laws was guaranteed by Faraday's belief in their divine source; indeed, science would be impossible without this guarantee. Moreover, providential design through universal, invariant laws guarantees our own personal safety and the very possibility of life itself.

E. *Consistency*. In his essay 'On the conservation of force', Faraday noted the consistency pertaining between one power and another.[19] The meaning of the term is not entirely clear, but Faraday seems here to be appealing to two rather different manifestations of the economy of nature. Firstly, nature is uniform and therefore any particular magnet, for example, will behave in a similar way to any other. Secondly, and more interestingly, there must be a consistency between all the different types of power, at least in as far as they are all governed by the law of force conservation (see F. below).

F. *Conservation*. Both matter and the forces by which it is known have 'undergone no change since man has been on earth, and, as we have reason to believe, none from the first moment of their creation until now'.[20] Elsewhere he referred to the 'stability of creation',[21] and in his lecture on mental education he also asserted that *we* can neither create nor annihilate matter or force. All that we can do is to transform one form of force into another so that an overall conservation is effected. It is important to notice that Faraday was here denying *our* ability to create, and twice in the same paragraph he dogmatically asserted that apparent counterexamples merely imply 'a creation of power, and that *cannot be*': he then rapidly dismissed these apparent refutations of this incorrigible principle.[22] Only God can create power and therefore forces must be conserved in the physical world.

G. *Inertia*. In a series of lectures delivered in 1847 Faraday rejected the idea that inertia is intrinsic to matter and instead argued that it is a law of nature, requiring that a cause (such as a moving body) should produce an equivalent effect (the inertia being conserved during collision). Thus inertia was merely a redescription of that '*great law*' of the conservation of force which required that no force can be created or annihilated.[23]

H. *Correlation*. Correlation, which will be discussed further below, is a relational term, expressing the correspondence or reciprocal dependence of two variables. In the early 1840s the term was popularised by Faraday's friend William Robert Grove, who endeavoured to show that any two powers of nature are correlates. Grove noted by way of illustration the example of the (inverse) correlation between up and down.[24] As this example makes clear, correlations are not causal relations (in the strong sense that Hume criticised) but merely represent a constant conjunction or lawlike relationship. There is thus a necessary lawlike relationship between travelling north and travelling south. Equally, cause and effect, action and reaction are correlates and Faraday even extended this principle to magnetism and electricity, such that the existence of a south pole or a positive charge must necessarily imply the simultaneous existence of a north pole or negative charge.

I. *Incoercibility*. Faraday used this term in a lecture of 1847 in which he stressed that the 'incoercibility' of force is 'a very important point in the philosophy of natural powers'. The meaning of the term is, however, far from clear. He seems to have equated it with the inexhaustibility of force, but also stressed that an incoercible force·is passive and intrinsic to the material body. His principal example is gravitation, a body's gravity being unchanging and acting through infinite distances. After instancing electricity and heat he claimed that it is probable 'that all force is *incoercible* except in those cases where it is convertible'.[25]

J. *Integrity*. In his memorandum on 'Matter' of 1844 Faraday refers to the close relation between the 'integrity [of the physical forces] & the integrity of the laws which govern the material universe'.[26] To attribute integrity to forces and the laws by which they operate is to conceive such forces as accounting for all natural phenomena and to make nature complete and self-sufficient. The term has also the strong moral association with perfection of design, and this is apparent in a lecture of 1846 in which Faraday dwelt on the interlocking forces of matter which bind 'worlds and suns together' and lead us to recognise 'the harmonious working of all these forces in nature'.[27]

K. *Unity in diversity*. Like the members of the Sandemanian community who work in harmony for the common spiritual good, so the different material bodies and the laws of nature cooperate with one another within the system of nature. Yet each type of force differed markedly from every other force in its mode of operation. Thus he considered that all the powers, except gravity, could readily be altered in the laboratory. Moreover, these powers are independent, each acting according to its own divinely ordained set of laws. However, within this considerable diversity the various powers are interrelated, or, as he expressed the principle before Prince Albert in 1849, there is probably *'unity in one'*.[28] The clear echo of the Christian tri-unity suggests both that the individual powers are mysteriously united and also that the different powers are the outward symbols of the invisible Godhead (Romans 1:20 again!). The point also applies to phenomena, for while there is great diversity in nature's appearances, this diversity is the result of a few simple laws cooperating.

L. *Stability*. For Faraday inorganic forces constituted not just an economy but one which naturally tends to stable equilibrium. From this rule he excepted 'vital or organic powers' which 'are for ever progressive, and have respect to a future rather than a present state'.[29]

M. *Simplicity*. Since the divine plan is perfect, nature manifests simple laws and not the complexity and confusion of a botched design. Indeed Faraday almost links truth with simplicity. This sheds further light on his rejection of mathematical complexities and his strategy of seeking simple laws requiring, at very most, the application of simple proportions. Thus in a letter of 1850 he expressed his wonderment at the 'simplicity of nature'.[30]

N. *Beauty*. As noted in the section 7.5, below, Faraday considered not only that nature is a major source for our own ideas of beauty but nature is beautiful in itself.

O. *No waste*. In this divinely ordained universe, there is a perfect match between ends and means, cause and effect. Since God constructed the world according to His plan, everything has a purpose and there is nothing useless or wasted in the whole of nature. As Faraday stressed to an audience in 1846, 'Nothing is unproductive in nature, there is no residue of action that is useless'. Never do we find, when we trace a cause, 'the least surplusage or deficiency in the amount of power or effect. All the power that God has infused into matter, He uses for various effects in creation'. A further example occurs in a letter to Schoenbein of 1852, in which Faraday claimed that 'there is nothing superfluous, or deficient, or accidental, or indifferent, in nature'.[31]

P. *Plenitude*. A classical expression of the economy of nature is found in the idea of plenitude; that God created nature full, without any wastage or gaps. This principle was frequently applied in the biological sphere to the great chain of being, which was envisaged as representing all possible species and all links in the chain were occupied by existent species.[32] Faraday held this principle in respect to the physical, and probably also the biological, realm. After reporting his discovery of the magnetic susceptibility of oxygen, he raised the 'great question': 'What is the final purpose in nature of this magnetic condition of the atmosphere ...? No doubt there is one or more, for nothing is superfluous there. We find no remainders or surplusage of action in physical forces. The smallest provision is as essential as the greatest. None is deficient, none can be spared'.[33]

Q. *Lines of force.* In his development of field theory in the 1840s Faraday distinguished between normal magnetic substances, which align themselves along lines of magnetic force, and diamagnetics, which orientate themselves across the lines. In considering the relation between lines of magnetic force and matter Faraday employed an economical principle such that they orientate themselves into positions involving a minimum of force. Thus, for example, the lines of force passing through an iron bar will become most concentrated and least disturbed when the bar is aligned with the lines. Again, the lines of magnetic force take the path of minimum resistance – they have a preference, as it were, for paramagnetics over empty space, but try to avoid passing through diamagnetics. Thus both lines of force and matter seek the most economical distribution.[34] It is also significant that Faraday conceived the lines of force associated with a current-carrying wire or a magnet as closed. A closed line is an economical one and symbolises the integrity of the universe. Indeed there is a long tradition which associates closed paths, particularly the circle, with perfection.

While all the above terms incorporate versions of the principle of the economy of nature, they also express Faraday's unswerving belief in a divine plan. All nature manifests God's wisdom, and in studying nature the scientist is trying to decipher parts of this plan. Yet teleology is a somewhat problematic notion, since Sandemanians have generally emphasised its role in the spiritual, not the material, realm. As already noted, they considered that Christians must strive to achieve perfection so that they may be granted eternal life. For Faraday there was also a teleological principle operating in nature, for the divinely ordained laws fulfil some purpose known only to God. As we shall see at the end of this chapter, one aspect of God's plan was that we should use the laws of nature to improve our technology. However, our more immediate concern is that Faraday's science was predicated on a teleological view of nature and that his science was constrained and directed by his belief that nature is a divinely ordained economical system.

7.3 THE UNIVERSE OF POWER

7.3.1 *Powers*

Throughout the eighteenth century the term power had acquired a rich variety of meanings. At the close of the previous century John Locke not only discussed the powers of the mind but he also considered that objects possess power; thus a magnet has the power to attract iron filings. Moreover, bodies possess the power to produce ideas in the mind; hence, to perceive a book as red or a fire as hot was, for Locke, the action of the book's or the fire's power affecting the mind. Reflecting on our knowledge of the external world, Locke claimed that we form the idea of power as an essential part of the complex idea of substance. Indeed, the complex idea of a substance, such as iron, includes a number of simple ideas specifying its powers – its power of being attracted to a magnet, of interacting chemically with oxygen, etc. The notion of power was thus central to Locke's account of mind and, more importantly, to his natural philosophy.[35] Locke's theory of powers was widely accepted during the eighteenth and early nineteenth centuries and Faraday can be located within this broad philosophical tradition.

Closely related to power was the concept of force. The key writer on this topic was Newton, who portrayed space as empty except for small inert particles of matter scattered through it. God activates these particles, and physical forces, such as gravitation, are a manifestation of His activity in the world. The term 'active principle' was used by Newton to refer either to a force or the cause of that force. In the case of gravity an attractive force acts between the centres of two particles such that the strength of the force decreases as the square of the distance separating them. Other centrally directed forces are repulsive, such as the one Newton postulated to account for the reflection of light at the surface of a mirror; that force is responsible for turning back rays of light at the mirror's surface. In Newton's picture of the universe phenomena are produced by particles of matter and their attendant centrally directed forces, both attractive and repulsive, acting across the intervening empty space.[36]

During the eighteenth and early nineteenth centuries there was a general move to interrelate Newton's forces and Locke's powers. Matter particles were clothed in fields of centrally directed forces which were attractive under some circumstances and repulsive under others. For example, the force would need to be attractive at great distances so as to account for gravitation but repulsive at an exceedingly small distance, in order to explain the reflection of light rays and the small volume occupied by condensed bodies. When more phenomena were included, the complexity of the picture increased. There was, moreover, an increasing tendency throughout the eighteenth century to exploit the notion of force as the principal explanatory tool and to downplay the role of matter particles.[37] Indeed, in the mid-century, Boscovich proposed that atoms do not require solid kernels and he reduced them to points which were merely centres of attractive and repulsive force. Many British writers, Faraday included, drew selectively on Boscovich's theory.[38]

Although Faraday can be located within this tradition linking British scientists and philosophers, it is important to notice that his notion of power, which was central to his natural philosophy, possessed profound religious connotations. Faraday's God was all powerful: 'Thine, O LORD, is the greatness, and the power, and the glory, and the victory, and the majesty: for all that is in the heaven and the earth is thine' (1 Chronicles 29:11). The act of Creation was the primary exercise of this power and through examining His Creation we see numerous manifestations of His power. Romans 1:20, which was twice quoted by Faraday in Royal Institution lectures, reminds us that in understanding the 'things that are made' we can perceive not only His power in the Creation but also 'his eternal power'. In the moral sphere the Bible and the life of Christ bespeak God's immense power. For example, in Romans 1:16 it is stated that Christ's gospel is 'the power of God unto salvation to every one that believeth'.

Sandemanians, like many other theologians of revelation, drew considerable attention to the biblical conception of God as all powerful. For example, John Glas, in rejecting John Hutchinson's arrogant claim that only his exposition of the Bible gave the reader a true vision of God's power, argued that Hutchinson had omitted the most important aspect of God's power, for the Bible proclaimed that God manifests his power through the acts of creation and destruction.[39] Likewise, in an exhortation preached on 29 June 1862, Faraday pointed out that happiness comes to those 'who are protected by the power of the Holy Ghost'.[40] Yet for Faraday power was not limited to spiritual power but was the very stuff of the physical world. At the commencement of his 1846 course of lectures on electricity and magnetism he defined his topic as the study of the electric and magnetic powers associated with matter. His subject was not a secular one, since he acknowledged that 'the

Creator had gifted matter' with these powers. Likewise, he concluded the course by declaring that all 'the power that God has infused into matter, He uses for various effects in creation'.[41] In other words, the powers associated with bodies were derived from God.

For Faraday physical nature was not constituted by atoms of inert matter moving in a void, since such a universe was incompatible with his insistence that the physical world manifested God's power; indeed, as will be discussed in the next section, his antipathy to mechanical conceptions of the physical world was rooted in his theology of nature. Instead, every particle of matter was an instance of God's power, and that power manifested itself in any of several active forms; namely heat, light, electricity, magnetism, gravitation, cohesion, chemical affinity and (in some of Faraday's lists) inertia. Absent from this list is motion, which Faraday considered to be a mere effect of one or more of the above powers. Faraday usually referred to the above-listed properties of matter as either 'powers' or 'forces'; indeed, these terms were taken as equivalent in many contexts. Rarely, however, did Faraday provide a definition of either term, although, on one occasion, he defined force as 'the *cause* of a physical action; the source or sources of all possible changes amongst the particles or materials of the universe'.[42]

Every material body contains a certain quantity of matter, which is indestructible. In addition, it possesses each of several powers to some degree. Unlike Newton's forces, which were conceived as acting at a distance, Faraday clearly attributed powers to material bodies, leaving open to investigation the manner in which powers act. Thus, for example, the body might be positively electrified and at room temperature. The amount of each power is not constant, but can be altered; thus 'we are enabled . . . to generate or throw into action certain powers of matter, and transmit them from place to place by certain laws'. We might heat the body by placing a hotter body in contact with it or by locating it in the focus of a burning glass.[43] The divinely created laws of nature govern the relation between the powers possessed by material bodies. A simple law would, for example, govern the gain in temperature of the aforementioned body and also the loss in heat power of the hotter body which came into contact with it.

For Faraday the Sandemanian the strands of power run from God through all aspects of our lives and through the physical world that He had created. Faraday therefore rejected the natural theologians' conception of God as a skilled artisan or craftsman. Instead of fashioning the world at the Creation, as a sculptor fashions a statue from a piece of stone, Faraday's God spoke the physical world into existence. The power thus imparted to the physical world at the Creation has subsequently been conserved and therefore the world Faraday experimented on contained as much power as God originally introduced into the system. Faraday's universe was a universe of power, and that power was of divine origin.

There existed, moreover, a scale of natural powers, with gravitation at the lower end and, usually, electricity at the top. Electricity is the most powerful and also the most esoteric of God's agents. As one of his nieces reminisced, Faraday was fascinated by thunderstorms, and she attributed this fascination to their religious significance:

> I shall never look at the lightening flashes without recalling his delight in a beautiful storm. How he would stand at the window for hours watching the effects and enjoying the scene; while we knew his mind was full of lofty thoughts, sometimes of the great Creator, and sometimes of the laws by which He sees meet to govern the earth.[44]

We can detect a biblical resonance here, since several passages, such as Psalms 77:18, likewise portray lightning and thunder as God's voice and His most potent signs. Moreover, the Book of Revelation predicts that there will be voices, great earthquakes, thunders and lightnings at the time of the Apocalypse.

While many scriptural exegetes have dwelt on the role of light in the Creation and on the metaphysical relation between the physical light, the spiritual light, and the person of Christ – 'I am the light of the world' (John 8:12) – the only evidence that Faraday came close to this tradition appeared in a very early lecture in which he referred to the sun as the 'great source of imponderable matter ... whose office it appears to be to shed these subtle principles over our system'.[45] Light metaphysics was totally lacking from his later writings. Moreover, light played a relatively small role in his research, partly because physical optics had become a highly mathematised subject and was therefore largely closed to him. Although he investigated the relation between light and magnetism in the mid-1840s and also speculated that lines of force might propagate light, electricity was far more central to his researches. Despite claiming that light and electricity were the 'two great and searching investigators of the molecular structure of bodies', he made relatively limited use of the former. Even in lectures where there was much scope for introducing the subject of light, such as his 1859–60 course for juveniles on the various forces of matter, Faraday usually omitted the topic.[46]

By contrast with the study of light, electricity was not only central to his scientific research, being the subject of numerous papers and lectures, but it was also for Faraday the most magnificent and potent of God's powers. In his view, electricity was the highest power known to man.[47] He stood in awe of electricity, its very power emphasising the vast difference between man and God. While Faraday devoted his religious life to studying the Bible and becoming perfect like Christ, he spent much of his scientific career studying the most potent physical force God had created.

7.3.2 Against atomism

Writing to his sister in June 1834, the Irish mathematician William Rowan Hamilton related a meeting with Faraday over breakfast at the Bilton Hotel. Hamilton was obviously impressed by Faraday but he was particularly delighted to discover that Faraday, 'who is the most distinguished *practical* chemist in England, has been led to almost as *anti-material* a view as myself. ... He finds more and more the [particulate] conception of matter an incumbrance and [a] complication in the explanation of phenomena, instead of an assistance'. Hamilton also noted that whereas his own anti-material-ism was largely based on philosophical arguments, Faraday had reached much the same conclusion by arguing from experimental evidence. 'He sees no proof from chemical facts, or from the phenomena of definite proportions, for the existence of those little bulks or bricks, of which so many fancy the outward world to be built up.' Moreover, he was delighted to note that Faraday had even abandoned the view that electricity consists of minute particles constituting one or more fluids.[48]

Over breakfast Hamilton and Faraday had joined in common cause against the prevalent view, based on Newton's writings, that matter consists of small, hard, impenetrable particles separated by void space. In 1808 John Dalton suggested that the particles composing material bodies were its indivisible atoms and that every chemical element was characterised by its own independent atoms, each type of atom possessing a different weight. Although Dalton's theory was widely criticised, it was consistent with the earlier view in that it postulated small particles (atoms) and void. It is not clear whether Faraday ever wholeheartedly adopted traditional atoms, for even in his early lectures to the City Philosophical Society, when he claimed to have followed Newton, he, like Davy, had emphasised the role of powers and ignored the notion of particles (or atoms).[49]

In the months before his meeting with Hamilton Faraday pursued electrochemical researches which were concerned with the funda-mental interaction between electricity and matter. In the course of these researches he had become progressively critical of the theory that matter consists of small, hard, impenetrable atoms. For example, towards the end of his January 1834 paper on electrochemistry he defined equivalent weights of different chemical substances as the ratio of their weights containing the same quantity of electricity. However, he added, '*if* we adopt the atomic theory or phraseology', then we can consider equal quantities of electricity associated with atoms which possess the same chemical action. 'But', he added, 'I must confess I am jealous of the term *atom*; for it is very easy to talk of atoms, it is very difficult to form a clear idea of their nature'.[50] A few months later, he wrote to a correspondent that he had 'thought long and hard about the theories of attraction and of particles and atoms of matter, and the more I think (in association with experiment) the less distinct does my idea of an atom or particle of matter become'.[51]

His major public attack on atomism was reserved until January 1844 when, in a Friday evening discourse at the Royal Institution, he not only criticised the speculative nature of atomism but pointed out that the atomic doctrine led to mutually contradictory views about the electrical conductivity of space. In the same talk he expressed his preference for Boscovich's theory, which eliminated the notion of hard, discontinuous particles and instead emphasised the role of continuous forces operating through space. As we shall see, one of Faraday's main criticisms of the atomic theory was that it required void space between the atoms.[52]

Although he occasionally used the term atom in his research papers, Faraday was much more at ease in referring to *particles of matter* or sometimes *molecules*. A particle was, for Faraday, merely a small part of a material body. Thus if you were to cut this book into very small pieces, each would be a particle but there would be no need to assume that those particles were the smallest possible components or the ultimate primary constituents of the book. Moreover, in his researches on electromagnetism he conceived induction as consisting 'in a certain polarized state of the particles'; it was impossible to conceive of an atom being polarised and thus induction could not be understood on the atomic theory.[53] By often engaging in particle talk he avoided atom talk, with its attendant assumptions about indivisibility and empty intervening spaces. Faraday could legitimately apply the 'analogy of nature' which Newton had used to argue that the 'least particles' of bodies possess the same primary qualities – such as extension, hardness, impenetrability and mobility – as we find in gross bodies.[54] Faraday refused to transfer these qualities to hypothesised atoms sunk in empty space, since the world we experience and experiment on is not populated by these discontinuous entities. Instead he considered that we should assume that particles possess the same properties and powers – such as the electric, magnetic and chemical powers – as the larger pieces of matter of which they were part and on which he was able to experiment. This was for him an important metascientific principle.

To this widely accepted picture of atoms, empty space and mediating forces has to be added the causes of heat, light, electricity and magnetism. For most writers of the period these were produced by ethereal fluids which were often conceived as consisting of exceedingly minute particles that repelled one another strongly. Heat was either the concentration of such a fluid (caloric), in the interstices of bodies, or its vibration. Light was either subtle particles projected at high speed from a luminous body or the vibration of an ubiquitous ether. Electricity (and similarly magnetism) was attributed to the distribution of one or two electric fluids; in the latter case one fluid would be positive and the other negative. The distance effects of electricity were often explained by the action of the electric fluid or by the influence of its associated forces.[55] Despite the considerable diversity among such theories, they all attempted to explain phenomena in terms of subtle matter and interparticulate forces, and many of the contributors to this tradition drew explicitly on Newton's speculations about a particulate, ubiquitous ether.

In the notes to a lecture delivered to the City Philosophical Society in 1819 Faraday discussed the four states of matter, namely, solid, liquid, gaseous and radiant. The theory of the fourth, radiant, state accounts for heat, light and possibly electricity as being composed of matter in this most subtle state. Yet against this topic Faraday added the assessment, 'Purely hypothetical'.[56] His scepticism about such issues remained throughout his scientific career. Thus, for example, he refused to commit himself either to a corpuscular theory of light or to the theory, which rose to dominance around 1830, that light is the vibration of a ubiquitous ether. There was much discussion about the structure of the ether and many different models were proposed, most of which assumed that it consisted of minute particles under the influence of very strong inter-particulate forces. Some writers, however, resisted these physical interpretations of the ether and instead sought to account for optical phenomena by solving the appropriate wave equations. Although Faraday did not engage the mathematics of wave optics, he was prepared in 1846 to entertain the view that light was a vibration, not of ether particles, but of the lines of force that he envisaged filling space. Yet, as he made clear, his strategy did not merely replace ether by lines of force, but eliminated atom-like ether particles and also dissolved the boundary between material bodies and the intervening space.[57] Similarly he rejected theories that postulated electric fluids comprised of minute atom-like particles.

Most of Faraday's explicit objections to atomism engaged methodological or scientific arguments which were, in turn, coloured by his religious conception of nature and of the scientist's role. More directly, atomism stood opposed to Faraday's conception of the universe filled with powers of divine origin. We see this opposition in an 1844 memorandum on matter in which Faraday favoured Boscovich's point centres of power over the traditional atomic theory. Might not God, speculated Faraday, 'just as easily by his word speak power into existence around centres, as he could first create nuclei [atoms] and then clothe them with power?'[58] This rhetorical question should be read as affirming that God was more likely to have created Boscovichian point atoms than hard billiard-ball atoms. The other theologically loaded argument against atomism was derived from Faraday's rejection of void space, which is the topic of the next sub-section.

7.3.3 Denying the void[59]

It has often been claimed that Faraday rejected the notion of action at a distance. This is certainly not true in the case of gravitation and only partially true in respect to the other powers in nature. While the gravitational power acted over considerable distances, the other powers could only be communicated locally. The meaning of local action becomes clear from Faraday's response to a well-known letter of 1840 in which the American scientist Robert Hare picked on an apparent contradiction in Faraday's eleventh and thirteenth series of electrical researches. In the first of these Faraday had denied that ordinary induction is produced by particles acting over 'sensible distances' and instead had attributed induction to 'an action of contiguous particles'. In the latter series he seemed to be saying that contiguous action might well imply a particle acting on another half an inch away. Surely, asserted Hare, half an inch is a sensible distance, and he therefore charged Faraday with inconsistency.[60] Since Faraday's subsequent remarks did little to clarify the objection, it has usually been taken that Faraday was confused and inconsistent. However, David Gooding has recently offered an insightful re-evaluation in which he shows that at worst Faraday did not make himself clear in public. Gooding's reconstruction turns on Faraday's distinction between the different ways in which gravity and induction act. Gravity is the paradigm case of action at a distance and it acts along the *straight line* drawn between the centres of the two masses in question. By contrast, the influence of induction depends on the particles comprising the intervening medium (or space), and therefore it can act along curves. Looking more closely at the process, one particle acts on its neighbour and only on its neighbour, which in turn acts upon the next particle in the sequence, and so on. Thus for Faraday contiguous action forbids a particle acting directly on particles which are further away than its immediate neighbour. Moreover, on this account contiguity forbids the kind of action at a distance involved in the gravitational example.[61]

Although this theory of local action still requires that one particle can exert an effect on another across the intervening space (which may be empty of gross matter), there is another way of interpreting this which lends itself more readily to Faraday's conception of the economy of nature. This reinterpretation is suggested by a paper on vaporisation, dated 1830, in which Faraday discussed, with obvious approbation, the view propounded by both Davy and John Dalton that particles of a chemical element possess an attendant 'sphere of influence'.[62] This operates in the locality of the particle and may be conceived of as an atmosphere reaching a small distance from the particle and thus affecting its neighbour. If this is what Faraday visualised, then void space had no role in his matter theory, since the power associated with one particle fills its immediate neighbourhood. By thus filling space with power, power becomes a property not only of particles of matter but also the interstices between these particles. In other words, powers are not localised but suffuse space. If this interpretation is correct, it shows Faraday trying to eliminate the void, a concept that was incompatible with his metascientific principle of plenitude and thus with the economy of nature.

This interpretation receives support from Faraday's frequent tendency to speculate that the space separating gross bodies is not empty but possesses various properties; thus it is not void. For example, in his 'Thoughts on ray-vibrations' (1846) he conceived lines of force extending through space so that an electromagnetic field is present at every point.[63] More than a decade later he confronted the problem of how the gravitational force might be formulated in a manner commensurate with the principle of force conservation and thus the economy of nature. Since the gravitational force increases as two bodies are moved closer to one another, this increase seems to demand a creation of force and is thus incompatible with force conservation. Some other force must be present and the magnitude of that force must decrease as the gravitational one increases. Space is a possible source for this balancing force, and Faraday envisaged that, far from being void, space might be conceived as a reservoir of power.[64]

When, in 1852, Faraday speculated on the reality of magnetic lines of force, he conceived that they indicate the existence of a magnetic medium extending into space beyond the confines of the magnet. Indeed, since the patterns adopted by the lines of force depend on the substances placed in their locality, the space surrounding the magnet is 'a part of the true and complete magnetic system'. Space is therefore not void but will be endowed with magnetic qualities by the presence of magnetic sources. As Faraday had shown in earlier experiments, lines of magnetic force can pass unhindered through the best vacuum he could produce mechanically. While acknowledging that space behaves differently from matter in respect to magnetism, he argued that the (apparent) vacuum should not be understood as a mere receptacle but as bearing a magnetic relation to ordinary matter. Although this space is devoid of particles of gross matter something must still be present, but what it is 'I cannot tell, perhaps [it is] the aether'.[65] It was a commonplace among eighteenth- and nineteenth-century scientists to appeal to the concept of an ether suffused through space. While for many such authors the ether was conceived as composed of particles with inter-particulate forces, a number of alternative models were available by the mid-nineteenth century. Although Faraday occasionally speculated on the nature and properties of the ether, sometimes likening it to a fluid in space, he avoided committing himself to any specific model of the ether. However, in line with the argument developed in this chapter, Faraday appears to have attributed to space (and thus to ether) magnetic and electrical properties, rather than the traditional mechanical ones.[66]

A further sustained discussion of space occurred in Faraday's 1844 paper, 'A speculation touching electric conduction and the nature of matter', in which he argued that, according to the atomic theory, the (void) spaces between particles of a (conducting) metal must be able to conduct electricity. This theory likewise required that in the case of an insulator, such as shellac, the space between particles must insulate. The theory was therefore inconsistent, since void was sometimes considered to be a conductor and sometimes an insulator. Sidestepping the problems of atomism, Faraday appealed to Boscovich's theory and suggested that in place of hard atoms and empty space we conceive of 'mere centres of forces or powers'. The nucleus of the atom thus disappears and so does the apparent contradiction concerning the electrical properties of space. All physical phenomena are explained by the actions of these forces or powers. Moreover the distinction between matter and intervening space also disappears and 'all matter will be *continuous* throughout'.

If matter and force are synonymous and forces stretch to the extremities of space, then this theory implies that 'matter fills all space'.[67] This view was by no means uncongenial to Faraday, not least because it removed void and thus more adequately exemplified the principle of plenitude and the economy of nature.

Faraday would not have read Boscovich's books, being in Latin, but he knew of Boscovich's theory from the writings of such British commentators as Joseph Priestley, some of whose chemical works he had read.[68] The connection with Priestley and others has led some historians to locate Faraday within the Newtonian–Lockean tradition that emphasised the primacy of power. Among the various possible influences must be included that of Humphry Davy, who repeatedly spoke of electrical powers and the powers of chemical affinity.[69] While much light can be shed on Faraday's views about powers and forces by locating him within the British natural philosophical tradition, there are facets of his views that make him peripheral to, if not outside, that tradition. Most importantly, he repeatedly questioned, and sometimes positively rejected, three of the fundamental tenets of that tradition: that there are hard, impenetrable, indivisible atoms; that there is void space between those atoms; and that centrally directed forces act at all distances from those atoms. Whatever the influences on Faraday from the various traditions in British natural philosophy, it is important to appreciate that, from the early 1830s to the late 1850s, he progressively developed his own views about matter, powers, space and fields. In reaching this synthesis his commitment to understanding the world as a divinely ordained economy played a prominent role.

Particularly during the 1840s Faraday sought ways of unifying matter and space so as to obviate the need for void (which lacks all properties except extension). For Faraday, all regions of space, irrespective of whether they were populated by material bodies, possessed electrical and magnetic properties. Lines of force likewise permeated all of space. In filling all of space with electric and magnetic properties Faraday was employing a principle of plenitude in his natural philosophy. However, the principle of plenitude has biblical roots and many writers, such as the Hutchinsonians, who sought to base their natural philosophy on Scripture, often cited Deuteronomy 33:16 and Isaiah 6:3 to justify their contention that the world was full of matter with no room for void space.[70] While Faraday did not subscribe to the Hutchinsonian theory of an ether circulating throughout the solar system, his mature view about space incorporated the principle of plenitude. Here, then, we can detect a resonance between the Bible and his natural philosophy.

7.4 CRITICISMS OF ENERGY PHYSICS

Faraday proposed a principle of force conservation which, at first sight, appears strikingly similar to the principle of the conservation of energy (the first law of thermodynamics) that became a central pillar of science in the second half of the nineteenth century. Indeed, Faraday has sometimes been hailed as one of the main formulators of that principle, alongside James Joule, Julius Robert Mayer and Hermann von Helmholtz.[71] Yet closer analysis shows that, while all these scientists expounded views about conservation, Faraday was strongly opposed to the concept of energy conservation, as propounded by Joule, among others, with its implied assumptions about the nature of matter and force. In this section we shall examine his reasons for this opposition, which shed considerable light on the theological underpinnings of his concept of power.

In a lecture series delivered in 1846, Faraday claimed that we cannot comprehend the origin of the powers in the universe or how those powers are united to matter; these are known to God alone. Instead, the role of the scientist is limited to discovering the effects of these powers, the laws by which they act and the relationship between them.[72] This last problem continually preoccupied Faraday and underpinned much of his research work. At the Creation God had made a perfect, providentially designed system, so there could be no loss of power by purely physical processes. The total amount of power must be preserved in the economy of nature and it can neither be created nor destroyed. However, we have no direct access to the absolute quantity of force – what he called the 'parent force' in one of his lectures[73] – associated with any piece of matter. All we can do is to determine the relation between specific observable powers, such as heat and electricity. Here we encounter one of the most important metaphysical principles in Faraday's science, for he firmly believed, as an implication of the indestructability of force, that a specific relation existed between detectable forces. However, it is difficult to interpret clearly Faraday's statements about that relation, since he used several different but closely related, terms.

In the winter of 1859–60 Faraday discussed in a series of juvenile lectures the 'various forces of matter and their relations to each other', and in the final lecture he used the term 'correlation' to specify the relation between the different powers or forces. As noted above, the term, popularised by Grove, meant that two forces are correlated if the increase in one corresponds to a decrease in the other.[74] Although Faraday used this term extensively in his lectures and illustrated the correlation of forces with many examples, there

were several places in the same lecture series where he appealed to
the stronger thesis that force is converted. Conversion, it should be
stressed, implies not just that the increase in one force correlates with
the decrease in another but that these various powers 'may be
changed one into the other', as in '*convert[ing]*' chemical affinity into
electricity or magnetism or any other force. Further evidence that he
did not distinguish these two terms clearly occurs on the final page of
the text, where he expressed the hope that he had proved 'the
universal correlation of the physical forces of matter, and their
mutual conversion one into another'.[75] It is not clear whether
Faraday distinguished these terms in 1859–60, although on other
occasions he wanted to demarcate them unambiguously.

Some 3 years earlier Faraday had entitled a Friday evening
discourse 'On the conservation of force'. Here he had argued a
thesis stronger than the correlation of forces. The crux of his
argument was that all theories about the physical world must be
consistent with the metaphysical but necessarily true principle that
force can neither be created nor annihilated. He argued that since
the gravitational force, as generally understood, appeared incompatible with this principle, the theory of gravitation needed to be
reformulated in order to remove this discrepancy. Faraday's views
on gravitation will be discussed in section 9.2, but for the present it is
important to notice that the only argument he used to support the
conservation of force was a theological one. He recoiled in horror
from the idea that powers might be created and thus able to increase
to an unlimited extent. Equally he was appalled by the prospect of
the '*annihilation* of force – an effect equal in its infinity and in its
consequences [unspecified] with *creation*'. While it is possible to
imagine universes where physical processes lead either to the
creation or destruction of forces (or a combination of the two), it
is clear that Faraday's revulsion against these possibilities stems from
his theology of nature. It is, he claimed, 'only within the power of
Him' to create or destroy force.[76] The amount of force in the
universe must therefore remain constant, or, in a slightly different
reformulation of the principle of the economy of nature, force must
be conserved.

Faraday used the terms correlation, conservation and conversion
almost interchangeably as formulations of the fundamental principle
that nature is a divinely ordained economy. Granted this termino-
logical difficulty, it will assist us to understand Faraday's response to
Joule if we can clarify the terms and some of the differences between
them. Correlation, which merely claimed a relation between observable properties, is the least rigorous of the three terms for the
following reasons. Firstly, in contrast to correlation, conservation

implies that some entity is conserved. Thus, together with the observable powers (which the theory of correlation addresses) there has to be some common stuff, call it power, energy, or what you will, lying behind the observable phenomena. Faraday was prepared to acknowledge that force could neither be created nor destroyed but was reticent about identifying force (or power) as an abstract entity that transcended the individual types of force. Although he was occasionally prepared to speculate on the existence of this substance, he generally erred on the side of caution and refused to proceed beyond the claim that force was conserved, since it could neither be created nor destroyed. Moreover, since 'we know nothing of [the] . . . origin' of powers or how they are united with matter, we are outside the legitimate bounds of science if we try to specify power or force as an abstract entity. Since all powers in the physical universe are manifested by God, speculations on the nature of force may be encroaching into the area of theologically dangerous hypotheses. Put simply, while correlation concerned the relation between observables, conservation was a metaphysically charged term.

Secondly, correlation needs to be distinguished from conversion. The difference can best be clarified if we examine Faraday's response to James Joule's paper, entitled 'The mechanical equivalent of heat', which Faraday communicated to the Royal Society in June 1849. In the letter accompanying the paper Joule requested Faraday to supply 'anything which requires addition or alteration . . . particularly with regard to the mechanical Doctrine'.[77] We do not know whether Faraday made any alterations when the paper was read at the Royal Society but he subsequently refereed the paper before its publication in the *Philosophical Transactions*. Faraday particularly objected to three phrases that Joule had used in the original version: these were Joule's claims (1) that *heat is convertible into force* and *force* into heat', (2) that friction consists 'in the conversion of Force into Heat' and (3) that a quantity of heat is 'equal to a mechanical force'. Faraday's objection to the first of these formulations becomes clear when it is set against his own claim in 1857 that under certain conditions heat is 'converted into mechanical force'.[78] Thus Faraday read Joule as denying that heat is a power or a force, a doctrine which, as we have seen, was part of Faraday's theory of power. Moreover, he interpreted Joule to be claiming that heat is merely the interstitial motion of particles. This disagreement (which will be discussed further below) gave rise to Faraday's complaint in his referee's report that Joule should only have reported the facts 'without any hypothesis as to their cause'.[79]

The above methodological criticism applied also to the second of Joule's claims cited above. The hypothetical nature of this claim becomes clear when set against the statement in one of Faraday's lectures that 'Friction will produce heat'.[80] Here Faraday treated friction as an observable form of action, while the heat developed by friction, i.e. the increase in the body's power of heat, was another observable. These two forms of power are correlative and no assumption need be made about the nature of friction or about whether friction is converted into heat. Thus, for Faraday, Joule was incorrect to think of friction, as it were, converting the body's interstitial forces into heat. The third phrase from Joule, (3) above, likewise involved the presupposition that heat (the interstitial motion of particles) is obtained from mechanical motion by the direct transfer of motion. Once again Joule had taken the mechanical theory of heat for granted.

Thirdly, unlike correlation, both conservation and conversion imply that there is a quantitative relation between the two forces. If you start with x units of force in the form of heat and convert it all to electricity, then you should end up with x units in the form of electricity. This procedure presupposes either that there is a common yardstick for measuring all forms of force or that the mathematical functions specifying each type of force can be related through a simple linear equation. Joule provided a classic example in his paper on the mechanical theory of heat, in which he determined the numerical constant relating mechanical work to heat (the so-called mechanical equivalent of heat). In brief, his apparatus consisted of a pair of weights which, in falling, turned a paddle-wheel immersed in a copper vessel containing water. The mechanical work produced (in ft-lbs) by the fall of the weights, i.e. the input to the system, was measured by the sum of the weights (W in lbs) multiplied by the distance through which they fell (h in ft). The heat produced in the water – the output – was measured by the product of the weight of water (M in lbs) multiplied by the temperature rise (T in degrees Fahrenheit). Thus Joule calculated the mechanical equivalent of heat (J) as the ratio between these numbers, viz.:

$$\text{Mechanical Equivalent of Heat, } J = \frac{\text{Mechanical Work}}{\text{Heat Produced}} = \frac{Wh}{MT}$$

Joule concluded his paper by noting that his experiments showed the heat produced to be proportional to the mechanical work expended. Moreover, 'the quantity of heat capable of increasing the temperature of one pound of water (weighed *in vacuo*, and taken at between 55° and 60°) by 1° Fahr. requires for its evolution the expenditure of a mechanical force represented by the fall of 772lb. through a space of one foot'.[81] The last figure specifies the mechanical equivalent of heat in ft-lbs.

Since Faraday was committed to the indestructibility of force, it would seem that he should have discovered the mechanical equivalent of heat or one of the other equivalents linking different powers or forces. What may seem to us a small step was an impossibly large one for Faraday. As noted above, Joule was prepared to consider the conversion of force, which made far more claims about the transfer of force (as an independent entity) from one form to another. By contrast, Faraday asserted both the empirically based correlation thesis and the metaphysical claim that force is indestructible, and therefore conserved.

There were, moreover, ways in which Faraday's approach to conservation was radically different from Joule's. For example, the calculation of the mechanical equivalent of heat required just the kind of manipulation of symbols and numbers which proved so inimical to Faraday (see section 8.4). There were, however, a few occasions when Faraday expressed the relation between forces in numerical terms or invoked the idea of force equivalents. For example, in a *Diary* entry for November 1837 he noted, 'Compare corpuscular forces in their amount, i.e. the forces of Electricity, Gravity, chemical affinity, cohesion, etc. and give if I can expressions of their equivalents in some shape or other'.[82] However, Faraday did not pursue this plan, and it is noticeable that of the several 'equivalents' known in his day he was predominantly interested in one, the relation between electricity and chemical affinity. For example, in a lecture on electrochemistry in 1846 he told his audience that one grain of water is decomposed by 800,000 discharges of a large collection of Leyden jars. However, it must be stressed that this numerically expressed equivalent does not involve the ratio between two different natural powers but rather the relation between two electrical actions. On the one hand, there was the electricity delivered from the Leyden jars, while on the other, there was the 'amount of electricity contained in a single drop of water, measured by chemical affinity'.[83] The chemical affinity displayed by the water was not a separate, incommensurable species of force but was dependent on the electricity within the water. As he had written in his *Diary* more than a decade earlier, '*affinity* IS *electricity and vice versa*'.[84] In other words, in comparing the decomposition of a grain of water by the repeated discharge of a Leyden jar he was comparing like with like: the two forces were of the same kind and could therefore be compared.

Thus Faraday not only equated electrical and chemical power, which he thought to be identical, but argued that the electrochemical decomposition of water provided an *'absolute measurer'* for electricity: one hundredth of a cubic inch of hydrogen collected (at a specified temperature and pressure) would define 1 degree of electricity.[85] Moreover, he sought to demonstrate that the electricities produced in different ways – viz. by electrochemical, electromagnetic and electrostatic processes – were merely varieties of the same species of power.[86] By contrast, Faraday avoided determining other so-called 'equivalents', since they involved two distinctly different and thus incommensurable species of power. In particular he ignored the 'equivalent' between heat and the mechanical power, the important relation researched by Joule and Thomson, among others.

7.4.1 *Opposing the mechanical philosophy*

The history of modern physical science has frequently been discussed in terms of the rise and elaboration of the mechanical philosophy.[87] This term has often been applied to those seventeenth-century writers who sought to explain physical phenomena in terms of particles of matter and their motions, all transfer of motion occurring by contact action. Newton, however, postulated action at a distance forces in place of collision and the term mechanical philosophy has therefore been extended to refer to theories which explain phenomena in terms of matter and force. Thus, Newton's account of the motions of the planets in the solar system, appealing to the three laws of motion and a centrally directed gravitational force attracting the planets to the sun, is the paradigm example of a mechanical system. Much of the progress, particularly in classical physics, has been attributed to the extension of this mechanistic programme to an ever-increasing range of phenomena. From this perspective, mechanics is the queen of the sciences.

Faraday is unusual, but not unique, in that he stood outside this pervasive tradition. As discussed above, Faraday's universe was a universe of power and not a mechanical system. Motion was not listed among the powers that bodies exhibit; indeed, it possessed a rather ambiguous status in Faraday's natural philosophy and was usually treated as the by-product of a power and not as a power in its own right. Thus, for example, when an electroscope was charged, the leaves diverged until they reached a maximum. For Faraday the divergence was due to the leaves possessing similar electrical states, which produced a stress in the intervening space, thus causing the leaves to take up new positions. Yet in pursuing this form of analysis Faraday paid no attention to the motions of the leaves *per se* and did not attribute power to that motion. Since he did not classify motion among the powers, the powers of matter (such as electricity, gravitation, etc.) could not be identified with or reduced to matter in motion. Mechanics was, for Faraday, not the queen of the sciences, and his natural philosophy was sharply opposed to the prevalent mechanical philosophy.

Although he omitted motion from his list of powers, Faraday sometimes included inertia. In a lecture of 1847 and in his force conservation essay of 1857 he briefly discussed inertia and its relation to the gravitational force. Inertia was, for Faraday, an internal principle of bodies which bore some relation to gravitation. Gravitation was the power responsible for moving two distant bodies towards one another, and as they approach 'each will have stored up in it, because of its *inertia*, a certain amount of mechanical force'. In this instance inertia, it appears, was not itself a force but was the principle which conserves mechanical force within a moving body, since it 'enables a body to take up and conserve a given amount of force until that force is transferred to other bodies, or changed into an equivalent of some other form'. Although Faraday acknowledged inertia as 'a pure case of conservation of force', it possessed an ambiguous status within Faraday's universe of powers.[88]

Faraday's non-mechanistic conception of science can be illustrated by his discussion of the excitation of electricity in 1838. The paradigm example of electrical excitation was clearly provided by the battery in which chemical forces produced electricity and, so he speculated, the intensity of the interaction should increase as the distance between the reactive particles decreases. Then he turned to the excitation of electricity by friction and sought to apply a similar model. When two bodies were rubbed, particles of different kinds are brought into close proximity and their interaction excited electricity. On this account the motion of the two bodies had no role, other than in bringing their constituent particles close together, and there was, moreover, no sense in which mechanical motion (or force) was related to electrical force. However, what principally concerned Faraday was the operation of the chemical and electrical forces (which he often equated) and these did not depend on any mechanical principles but only on the spatial proximity of the particles.[89]

A further example is provided by Faraday's conception of heat, and here the contrast with Joule will again prove helpful. Joule must be counted as a mechanical philosopher, since he considered heat to be the vibration of the particles constituting a body. Thus when water is heated by a mechanical process, work is performed, and the (apparent) consumption of work involves the transfer of that quantity of motion to the particles of water, whose motion (or more precisely 'vis viva') thereby increases. Joule was greatly concerned with expressing this process in terms of measureable parameters. The work was expressed in terms of a weight and the distance through which it fell, while the increase in the 'vis viva' of the water particles was measured by its perceptible effect, namely the mass of water multiplied by its temperature increment.[90] Crosbie Smith has rightly stressed the importance of the concept of work for Joule and others, most especially William Thomson. For Thomson not only heat but also electricity had to be expressed in terms of its mechanical effect, and that effect could be measured and thus related directly to the work performed or consumed. Thus the work performed by a heat engine or an electric motor offered an absolute measure of physical quantities. Utilising this approach, Thomson had developed his absolute scale of temperature.[91]

By contrast, Faraday dismissed both the dynamic theory and the caloric theory (according to which heat is a fluid, caloric, obeying mechanical laws). Instead he considered heat to be a power possessed by the body, and although that power might manifest itself in the form of motion (perhaps in driving an engine), heat was *not* motion. Even when the expenditure of heat led to the production of motion, as in a steam engine, Faraday did not conceive this motion as a measure of the heat power.

One reason why Faraday was so strongly opposed to the mechanical philosophy adopted by many of his contemporaries was his lack of a formal scientific education. Those contemporaries, such as Whewell and Herschel, who had attended Cambridge University and sat the Mathematical Tripos examination, were trained to solve problems in mechanics with the aid of mathematical methods. By contrast, having sat at Davy's feet and been trained principally in practical chemistry, Faraday was not exposed to mechanics and doubtless imbibed some of Davy's strong preference for talking about the activity of chemical and electrified bodies.[92] However, lack of education in mechanics does not seem a sufficient cause for Faraday's persistent and dogmatic rejection of the mechanical philosophy.

Faraday's rejection of the mechanical philosophy takes on further theological significance if we extend the contrast with Joule into their respective theologies of nature. Joule, from Congregationalist stock, argued in a public lecture in 1847 that living force ('vis viva') cannot be created or destroyed 'because it is manifestly absurd to suppose that the powers with which God has endowed matter can be destroyed any more than that they can be created by man's agency'.[93] Although Faraday could equally well have written these words, it is important to notice that such a belief in the economy of nature was subject to a number of different and even incompatible interpretations. For example, Joule and Thomson, who subscribed to latitudinarian principles and to natural theology, envisaged their God conserving the mechanical operations of nature in accord with the principle of conservation of work and 'vis viva'. Joule could speak of 'the wisdom and beneficence of the *Great Architect* of nature' and of 'the *entire machinery*' of the universe working 'smoothly and harmoniously' – terms from natural theology that connote God as the designer of a mechanical universe.[94] By contrast, as a Sandemanian Faraday interpreted the Bible in a more literal manner and portrayed the God of Genesis as the source of *power*, which is conserved. The powers we observe in matter are ultimately derived from that power but they are not reducible to mathematics or mechanics. Faraday's God was no artificer or engineer.

7.4.2 *Faraday and utilitarianism*

A very different explanation for Faraday's antipathy towards the mechanical philosophy is provided by Crosbie Smith, who has noted Faraday's lack of contact with industry when compared with the industrial contexts in which both Joule and Thomson worked, the former in Manchester and the latter in Glasgow. The efficiency of steam and electrical engines was an issue of crucial industrial and economic importance in those two mighty engineering and manufacturing centres. Joule and Thomson were thus imbued with the social and economic values that placed a premium on efficiency, and much of their work was aimed at determining and improving the efficiency of machines. To achieve this they needed to develop theories that predicted the maximum work that could be obtained from a given input, since only then could the amount of wasted power be ascertained and steps taken to reduce that loss and the attendant financial losses. Smith argues that many aspects of their scientific work can be understood as deriving from these industrial concerns. For example, they placed great emphasis on devising ways of measuring the input and output of machines, and Thomson in particular was concerned with electrical standards that could easily be replicated in the artisans' workshop. Again, they made extensive use of the terminology used by engineers, such as 'work' and 'vis viva'.[95]

Several significant contrasts can be drawn between Faraday and both Joule and Thomson. Faraday became an ornament of metropolitan science who was associated with the capital's scientific institutions and not the provincial literary and philosophical societies.[96] Working in London, he was geographically isolated from the main industrial centres, where engines were put to industrial uses and where their efficiency was of concern both to manufacturers and to the whole community. He neither used the engineers' vocabulary nor did he centre his research on the engineers' problems, arising especially from mechanics and heat theory. Ironically, while he discovered the scientific principles underlying the electric motor (even inventing a primitive form of electric motor) and the laws governing electrochemistry, he left to others the industrial application of his work in both these areas.

Although such contrasts can be drawn, they should not be overstated, and we must not portray Faraday as totally dissociated from applied science. His father was, after all, an ironsmith and the young Faraday grew up not only near the smithy but also in an area teeming with workshops. Moreover, as Berman has forcefully argued, he frequently performed (small-scale) chemical analyses, experimented on improving optical glass, advised Trinity House and had extensive contact with manufacturers.[97] While London offered different social and industrial experiences from Manchester, there are some far more important issues that set Faraday apart from both Joule and Thomson. To appreciate these we need to examine Faraday's views about the application of science.

As discussed in section 7.2, Faraday held a teleological view of nature. A further aspect of his teleological commitment was that the history of mankind also forms part of God's plan. Since all parts of His plan are interconnected, the laws of physics and chemistry must be of use to us. As Faraday told an audience in 1846, 'the Creator has conferred' the power of gravitation on all matter 'so that *it may serve our purposes*'. The uses of the power of gravitation are numerous, including the operation of most types of machine. More generally, the powers of nature operate 'always for our good'. The scientist should therefore understand that the discovery and subsequent utilisation of natural processes confers great benefits on mankind. Thus in the applications of electricity, such as Wheatstone's electric telegraph, the 'gifts of God' are conveyed to man.[98] Faraday's numerous ventures in applied science – from seeking stronger, less brittle steel to improving the illumination of lighthouses – need to be understood as instances of this ideal which arose from this vision of God's providence. Through the applications of science scientists can help mankind to advance and they therefore have a duty to apply scientific knowledge.

Thus far Faraday appears to have been articulating a utilitarian philosophy, and Berman has even argued that Faraday should be firmly bracketed with the Utilitarians.[99] However, it must be stressed that Faraday was no secular utilitarian nor did he consider that the primary aim of science was to improve the physical condition of mankind via technology. Instead, the foremost rationale for science was that it displayed the structure of the Creation and thereby glorified the Creator. The justification for science was therefore moral not utilitarian. Faraday's experimental researches were aimed primarily at disclosing the divinely ordained laws that govern the physical universe and he paid relatively little attention to how these laws could be used to improve technology and thus society. Whatever happened in this world rested in the hands of God and would occur according to His plan. Like other Sandemanians, Faraday placed far greater emphasis on the eternal life than on the present. Moreover, the industrial and entrepreneurial ethos, with its emphasis on efficiency and profit, was foreign to him as a Sandemanian.

While Faraday sometimes stressed the importance of science in improving our physical conditions, there was another rationale for science to which he attributed greater weight. This was the role of science in improving the intellect of the individual. Intellectual improvement figured far more prominently in his writings than the technological utility of science. Thus, although he stressed the importance of improving our physical conditions through the application of science, he attached greater weight to improving the rational and judgemental powers of the mind. Far above both of these was his insistence that science inculcated the correct moral and spiritual values.

8 *Faraday on Scientific Method*

8.1 READING THE BOOK OF NATURE: FARADAY'S EMPIRICISM

That Faraday expounded an empiricist view of science all commentators agree. Yet there are many incompatible strands within empiricism, and this chapter sets out to specify in a much more precise manner Faraday's rather complex, and apparently inconsistent, views on scientific method, and also to examine why he held those views. Questions about the proper method for pursuing science were very important to him and he felt obliged on numerous occasions to expound his views on this subject so as to defend himself and his scientific work against alternative, but false, conceptions of science. In this chapter we shall be concerned with his methodological pronouncements rather than the actual methods he employed in the laboratory (a topic which will be discussed in chapter 9), although the two were closely related. An appropriate starting point for this discussion is his views about the proper role of the imagination.

In 1858 he related to his close friend Auguste de la Rive his initial encounter with science half a century earlier through reading the books which he bound in Mr Riebau's shop, especially Jane Marcet's *Conversations on chemistry* (1806) and the *Encyclopaedia Britannica*. Yet, he admitted to de la Rive he was not a particularly bright youth or a deep thinker. He was, however, 'a very lively, *imaginative* person, and could believe in the Arabian nights as easily as in the Encyclopaedia'. It is significant that, looking back over some 50 years, Faraday recalled that he was forced to struggle with his overactive imagination in order to keep it in check. In contrast to flights of imaginative fancy, he turned to facts – 'facts were important to me & saved me. I could *trust a fact*'.[1] Security was to be found in solid facts, while danger lurked in the world of fantasy. This theme also appeared in his juvenile lectures for 1853, following his attack on the table-turners: he warned his audience, 'Keep your imagination within bounds, taking heed lest it run away with your judgment'. An unbounded imagination was therefore the source of prejudice and the subversion of the discriminatory faculty. Yet the imagination was not to be stifled but played an essential role in science, which would otherwise degenerate into rote learning and all advance would be retarded if not halted.[2] For Faraday, whose imagination was particularly lively, a strong dose of empiricism was needed to keep his flights of fancy in check and to direct and guard the imagination, which could not be trusted if granted free rein. If an excitable imagination posed dangers for science, it was also, in Faraday's opinion, the source of false religious beliefs. Thus he censured revivalist meetings at which the unwary and credulous readily lost their reason and judgement and instead gave way to flights of fancy.[3]

This point was frequently emphasised by other Sandemanians when criticising religious enthusiasts and popular preachers, who, in their opinion, threw safe judgement to the wind and engaged in their personal interpretations of Scripture. To allow human weakness and fallibility into scriptural exegesis was to debase the text's divine authority and certitude. Instead, the Sandemanians insisted not just on the truth of Scripture but also on the need to interpret the Bible as literally as possible. This point had been repeatedly stressed by John Glas, who chided those divines who indulged their private self-serving interpretations, and thereby perverted God's message. According to Glas, the true Christian should proceed by interpreting the Bible as literally as possible and thereby preserve its true meaning. Every biblical passage must be interpreted in its proper context so as to preserve the meaning intended by its Author. Moreover, difficult passages should be resolved by comparison with 'plain and certain ones'.[4] To obtain a pure and undefiled religion Glas recognised the need to employ a mode of exegesis which avoided all distortions of human origin. Sandemanians described their form of exegesis as 'pure style' or 'plain style', and they particularly eschewed the notion that the Bible (excepting perhaps prophetic passages) is written in a mysterious, or even a metaphorical language. Sandemanian congregations did not employ university-trained clergymen, since any man or woman of common sense could understand God's unembroidered word. Indeed, a trained ministry was likely to learn, and thus propagate, error. Instead, any Sandemanian could construct an exhortation using passages from the Bible with a minimum of linking material. Thus the Sandemanians shared with John Locke and the deists a concern with the common sense understanding of plain language, but, unlike these writers, they founded their Christianity not on reason but on a plain, literal understanding of the Bible.[5]

It is a commonplace among theological writers that while the Bible is the word of God, nature is His handiwork. Robert Sandeman was one of the few Sandemanian writers who took an interest in scientific matters, and it is instructive to see how he drew the relationship between God and His creation. The crucial text on this issue is Romans 1:20: 'For the invisible things of him from the creation of this world are clearly seen, being understood by the things that are made, even his eternal power and Godhead'. Although Glas had briefly glossed this passage in his critique of John Hutchinson's 'most childish' attempt to understand the natural philosophy hidden in the Hebrew text of Genesis, he did not elaborate on 'the things that are made'.[6] Sandeman, however, subjected Romans 1:20 to a more searching analysis in his *The law of nature defended by scripture* (1760). He particularly praised John Locke's *A paraphrase and notes on the epistle of St Paul to the Romans* (1709), and then proceeded to argue that since 'every machine and contrivance suggests to us an artist or contriver', the book of nature likewise 'sufficiently evinces its author'. This is the argument from design as widely employed by natural theologians, who placed reason above revelation. However, it is important to note that Sandeman was staunchly opposed to natural theologians who turned their backs on Scripture and instead deployed this rational argument to demonstrate the power, wisdom and goodness of God. While Sandeman had no doubts that the physical world was of God's design, his legitimation of the argument from design lay not in the power of reason but in God's explicit statement in Romans 1:20. This passage provided the scriptural foundation for his assertion that the natural world is necessarily the reflection of the divine.[7]

Romans 1:20 had particular importance for Faraday, who cited it on at least two occasions before audiences at the Royal Institution. When addressing Prince Albert and other dignitaries on mental education in May 1854, he stressed the need for man to walk humbly with God, since the truth about our future life can be known only from revelation and not by the power of reason. In this rare, if not unique, public statement of his religious views, he attacked not only the presumptions of table turners but also, and more importantly, those natural theologians who claimed 'by reasoning' to 'find out God'. It was necessary, he claimed, to draw 'an absolute distinction between religious and ordinary belief'. While reason ought to exercise a major role in the latter sphere, its applicability to religion was severely limited. Then, quoting Romans 1:20 as an article of his own faith, Faraday claimed that the physical world manifests its divine origin as follows:

> . . . even in earthly matters, I believe that the invisible things of HIM from the creation of the world are clearly seen, being understood by the things that are made, even His eternal power

and Godhead; and I have never seen anything incompatible between those things of man which is within him, and those higher things concerning his future, which he cannot know by that spirit.[8]

Seven years earlier Faraday had concluded a course of lectures by surveying the role of centrally directed forces in the economy of nature. He conceived that surrounding each particle of matter was 'a *centre* of force reaching to an infinite distance, binding worlds and suns together, and unchangeable in its permanency'. All the phenomena observed in nature appeared to be referable to these forces, including the stability of the heavens, the role of heat, the lightning discharge, tidal motions and even life itself. In contemplating the dependence of these diverse phenomena on central forces we are led, claimed Faraday,

> to think of Him who hath wrought them; for it is said by an authority far above even that which these works present, that 'the invisible things of Him from the creation of the world are clearly seen, being understood by the things that are made, even His eternal power and Godhead'.[9]

As these uses of Romans 1:20 make clear, Faraday did not view the physical world as an inanimate object to be subjected to scientific analysis but instead as the work of God that manifested its divine origin. While he insisted that science could provide but a very partial appreciation of God's powers, nevertheless science reveals 'the evidences in natural things of his eternal power & Godhead'.[10] The scientist is uniquely placed in being able to appreciate God's handiwork, and even the study of some outwardly uncongenial phenomena, such as putrefaction, could nevertheless reveal the beauty, power 'and evidences of a wisdom which the more a man knows the more freely will he acknowledge he cannot understand'.[11] While accepting that the wisdom and goodness of God is manifested in the creation, Faraday, like Sandeman, nevertheless parted with those natural theologians who paid exclusive attention to reason and he instead insisted that the Bible provides a sufficient warrant for the belief that the physical and moral worlds manifest the divine plan. In studying nature the scientist thereby examines the works of God, and it was of the utmost importance to Faraday that he perceived nature clearly and distinctly without prejudice or distortion. Thus empiricism was the method by which the scientist of integrity must study nature since observed facts provide the scientist with the only means of access to the world that God had created.

It is no coincidence that in his science Faraday should have employed the same hermeneutic strategy that the Sandemanians used in interpreting the Bible, since both the Bible and nature are God's creations and, according to Romans 1:20, nature bears the marks of its Creator. In a revealing sentence in his lecture on mental education Faraday vividly portrayed the relationship between God, man and nature in the following terms: 'for the book of nature, which we have to read, is written by the finger of God'.[12] The phrase 'finger of God' may appear misplaced in this context, since in Exodus 31:18 Moses received the ten commandments 'written by the finger of God'. Yet it is significant that in his lecture Faraday conceived the author of the ten commandments also using his authorial fiat to write the book of nature in accordance with similar syntactical rules. This evidence suggests that Faraday adopted the two books metaphor not merely as a literary device but also as embodying a profound truth about the world.[13] Moreover, as he emphasised in the above quotation, the investigation of nature is like reading a book, and just as the Sandemanians sought to understand every word in the Bible without introducing distortions of human origin, so Faraday aimed to comprehend God's other book with integrity and without giving reign to imagination or prejudice. In pursuing the empirical method great care had therefore to be taken to read correctly the God-made signs that are accessible to the scientist.

In order both to curb his overactive imagination and to read God's works correctly, Faraday nailed his flag firmly to the empiricist mast. He believed that facts, and only facts, are the basic signs of nature and the foundation on which the whole edifice of science has to be constructed. As he told the audience at his lecture on mental education, a 'fundamental fact . . . never fails us, its evidence is always true', and he portrayed science as dependent 'upon carefully observed facts'.[14]

Repeatedly expressed throughout Faraday's discussions of facts was his concern to differentiate unambiguously between facts, on the one hand, and opinion, imagination and conjecture, on the other. In 1850 he complimented Adolphe Quetelet for reporting certain facts (relating to atmospheric electricity) 'without any addition of imagination or hypothesis. They are FACTS and ought not too hastily to be confounded with opinion; for the *facts are for all time, whilst opinion may change as a cloud in the air*'.[15] Facts are therefore totally incorrigible and provide the stable foundation on which science is to be constructed. Moreover, it would seem on this view that no two facts could be inconsistent, and any apparent inconsistency would have to be resolved experimentally, with the elimination of the false claim(s). Nature is ultimately consistent and, because it is God's book, it does not deceive the honest experimenter.

It is important to notice that Faraday employed a simplistic dichotomy differentiating facts from opinions. Moreover, as will be discussed further in section 10.3, his appeal to facts possessed clear psychological overtones. Thus, writing to Whewell in 1835 he admitted that 'I feel that my *safety* consists in facts; and even these I am but too anxious [not] to pervert through the influence of preconceived notions'. Likewise, when he recounted to Auguste de la Rive his youthful fantasies concerning the Arabian Nights, he claimed that 'facts were important to me & *saved me. I could trust a fact*'.[16] His fears of losing control and of being carried away by flights of imagination (or by prejudice) were overcome by the safe shelter that firmly based facts could provide. His persona, and with it his science, could be saved by facts and facts alone. As he wrote to Schoenbein in 1858, 'without experiments [which produce facts] I am nothing'![17]

A letter to Schoenbein dating from February 1845 indicates a further significance of Faraday's cult of facts. This letter comes from the period when his memory was most treacherous, and he was frequently unable to think constructively or understand the work of other scientists. His thoughts were unstable and tended to evaporate unless 'there is some visible body before my eyes, or some large fact approaching with force to the external senses, and easy to be produced, to sustain, by a sort of material evidence for the existence of a thought'.[18] At this dark period of his life Faraday could stabilise his mind by concentrating on 'some large fact' that also provided assurance that he was still in contact with physical reality. If in his philosophy of science Faraday appears to have adopted a form of empiricism found among many eighteenth- and early nineteenth-century writers, it should be remembered that his emphasis on facts bears not only the marks of his religious beliefs but also possesses psychological significance.

8.2 NATURAL LAW AND MORAL LAW

If facts were the signs comprising the language of nature, Faraday considered that laws were its syntax. On numerous occasions he emphasised that the aim of science is to discover the laws of nature: 'The *laws of nature*, as we understand them, are the foundation of our knowledge in natural things', he told the audience at his mental education lecture, while the notes for a much earlier lecture contain the strident injunction, 'Search for laws'![19] Moreover, much of his research work was directed towards the discovery of laws, including such successes as the law of electromagnetic induction, the general law of electrical conduction and the laws of electrolytic action. In his opinion the discovery of a law was the greatest acquisition to scientific knowledge. Indeed, he sometimes equated science with lawlikeness, and he explicitly denied that table-turning was a science, since the phenomena it dealt with were indiscriminate and were thus not governed by regular laws. Nature is consistent and every natural phenomenon is the result of laws, which determine what must occur. There is no room for chance and every effect is the result of an appropriate cause. Thus he criticised those who considered electricity to be a mysterious power, instead asserting that the 'beauty of electricity . . . [is] that it is under *law*'.[20]

For Faraday God was the lawmaker who had imposed laws on matter at the Creation. As he argued in his 1844 memorandum on the nature of matter, 'God has been pleased to work in his material creation by laws', and he went on to state that 'the Creator governs his material works by *definite laws* resulting from the forces impressed on matter'. In his evidence before the Public School Commissioners he likewise claimed that God had 'impressed [laws] on all created things', and he also asserted that there could be no better subject for educating the young mind than the study of those laws.[21] In the beginning God had created not only matter but also the laws to which all physical bodies are subservient. Thereafter the laws had remained unchanged. Faraday did not require an immanent God but only the God of Genesis, who created the world and found that it was 'very good'.

There was no doubt in Faraday's mind that the physical world that God had created is perfect. Since nature is designed according to a divine plan, it is a perfectly regulated system and the regulation is effected through laws. The original laws still operate and, owing to the perfection of God's plan, the physical world has not decayed but instead it manifests signs of both design and economy. By observing nature and judiciously collecting the facts, scientists are able to ascertain the small number of laws governing phenomena. The laws are consistent and universal, so that under similar circumstances similar objects behave in similar ways. Moreover, there is simplicity in the design, so that even apparently complex phenomena can be accounted for by very few laws. 'How wonderful it is to me the simplicity of nature when we rightly interpet her laws' asserted Faraday in a breathless tone.[22] Yet the consistency and simplicity of nature were not only conclusions that Faraday drew from his scientific work but they were also metaphysical presuppositions that directed his research. As discussed in section 7.2, Faraday accepted that nature is a divinely planned economy and that the total amount of force is conserved.

Since natural laws underpin that economy, we need to investigate their significance for Faraday. While many of his contemporaries accepted that the search for laws was a major goal for science, Faraday's dogmatic insistence on the central role of laws and his rejection of hypotheses shows that the concept of law was of particular and peculiar importance to him. In part this emphasis on laws derives from the widely shared belief that at the Creation God impressed laws on matter, but in Faraday's case the lawlikeness of the physical universe was further sustained by his close adherence to the Word. The prevalence of the word law(s) in the Bible – it appears over 300 times – suggests that it was an important term for Faraday. God's laws were also frequently discussed in the exhortations and writings of Sandemanians such as John Glas and Robert Sandeman. For example, in discussing Romans 10:5–10 – 'Moses describeth the righteousness which is of the law . . .' – Glas asserted that we attain righteousness by choosing obedience to God's law, and that by living by His law we are thereby granted life.[23] The world of the Sandemanians was circumscribed by laws of divine origin.

While Glas occasionally employed the term 'law of nature' in his essays, this term was more widely discussed by Robert Sandeman, who even wrote a small book appropriately entitled *The law of nature defended by scripture* (1760). As he emphasised in his opening remarks, although deists frequently appeal to natural laws, Christians should not react by denying the existence of laws of nature but should, instead, utilise them for their own purposes. Indeed, as he stressed, the notion of natural law bore a close connection with the divine laws which God had explicated in the Bible and which the true Christian is bound to obey. Since the world was created by God, the laws of nature are of divine origin and universal applicability. Sandeman defined natural laws as those laws which are consonant with the moral inclinations of all mankind. Thus, for example, mothers feel affection for their children and 'the common propensity between the sexes . . . is [also] natural'. By contrast, certain other actions, such as 'a similar propensity between those of the same sex [!], is unnatural or against nature – so, highly criminal'.[24] To differentiate between the natural and the unnatural we have only to appeal to our heart, our conscience. Like the third Earl of Shaftesbury earlier in the eighteenth century, Sandeman believed that we possess a moral sense which provides us with clear, incorrigible judgements on moral questions.[25] Yet Sandeman considered that the moral sense operates in accordance with God's moral law and with the laws expounded in the Bible.

Sandeman's defence of moral law was based on the first three chapters of St Paul's Epistle to the Romans, especially Romans 1:20. In discussing this passage he praised John Locke's paraphrase as offering 'Paul's sense in a plain and easy manner'. Locke understood Romans 1:20 to mean that God's invisible being 'might be clearly discovered and understood from the visible beauty, order, and operations observable in the constitution and parts of the universe'.[26] Sandeman likewise accepted that 'the great frame of nature sufficiently evinces its author'. This is the argument from design so extensively employed by natural theologians in the eighteenth and early nineteenth centuries, but it must again be stressed that for Sandeman the light of nature offered only partial illumination. Romans 1:20 makes clear that heathens have access to nature and are thus able to read God's moral laws from the book of nature; from an understanding of these laws they can thus gain a partial appreciation of God. However, although natural religion speaks this message, it is not sufficient, and the deists are thoroughly mistaken in turning their backs on revelation, which contains the full (not partial) account of God and His moral laws. The true Christian cannot therefore be satisfied with the study of the laws of nature alone or with the voice of his conscience, but must also pay close attention to God's word as written in the Bible. Yet it is important to note that, unlike some theologians of revelation, Sandeman did not eschew natural religion and natural law, but saw them as necessary for the non-Christian and as non-trivial but incomplete statements of truths known to Christians through the Bible.[27]

Sandeman's concern with moral and physical laws indicates that he was well-disposed to the study of science, and we would also expect other Sandemanians to view science in a similar light. This is certainly true in Faraday's case, and it is important to notice that he considered that God works through laws in both the moral and the physical spheres. In one of his few extant exhortations Faraday reiterated Sandeman's concern with God's moral law as expressed in Psalm 19 – 'The law of the Lord is perfect, converting the soul'. Yet, argued Faraday, if we fail to obey or refuse to obey God's law, then we will have to bear the consequences and accept a life of suffering. The true Christian, on the other hand, tries to fulfil God's law in every deed and thought, taking Christ as his exemplar, since Christ came into the world to show men how to live righteously and thereby be granted eternal life. Although we are constantly aware that we fail to attain the standard set by Christ, 'we ought to value the privilege of knowing God's truth far beyond anything we can have in this world'.[28] Notes also exist for another exhortation entitled 'To the *law* & *testimony*' (Isaiah 9:20), which deals with the Mosaic law and likewise stresses that 'the law of the Spirit of life in Jesus Christ hath made me free from the law of sin and death'[29] (Romans 8:2).

The crucial point is that for Faraday God rules both the moral and the physical world by laws. Yet the two types of law are rather different in their modes of action. The moral law is binding on true Christians but any individual is able to break that law, although he then slips into sin and is denied eternal life. By contrast, electricity possesses no choice but simply obeys the laws of electromagnetic induction. Both types of law are, then, deterministic, except that the former brings in the will; Christ alone exemplifies God's moral laws, while the rest of us can but try to emulate Christ in this respect. The scientist is likewise responsible for trying to discover the laws that God used in framing nature. However, while God certainly deployed immutable laws, the laws currently accepted by scientists are provisional, may be proved false and thus need to be replaced by other, better ones. Yet the application of the correct scientific method has nevertheless enabled scientists in many instances to grasp laws of 'assured and large character' which Faraday considered to be the ones which God had in fact used.[30] Newton's inverse square law of gravitation would be an example of such a law.

The importance that Faraday attributed to laws (both moral and physical) suggests that their significance is not confined to his science and his religion. The very existence of laws of divine origin emphasised that all is not chaos and confusion but that there is a meaningful plan and a higher level of order behind apparently chaotic events. Through his Sandemanianism he discovered the way to live in obedience with God's moral law and with the promise of eternal life. Through his science he came into close contact with the physical laws that God had chosen to govern the universe. Since Faraday could not abide confusion, the existence of laws was a firm assurance that order reigns supreme. He thus created for himself a protective haven of order.

8.3 SCIENCE AND SPECULATION

Prejudice in science took a number of different forms; it was manifest in those who sought patronage, rewards, or riches from their scientific endeavours and who were thereby liable to be biased in their judgement. Again, the use of language which lacked clarity or was weighed down by theoretical baggage was liable to prejudice the investigator. In the use of scientific hypotheses Faraday, like Francis

Bacon, recognised a further stubborn and pervasive form of prejudice.[31] Even in his first surviving letter to Benjamin Abbott, dating from July 1812, Faraday claimed that 'in [natural] philosophy we do not admit supposition'.[32] By the time he wrote his 'Historical sketch of electro-magnetism' (1821–2) Faraday found his scepticism fully justified when he witnessed the plethora of competing hypotheses evoked by other scientists to explain electromagnetism.[33] Such hypotheses, he claimed, are very likely to be shown, in time, to be false. Indeed the historical record indicated that hypotheses are transitory and that the pet hypothesis of one generation is rejected by the next and another temporary hypothesis is substituted in its place. Thus, in 1854, at the end of a lecture reviewing the different hypotheses about the nature and operation of magnetism, he stated: 'Our varying hypotheses are simply the confessions of our ignorance in a hidden form; and so it ought to be, only the ignorance should be openly acknowledged'.[34] Hypotheses, in other words, are dishonest subterfuges masking our ignorance.

Throughout history hypotheses have been refuted by correctly ascertained facts; indeed, argued Faraday, 'the progress of experimental philosophy will show you that it is a great disturber of preformed theories'.[35] In his mature writings he repeatedly stated that in science hypotheses had to be clearly distinguished from facts. Facts, if properly ascertained, were incorrigible, whereas hypotheses go beyond facts, express mere opinion, and are always liable to be refuted by experimentally determined facts. The contrast between experiment and hypothesis was neatly summarised in an entry in Faraday's *Diary* for 19 December 1833: 'I must keep my researches really *Experimental* and not let them deserve any where the character of *hypothetical imaginations*'.[36] Hypotheses (unlike facts) are the result of an overheated imagination and therefore possess no legitimate place in science. In a footnote added to the published exchange with the Berlin electrical researcher Peter Riess, Faraday claimed that it is 'not the duty or place of a philosopher to dictate belief, and all hypothesis is more or less matter of belief'.[37] Not only that, but, when taken seriously, hypotheses are likely to corrupt the scientist's judgement. Moreover, even facts are liable to be perverted 'through the influence of [such] preconceived notions'.[38] Hypotheses, then, are thoroughly pernicious and must be kept separate from facts. A previously quoted passage from Faraday's 1850 letter to Adolphe Quetelet bears repetition, since it clarifies Faraday's position on the issue at hand. Here he praised Quetelet's papers on atmospheric electricity, which were 'without any addition of imagination and hypothesis. They are FACTS and ought not too hastily to be confounded with opinion; for the facts are for all time, whilst opinion may change as a cloud in the air'.[39] In these and numerous other quotations Faraday incisively contrasted facts with prejudices, hypotheses, opinions and imagination.

In the above quotations Faraday emerges as a committed positivist, who wished to base science firmly on facts and to exorcise all hypotheses. Yet in his adoption of this position we can detect metaphysical, if not religious, overtones. Just as the Sandemanians were contemptuous of theologians who, in their interpretation of Scripture, forsook God's word and instead imposed their own prejudices, Faraday was equally dismissive of those scientists who paraded their prejudices and failed to read the book of nature correctly. To commit oneself to a theory and hold that theory as if it dictated the course of natural phenomena was what Faraday consistently rejected, since to raise man's conjectures above God's works was the very essence of prejudice. He would have agreed with the claim in Roger Cotes's preface to the second (1713) edition of Newton's *Principia* that the 'business of true philosophy is to . . . inquire after those laws on which the Great Creator actually chose to found this most beautiful Frame of the World'. This argument was used to justify Newton's dismissal, as mere opinion, of the systems of the world invented by Descartes and others.[40] In a similar vein Faraday rejected the major theories that dominated early nineteenth-century science, such as mechanistic interpretations of the wave theory of light, the one and two fluid theories of electricity and the numerous hypothetical schemes which sought to explain the simplest electromagnetic phenomena. Hypothesisers had committed the unforgivable sin of pride. In his pronouncements against hypotheses Faraday adopted the tone of a preacher trying to eradicate sin.

There is a further aspect of Faraday's rejection of hypotheses that requires brief exploration. The advocates of hypotheses invariably form parties – in other words, they group together to expound, expand and defend their view of nature against the proponents of alternative hypotheses. Thus the history of science is littered with controversies between the supporters of incompatible hypotheses. Thus, for example, the proponents of the wave and particle theories of light formed clearly demarcated factions that engaged in particularly fierce battle during the 1830s. Not so Faraday, who carefully avoided aligning himself with any intellectual or social faction in the scientific community. In so doing, he adopted a typical Sandemanian strategy, since members of the sect sought (not always successfully) to avoid theological controversy, considering that the protagonists had failed to pay adequate attention to God's word but instead had behaved like political parties.

Faraday's views on scientific method are not, however, as straight-forward as the above discussion would seem to imply. In his book *Faraday as a natural philosopher* (1971) Joseph Agassi sought to invert the traditional account of Faraday and show that he was not a plodding empiricist but a bold theoretical speculator in the mould of Karl Popper. Agassi is certainly right to dwell on this aspect of Faraday's scientific method, but his account fails to engage Faraday's deep antipathy towards hypotheses and also the important role of laws in his science. Faraday was an experimentalist committed to building science on a firm, safe foundation of facts. Yet as keenly as he opposed the use of hypotheses, with their associations with prejudice and sin, he was also one of the most brilliant speculators in the history of science who was able to transcend empirical data and offer deep insights into the structure of the physical universe. For example, his conception of lines of magnetic force and his hypothesis of ray vibrations proved particularly successful.[41] Indeed, a strong case can be made for Faraday's speculative writings having been more important to the history of science than his empirical discoveries. Therefore against his strident opposition to hypotheses must be set both his own deployment of hypotheses and also a number of passages (to be discussed below) in which he argued for hypotheses.

There may be some inconsistency in Faraday's methodological pronouncements but it is also clear that his position on these issues changed significantly, and we have to wait until the 1840s for his more speculative papers and for his most explicit statements in favour of hypotheses. Moreover, we can help clarify the apparent inconsistencies if we recognise that Faraday sought to discriminate between the legitimate and illegitimate uses of hypotheses. In the following discussion it will be helpful to divide his speculations into three categories and subject each to analysis.

The first type consists of what we might call working hypotheses, in that these were generated and tested in the laboratory. David Gooding and Elspeth Crawford have recently concentrated on Faraday's *Diary* in order to analyse the complex processes in his experimental procedures. As these historians have emphasised, Faraday was not simply testing hypotheses or working by a trial and error method, but there was a constant, constructive interplay between his thoughts, the evolving structure of his apparatus and the results he obtained with it. Moreover, he was seeking ways of isolating classes of phenomena and of displaying them before the audience at the Royal Institution. Yet within these processes he was forever testing low-level conjectures about the structure of the world.[42] For example, in an electrostatic experiment in which he held one end of a piece of gutta percha and brought the other into contact with a charged electrometer, he noticed the leaves collapsed, but returned to their original positions when the gutta percha was removed. 'Suspected [that] this was due to the presence of water in the inner part of the [gutta percha] strap piece', speculated Faraday, 'and so kept it in a warm place for several days; yet still produced no improvement in its qualities'. Another example occurred in the middle of a series of electromagnetic experiments in which he speculated, 'Ought not two wires or currents at right angles tend to move in the direction of *their length*?'[43]

A second type of hypothesis concerned more general theoretical structures, including Faraday's incisive original speculations – such as his lines of electromagnetic force – which went far beyond the facts and offered new ways of understanding the universe. In 1844 he began a series of lectures on the phenomena and theories of heat by pointing out that speculations offer 'dangerous temptations', and he cautioned against employing them, but he added, paraphrasing Ecclesiastes 3:1–8, there is 'a time to speculate as well as to refrain, all depends upon the temper of the mind'.[44] Faraday certainly possessed such a temper of mind and in a letter of 1846 he wrote:

> I keep dreaming away with views of matter & its powers that I do not think it wise or philosophic to put forth because I hold them so that they may change with the evidence of experiment: but I use these views *as stimulants & guides* in some degree into the course of new enquiries . . .[45]

Here he attributed a specific and necessary role to speculation in the realm of matter theory. This quotation also indicates his reticence in placing conjectures before the public; indeed, when he did publish them, he issued a warning that he was leaving the realm of fact and entering the world of opinion.

Thus he opened his 1844 paper, 'A speculation touching electric conduction and the nature of matter', by stating that he merely wished to record his 'opinion and views'. He went on to complain that students were taught the atomic theory – that matter consists of small particles separated by empty space with inter-particulate forces – as if it were a fact, whereas it, too, was only an opinion. Moreover, he noted that the use of the word *atom* necessarily includes 'much that is purely hypothetical'. He next emphasised the need to divide facts (and laws) from hypotheses:

> But it is always safe and philosophic to distinguish, as much as in our power, fact from theory; the experience of past ages is sufficient to show us the wisdom of such a course; and

considering the constant tendency of the mind to rest on an assumption, and, when it answers every present purpose, to forget that it is an assumption, we ought to remember that it, in such cases, becomes a *prejudice*, and inevitably interferes, more or less, with a *clear-sighted judgement* . . . [Therefore the scientist must] distinguish that knowledge which consists of assumption, by which I mean *theory and hypothesis*, from that which is knowledge of *facts and laws*; never raising the former to the dignity or authority of the latter, nor confusing the latter more than is inevitable with the former.[46]

This is a very cogent statement of Faraday's primary methodological distinction. Yet this distinction did a lot of work for Faraday, since he then used the incorrigibility of certain facts about electrical conduction to undermine the atomists' conception of empty space. Indeed this criticism of the atomic theory led Faraday to propose an alternative speculative theory, and he argued that facts are far more conducive to the view espoused by Roger Boscovich in the mid-eighteenth century. Boscovich had rejected the traditional view that matter is composed of small, hard, impenetrable atoms and had instead suggested that there are point atoms surrounded by fields of force which permeate space and produce the observable phenomena. Faraday argued that the scientist working in this area must hold some theory about the nature of matter, but he should nevertheless assume as little as possible and 'in that respect the atoms of Boscovich appear to me to have a great advantage over' the traditional atomist doctrine. Later in the paper he praised one aspect of Boscovich's theory as 'a more beautiful, yet equally probable and philosophic idea' when compared with the alternative theory.[47]

It would be incorrect to claim that Faraday was a Boscovichian, since both before 1844 and after he entertained ideas about matter and space which were incompatible with Boscovich's theory.[48] A further objection to this label derives from Faraday's methodological distinction which placed Boscovich's theory among hypotheses, not facts, although he considered it to be more consistent with the facts, and thus more probable than traditional atomism. What is less easy to ascertain is whether Faraday was consistent in holding Boscovich's theory lightly – as an hypothesis and not as a prejudice. He certainly utilised Boscovich's theory and many other similar hypotheses as conceptual aids in trying to understand the workings of nature, but this use of hypotheses does not constitute *prima facie* evidence that Faraday's methodological pronouncements were incompatible with his various speculations.[49] He saw (probably more clearly than we see) facts and speculations as different forms of discourse. For example, he emphasised at the beginning of his 'Thoughts on ray-vibrations' paper of 1846 that he 'merely threw out [these ideas] as matter for speculation, the vague impressions of my mind, for I gave [in his recent Friday evening discourse] nothing as the result of sufficient consideration, or as settled conviction, or even probable conclusion at which I had arrived'.[50] For all his equivocation he recognised that, whether he should ultimately be proved right or wrong, both Boscovichian atomism and his hypothesis of ray vibrations offered deep, if insecure and partial, insights into natural phenomena and ones which the scientist could employ. The methodological distinction was not only reflected in Faraday's hesitant style of presenting speculative material but he also reserved his more speculative papers for the *Philosophical Magazine*, while the positive results of experimental investigations were generally published in the Royal Society's *Philosophical Transactions*. Yet even these latter papers contained much speculative material, such as his brilliant excursion into the *'electro-tonic* state' in the first series of his 'Experimental researches in electricity' of 1831.[51]

One of Faraday's strongest defences of this second kind of hypothesis appeared near the beginning of a highly speculative paper of 1852 concerning the physical reality of magnetic lines of force. Choosing to publish his speculations in the *Philosophical Magazine*, he argued that while hypotheses are liable to change and should be demarcated from facts and laws, they were nevertheless useful to the scientist. In particular, hypotheses help clarify vague ideas so that these can then be incisively tested against experiment. Moreover, through their suggestibility they can pave the way to the discovery of new phenomena and thus advance the cause of truth; once an incorrigible truth had been obtained, he added quickly, the hypothesis could be removed like a temporary scaffolding. Both the wave and emission theories of light had proved fruitful in this manner.[52]

In respect to the first and second kinds of hypothesis Faraday insisted not only that they were not to be claimed as scientific truths but also that the sensitive scientist must be constantly aware of their dubious epistemological status. In a letter to Auguste de la Rive he spoke of 'the necessity one is under of holding them loosely & suspending the mental decision'.[53] Yet Faraday admitted that he personally experienced considerable difficulty in adhering to this prescription. For example, in his first paper on electromagnetic induction (1831) he suggested that the particles constituting the secondary coil assumed a peculiar state, which he called the 'electro-tonic' state, while current flowed in the primary circuit. Although he was subsequently forced to relinquish this conjecture for want of empirical proof, he confided to Whewell in September 1835 that he still clung to it 'in fancy or hypothesis from general impressions produced by a whole series of results'.[54] The conjecture sprang from his fertile imagination and, as he admitted in a subsequent discourse at the Royal Institution, 'every one is of course partial to the child of his own imagination'.[55] Although in his letters and papers Faraday could often draw a clear distinction between fact and hypothesis, there was also an internal battle to free himself from prejudice, both scientific and religious. He was never quite sure whether he was entering the land of fantasy and losing his grip on the reality that God had created by the Word. He was frequently carried along on a wave created by his own imagination and he had to stop and check to see where truth lay, fearing all the time that he might not be able to distinguish fantasy from reality – like the Taoist Chuang Tzu, who dreamt he was a butterfly and not Chuang Tzu. He suddenly woke to find that he was Chuang Tzu and not the butterfly. However, he could not decide who he really was, whether he was Chuang Tzu who dreamt he was a butterfly, or a butterfly dreaming he was Chuang Tzu.[56]

A particularly revealing moment occurred in 1833 when Faraday was one of several speakers at a meeting held at the Freemason's Hall to celebrate the centenary of the birth of Joseph Priestley. Faraday was especially impressed by Priestley's 'freedom of mind' and his 'independence of dogma and of preconceived notions, by which men are so often bowed down and carried forward from fallacy to fallacy'. While antipathetic to Priestley's Unitarianism, Faraday considered his scientific methodology as morally praiseworthy, since he sought to eradicate prejudice, 'which not only influences our judgement, but even the perceptions of our senses'. Although some commentators had chastised Priestley for changing his mind, this characteristic was, for Faraday, a strength, and it had enabled him to make such important discoveries as oxygen. Indeed, the way forward in science was to forsake dogmatism and to follow Priestley in being prepared to reject our accepted views and instead to be prepared to receive new, even uncongenial, ideas.[57] This openmindedness, which Faraday found so attractive, was an essential component of his own attitude towards hypotheses, which, he believed, were necessary for stimulating and directing research but which must not be maintained in a rigid manner.

While Faraday readily admitted the importance of hypotheses of this second type and utilised them as aids to his research, there was a third type of hypothesis which he employed but did not acknowledge as conjectural. These hypotheses were, instead, necessary truths about the world and about how the scientist understands nature. As we have already seen, Faraday accepted that God had written laws into the book of nature and that the aim of science was to uncover these universal laws. Again, he appears to have held the deep and unassailable conviction that space (between atoms, on the atomists' account) is not empty. A third example is that nature is an economical system and therefore force is conserved. All three examples illustrate the kinds of metaphysical assumption that underpinned much of Faraday's research. He did not usually treat these assumptions as hypotheses but rather as necessary truths, and would not have countenanced any natural philosophy that denied these truths; for example, it was inconceivable to him that forces could be created or destroyed by purely physical processes.

Before concluding this section I would like to suggest why Faraday was more prepared to speculate in the 1840s and 1850s. This change in attitude may have resulted from his increasing maturity and self-confidence, or it may have been a natural extension of his successful empirical work of the 1830s. However, the coincidence between this change and his extended illness suggests that after about 1840 his altered mental state played some role in his accommodation of speculative thinking. In several letters of the 1840s and 1850s Faraday complained that he could not work because loss of memory prevented him from remembering scientific facts, and in 1858 he told Schoenbein that he was 'terrified' at the thought of lecturing at the Royal Institution in case he forgot what he intended to say.[58] Although he appears to have recovered his memory for considerable periods, the recent experience of severe memory loss may have caused him not only to emphasise the importance of facts but also to be more prepared to speculate in order to perceive how facts are interrelated in nature. Thus he wrote to Schoenbein, 'You can hardly imagine how I am struggling to exert my *poetical ideas* just now for the discovery of analogies and remote figures respecting the earth, sun, and all sorts of things – for I think that is the true way (corrected by judgement) to work out a discovery'.[59] This acknowledgement of the importance of speculation was written in November 1845, when he had largely recovered from his extended illness and a few months before his most celebrated speculative paper entitled 'Thoughts on ray-vibrations'.

8.4   THE LANGUAGE OF SCIENCE

8.4.1   *The language of facts*

In book three of the *Essay concerning human understanding* (1690) Locke articulated the theory that words are signs for ideas and that in philosophical (i.e. scientific) discourse ideas have to be clear and distinct so as to evoke accurately the ideas for which they stand. This theory gained considerable popularity and was reiterated in many subsequent works. While experiments were, for Faraday, the source of ideas, those ideas had also to be expressed clearly and accurately in words, and every care had to be taken not to pollute a fact by misrepresenting it or expressing it in an imprecise way. This concern with the proper use of language is to be found in his earliest letter to Benjamin Abbott, in which he praised the exercise of letter writing because 'it tends I conceive to make ideas clear and distinct'.[60] Likewise, when lecturing on mental education nearly four decades later, Faraday argued that in order to sharpen the judgemental faculty we should form the habit of framing *clear and precise ideas*. Similarly we should 'accustom ourselves to clear and definite language, especially in physical matters; giving to a word its true and full, but measured meaning'. By purifying ideas and words the link between experiment and language was complete and facts could be expressed and communicated without distortion by the linguistic medium. While he argued for this ideal in respect to scientific communication, he also recognised that a scientific term is not very precise when first used but he considered that through regular use its referent becomes increasingly exact.[61]

Faraday was greatly concerned with maintaining clear and accurate communication in science. In his public lectures he sought to convey the principles of science through a mixture of polished prose and experiment – nature speaking directly. Judging from reports of his lectures he achieved his ideal to an impressive degree. However, he was not always so successful when addressing other scientists, particularly those whose terminology was impregnated by alternative theoretical notions and those who found clarity in mathematics, rather than in ordinary language, however refined. For example, in an exchange with the American scientist Robert Hare in 1840, some of Faraday's papers were criticised and he was forced to qualify what he had written in order to remove ambiguities.[62] Given his concern with clear expression, Whewell touched a raw nerve when he admitted in 1853 that he had failed to understand some of Faraday's papers. 'You frighten me', replied Faraday, 'how *obscure and confused* my last three papers must have seemed'.[63] His use of the word confusion is telling, since it would have possessed for him strong biblical overtones. For example, the confusion of tongues occurred at Babel and 1 Corinthians 14:33 states that 'God is not the author of confusion, but of peace'. These resonances suggest that Faraday was accusing himself of committing a far more serious offence than merely failing to communicate with another scientist. He had misused language and thereby distorted the truth. It was as if he had misread the Bible and instead substituted his own confused account.

In a famous exchange with Whewell (wearing his classicist's hat) dating from 1834, Faraday pointed out that theoretical notions are often smuggled into science through its terminology and thus the medium of language distorts scientific facts. He therefore asked Whewell to advise him on adopting terms that would accurately convey an experimental situation but would not presuppose any particular theory. What he sought was a pair of terms to refer to the surfaces in an electrochemical cell by which the electricity enters and leaves. 'Poles' was a widely used term but it definitely suggested that chemical substances were attracted, like small magnets, to the surfaces of the electrodes – hydrogen to one pole and oxygen to the other in the case of water. In rejecting this view of electrochemical interactions Faraday looked for an alternative nomenclature. Eisode (entry) and exode (exit), he suggested. Dexiode (path to the right) and sceode (path to the left), even orthode (the straight way) and anthode (the road opposite), responded Whewell. Yet Whewell clearly preferred anode (up road) and cathode (down road), the terms that Faraday adopted.[64] Facts had to be described with integrity and without being unduly adulterated by theory. Just as Galpine's *Botany*, which Faraday carried on country walks, gave the correct Linnaean name for each species of plant, so Faraday saw himself as the Linnaeus of electrochemistry, attributing to each entity its proper theoretically neutral name. This aspect of Faraday's work may be related not only to the Sandemanian emphasis on a plain, true reading of the Bible but also to Genesis 2:20, where Adam attributes to each type of animal its true and accurate name. Correct naming was therefore a necessary part of correct observation.

8.4.2   *The language of mathematics*

While Faraday allowed specific roles to hypotheses, he consistently rejected the claim that hypotheses, such as the two fluid theory of electricity or the wave theory of light, were truths about nature. Another aspect of his criticism of such hypotheses was that they were dressed in the language of mathematics, which further increased the impression that they were unassailable truths. In the 1830s, when Faraday was at the height of his powers, large areas of the physical sciences – most noticeably optics, electricity and heat theory – were rapidly becoming dominated not just by mathematics but by its higher branches. No longer would simple geometrical and algebraic methods suffice: calculus and the higher branches of algebra were increasingly necessary tools for the scientist. Those who could not handle the new rigorous symbolism found much of the journal literature inaccessible, and they were sometimes even edged out of these technically demanding branches of science. Thus in 1843 Richard Potter, the recently appointed Professor of Natural and Experimental Philosophy at University College, London, was suspended from his post following an investigation by a committee of the College Senate which found him unable to keep abreast of developments during the previous 15 years owing to his lack of mathematical knowledge.[65]

Faraday was far less competent at mathematics than was Potter, and he frequently complained to his correspondents that he was 'quite unable to enter into the mathematical part' of their arguments.[66] The mathematician Augustus de Morgan offered an even more gloomy view, claiming that Faraday 'is more ignorant of mathematics than any one would imagine', and he related an incident in the early 1830s when he was asked by a friend to obtain from Faraday a diagram that the friend needed for a lecture at the Royal Institution. 'I did so', added de Morgan, 'and found I could not trust F., nor could he trust himself to give directions for one of the simplest astronomical diagrams possible. It was just a comet's orbit, with the sun in the proper place.'[67] This diagram demanded not just mathematics but mechanics, towards which, as noted above, Faraday was deeply antipathetic.

Although Faraday was thus debarred from entering into an increasing proportion of contemporary science, especially physics, his work (unlike some of Potter's) was not rendered useless or *passé* but hailed as of first-rate importance. Paradoxically, at the time when mathematical physics was on the ascendency, some of the most sublime discoveries were made by an experimentalist who was unable to comprehend the language in which these new developments were expressed. Yet Faraday was not completely ignorant of mathematics, for he claimed to Whewell that 'all my mathematics consists in [is] that rough natural portion of geometry which everybody has more or less'.[68] He was certainly able to perform simple arithmetical operations, and in his Commonplace Book he expressed considerable interest in the methods used by Zerah Colburn, a prodigious 13 year-old who was able to perform high-speed manipulations.[69]

What little mathematics he used in his scientific papers was confined to simple ratios. In his early chemical work he frequently calculated the ratios of combining weights of chemical substances, and he appears to have experienced no difficulty in expressing such proportions in terms of weights totalling 100 grams. Likewise in his electrochemical researches he expressed the ratios between the quantity of electricity and the amount of chemical decomposition obtained. Moreover, he calculated from this the electrochemical equivalents for a large number of elements, taking hydrogen as unity. Thus direct ratios entered not infrequently into his research. However, such ratios called for the minimum of mathematical manipulation and they were, moreover, an expression of the law-likeness of nature and thus the economy of nature (discussed more fully in section 7.2). If nature is lawlike, then two chemical substances must always combine in the same ratio. Viewed from this perspective, the above examples do not show Faraday entering the world of mathematics but rather they are merely direct numerical expressions of the consistency and lawlikeness of nature. Totally lacking from Faraday's *Diary* and his scientific papers were mathematical formulae and algebraic expressions. Unlike so many of his contemporaries, Faraday did not use mathematics as a language and a manipulative tool.

Whatever the source of Faraday's difficulties with mathematics, in his later years he experienced particular difficulty in following long algebraic arguments, since, as he admitted, he could not mentally retain the meanings of the symbols. While shortness of memory may have exacerbated his difficulty in coping with mathematics, he seems to have experienced a much longer-term and deeper antipathy towards the application of algebraic symbols in the physical sciences. As Tyndall later wrote, 'No man felt this tyranny of symbols more deeply than Faraday, and no man was ever more assiduous than he to liberate himself from them, and the terms which suggested them'.[70] Faraday believed that mathematics is not the appropriate language for understanding God's nature. Although mathematical physics was closed to him, he generally adopted an unaggressive tone when corresponding with mathematical physicists, particularly those of the younger generation. Thus in writing to James Clerk Maxwell in 1851 Faraday expressed the hope that mathematicians would take the trouble of translating their conclusions 'out of hieroglyphics that we might work on them by experiment'.[71] However impressed Faraday was with the fruits of Maxwell's incisive researches, this quotation betrays an implicit opposition between mathematical physicists and experimentalists. He frequently portrayed mathematicians as failing to reach a consensus and therefore being unable to grasp truth, whereas he believed that experimentalists would arrive at consensus through interrogating nature.[72] Again, as he confided to Isambard Kingdom Brunel, while mathematics is sure, the conclusions of mathematicians cannot be trusted when compared with the solid results obtained by experimentalists.[73]

The contrast between these styles of science also appeared in a letter of 1831 to Richard Phillips, in which Faraday claimed that it 'is quite comfortable to me to find that experiment need not quail before mathematics, but is quite competent to rival it in discovery'.[74] In 1857 he adopted an even more strident position in public in claiming that he could not see any advantage in a mathematical mind over a non-mathematical one (such as that possessed by an experimentalist) 'in perceiving the nature and power of a natural principle of action'. Indeed, the mathematical mind 'cannot of itself introduce the knowledge of any new principle'. In other words, experimentalists and experimentalists alone were capable of making scientific discoveries – such as the discovery of the law of conservation of force or the laws of electromagnetism – and mathematicians were therefore relegated to the subservient role of explicating, though in their own terms, the discoveries made by experimentalists.[75] Likewise, in commenting on the proof sheets of Mary Somerville's *On the connexion of the physical sciences* in 1833, Faraday remarked:

> 'Predicted'[:] I do not remember that Mathematic[ian]s have *predicted much*[.] Perhaps in Ampere[']s theory one or at most two independent facts. I am doubtful of two. Facts have preceded the mathematics or where they have not the facts have remained unsuspected though the calculations were ready as in Electro magnetic reactions & Magneto electricity generally and sometimes when the fact was present as in Arago[']s phenomenon the calculations were insufficient to illustrate its true nature until the facts were in to help.[76]

From this evidence it would appear that Faraday consistently asserted the primacy of experiment over mathematics and the inappropriateness of the language of mathematics for reading nature's book.

From two letters written 35 years apart we glimpse the deeper currents that impelled Faraday to reject hypothetical systems in general and the deployment of abstract mathematical systems in particular. Writing to Ampère in 1822, Faraday noted his inability to leave experiment far behind and then compared himself to 'a timid ignorant navigator who though he might boldly and safely steer across a bay or an ocean by the aid of a compass which in its action and principles is infallible[,] is afraid to leave sight of the shore because he understands not the power of the instrument that is to guide him'.[77] For a man whose safety lay in facts the use of mathematically expressed hypotheses was fraught with immense danger; the danger not just of being lost but also the terror of self-annihilation. Fear also crept into a letter to Maxwell of 1857 when Faraday claimed he 'was at first almost frightened when I saw such mathematical force made to bear upon the subject' of electromagnetism.[78] Although he was intellectually impressed by Maxwell's mathematisation of his lines of force, Faraday clearly manifested a primitive insecurity on finding his ideas translated in the powerful but obscure language of mathematics.

Faraday's rejection of mathematical physics cannot be related directly to any specific part of the Bible. While a number of biblical passages deal with weighing and measuring, they do not offer a coherent picture. For example, although Isaiah 40:12 speaks of weighing 'the mountains in scales, and hills in a balance', Jeremiah 31:37 states that 'If heaven above can be measured, and the foundations of the earth searched out beneath, I will cast off all the seed of Israel for all that they have done, saith the Lord'. Thus different biblical passages could be recruited both for and against the use of measurement and thus mathematics in understanding the physical universe. The Hutchinsonians, for example, emphasised Jeremiah 31:37 to the exclusion of other such passages. John Hutchinson then proceeded to dismiss the diagrams in Newton's *Principia* as almost useless to science – 'cobweb[s] of Circles and Lines to catch Flies in'. He also claimed that Newton had erred in calling his book the *Mathematical principles of natural philosophy* (1687), since, 'when 'tis examined, it will be found, that . . . [mathematics] have no Place, but the last in Science'.[79] Faraday would have agreed with Hutchinson's opposition to mathematics (though in a less strident tone) but it is unclear whether his opposition was in any way influenced by his reading of Jeremiah 31:37.

There is, however, one instructive comparison to be drawn between the attitudes of Faraday and Hutchinson towards the use of mathematics in understanding the physical world. Both considered mathematics to be a language of human invention and not the language which God had deployed in writing the book of nature. Therefore mathematics was not the language in which we could understand nature. At this point their agreement ends, for while Hutchinson turned to revelation as the principal source for natural knowledge (although he also accepted empirical evidence), Faraday turned to experiments and observations. These provided for him the signs to help him read the book of nature.

As argued in section 8.1, above, Faraday's way of reading the book of nature bears a close similarity to Sandemanian hermeneutics. From this hermeneutical perspective there is a further Sandemanian practice which helps illuminate Faraday's attitude towards the application of mathematics in science. In the Bible there are many references to the use of lots: for example, Moses was instructed by God to divide the Promised Land and then to use lots to assign the different parts of it to different families.[80] The lot was extensively used by the Sandemanians, who considered it to be a sacred sign from God indicating His will. Thus lots were drawn to determine where members of the congregation should sit at the Love Feast after the Sabbath morning service. Since the lot is of divine origin, Sandemanians were prohibited from profaning it by playing cards, dice or other games of chance.[81] This aspect of Sandemanian doctrine indicates their profound respect for natural signs and, as we have already seen, Faraday manifested just this attitude in his experimental science. If facts are signs of divine origin, then to operate mathematically on any sign by, say, squaring it, is to destroy that sign and to make it unrecognisable. This seems to be at the heart of Faraday's criticism of mathematics, especially algebra, for he viewed empiricism as allowing an undistorted reading of the book of nature and hypotheses, particularly mathematical ones, as distorting and thus impugning God's signs.

8.5 FARADAY AND NATURE

To interpret Faraday's views on scientific method solely in terms of his stated attitudes towards facts, laws and hypotheses is to ignore his feeling of intimacy towards the physical world. Nature was not an external object which he, with superior intellect, was able to experiment on, analyse and rob of its secrets. Instead nature possessed a benevolent personality, a reflection of the Creator. As Faraday told John Tyndall in 1851, 'Nature is our kindest friend'. In similar vein William Barrett wrote, in a review of Gladstone's biography, that to Faraday 'the Universe was no machine. His was "a face-to-face, heart-to-heart, inspection of things"'.[82] For Faraday nature possessed not just personal but distinctly feminine qualities. Thus we find him castigating the (male) scientist for 'substituting the whisperings of his own fancy [i.e. his prejudices and hypotheses] for the revelations of the goddess'.[83] As will be discussed in section 10.2, Faraday felt more comfortable in the company of women than men, and it may connote more than convention that he referred to nature not just as a woman but as a 'goddess' who would reveal her secrets to the humble experimenter.

Faraday's response to nature was not limited to the intellectual faculties, for he insisted that the experimenter had to adopt the correct emotional state when pursuing investigations. Most importantly, the scientist must stand before nature with humility, and not burdened with preconceived opinions.[84] In addition, as he frequently expressed in his letters and especially in his juvenile lectures, the scientist must feel wonder, joy and excitement in the contemplation of nature. The scientist must therefore feel a positive emotional response towards the natural world and must neither maltreat nature nor view her with indifference. As one of his nieces recalled, even towards the end of his life Faraday delighted in observing nature during country walks and enjoyed watching thunderstorms.[85] Faraday manifested a disarming simplicity and lack of sophistication but also a deeply religious empathy with nature, God's creation.

This emotional bond makes Faraday's scientific work so personal. As David Gooding suggested, nature was Faraday's 'instructress' or, better still, his 'collaborator', in that he seemed to work effortlessly in concert with nature, rather than in prising her secrets from her.[86] Elspeth Crawford has likewise emphasised the emotional basis of Faraday's science as manifested both in his learning processes and in his immense creativity, which sprang from and was channelled by his close empathetic relationship with nature. His *Diary* entries often convey a sense of intimate dialogue through which Faraday learnt nature's secrets by posing questions and listening attentively for the answers. This dialogue was not conducted solely at an intellectual level, since Faraday immersed his whole being in this empathetic discourse with nature. Only once this emotional, if not spiritual, bond had been created would nature reveal herself to Faraday.[87] Yet nature was not always so amenable, and she could occasionally hold back. Particularly at times of illness and during old age Faraday's research failed to progess, and he then described himself as engaged in a 'struggle' with nature which resulted in him feeling tired and worn out.[88]

Faraday's affective (rather than intellectual) relationship with nature is evident in the highly personal way he dealt with facts. Although facts are objective, in the sense that they pertain to nature and not to the observer, Faraday frequently emphasised that he could not form a clear notion of a fact from reading an experimental report written by someone else. 'I was never able to make a fact my own without seeing it', he claimed to one correspondent. Thus direct witnessing played a major role for Faraday and he relied far less on facts gleaned by others and merely transmitted through the communication channels of the scientific community.[89] Nevertheless, he often greatly appreciated the facts reported by other scientists: for example, he praised the 'beautiful facts' which Julius Plücker reported concerning the magnetic action exhibited by the optical axes of certain crystals.[90] Yet such facts were only meaningful to Faraday when he could internalise them through his own direct experience in the laboratory or by linking them with significantly similar observations of his own. A further example was provided by Robert Mallet, an Irish engineer, who related how he had performed an experiment before Faraday, who then insisted on repeating it himself as if he 'needed absolute satisfaction that he had grasped a *fact*'.[91] The study of nature was only meaningful through close, personal and empathetic interaction.

One investigation dramatically symbolises this point. In December 1835 Faraday was conducting experiments to map the intensity of the electrostatic field inside a charged hollow metal container, such as a copper bucket or boiler. But, he asked, what would be the electrical charge inside a completely enclosed charged container? If, as he suspected, an opposite charge was induced on the inner surface, the net result within the container would be zero; a conclusion which was denied by contemporary mathematical theories of charge intensity. The experiment could not be carried out with a small metallic container, since metals, being opaque, prevent observations from being made from outside the container. Faraday therefore constructed in the lecture theatre of the Royal Institution a hollow 12 foot cube consisting of copper wire wound on a wooden frame. It stood on insulating legs and he connected its conducting wire surface to an electrical machine. 'I went into the cube and lived in it,' he reported. 'I now went inside the cube [with an electroscope] standing on a stool and Anderson worked the machine' but Faraday could not 'find any traces of electricity in myself or the surrounding objects'.[92] Inside the cube Faraday was a participant, rather than an external observer. If the cube can be treated symbolically, then so too can space. While most of his contemporaries conceived of space as empty, with a few atoms dispersed through it, Faraday filled space with fields of force. Hence, like the cube, he was enclosed in and part of space. From a field theoretic perspective he was merged with nature.

As noted in the preceding sections, Faraday possessed a highly ambivalent attitude towards hypotheses and the imagination. While on many occasions reticent about allowing his imagination free rein, we occasionally catch Faraday in full flight and capture the emotional charge behind his creativity. Faraday's *Diary* entry for 19 March 1849 contains a remarkable train of thoughts and feelings about how the force of gravity might be conceptualised and related to other forces in nature. If in its operation gravity could produce electricity, then an electric current should be produced by water falling in a pipe and even by the motion of a planet round the sun. The last excited entry for the day reads, 'ALL THIS IS A DREAM'. At this point reality began to intrude and he added, 'Still, examine it by a few experiments. Nothing is too wonderful to be true, if it be consistent with the laws of nature, and in such things as these, experiment is the best test of such consistency'.[93] A second example occurred in the middle of a series of magnetic investigations dating from the following year. He wrote to Whewell: 'At present my head is *full of visions*: whether they will disappear as experiment wakens me up or open out into clear distinct views of the truth of nature is more than I dare say. But my hopes are strong'.[94] These quotations offer remarkable insights into Faraday's mental processes. When he was standing at the laboratory bench, dreams, visions and hypotheses welled up in his mind. He was, at one level, afraid of losing control by allowing himself to be borne along on a wave of fantasy. At the same time he recognised that the imagination was the source of all insight and must not be stunted. Instead it must be directed by self-discipline and that direction takes place by applying the touchstone of experimental fact.

While historians of science have generally viewed scientific method in purely intellectualist terms, the evidence offered in this chapter strongly suggests that methodological categories are as much emotive and psychological as logical. Some writers have wanted to go even further and suggest that Faraday possessed some preternatural ability. For example, Elspeth Crawford has been greatly impressed by Faraday's uncanny ability to make correct judgements in an extraordinarily high proportion of cases, when compared with other scientists. Similarly the German chemist Rudolph Kohlrausch asserted that Faraday 'smells the truth' and a number of his biographers have pointed to his rarely failing intuition. Crawford has also claimed that Faraday believed that the truths of nature are revealed to the honest enquirer, the word 'reveal' having, of course, theological associations.[95] Scientific discovery then becomes not the application of a method but a revelation to the scientist, who must be in a suitably receptive state. For Faraday such revelation required not only imagination but also a close, even intimate, rapport with nature, especially in the laboratory.

By its very nature creativity cannot be reduced to the confines of scientific method and Faraday's accounts of his research, particularly his *Diary*, will continue to fascinate those concerned with the processes of scientific innovation. For the present we can perhaps proceed no further than to point to Faraday's remarkable ability to combine a lively imagination with rigorous experimental work. In a short note dated June 1858 Faraday reasserted his commitment to the principle of force conservation (to be discussed in the next chapter) which he considered to be a necessary truth concerning the physical world. The way forward for the scientist, he argued, was to accept that principle and to 'permit its guidance in a *cautious yet courageous* course of investigation'.[96] In this italicised phrase he succinctly expressed the dichotomy which bounded his practice of science. On the one hand, he was the cautious experimenter who sought safety in facts and saw danger in irresponsible hypothesising; on the other, he was the courageous speculator who tested his insightful conjectures against experiment.

The phrase 'cautious yet courageous' also expresses two contrasting aspects of Faraday's personality. Both courage without caution and caution without courage were inadequate qualities for the scientist. Likewise neither speculation nor experimentation was sufficient, but the scientist had to combine the two in a constructive manner. That there existed a tension between Faraday the cautious, careful experimentalist (who was concerned with facts and the laws connecting them) and Faraday the courageous speculator may provide an important clue to his scientific creativity. He could not rest content in pursuing either mode, but instead interrelated constructively his experimental work with his speculations. Moreover, he allowed considerable rein to his feelings and, in particular, insisted on feeling at one with nature. Thus we see in Faraday's scientific investigations the harmonious interplay between his thoughts, feelings and actions.

9 Scientific Investigations

In the previous two chapters I have argued that Faraday pursued his science within the context of a theologically informed conception of nature and that he advocated a scientific method that can likewise be related to his religion. However, the characterisation of Faraday developed in these chapters is far from complete. We have, as it were, led him to the door of his basement laboratory at the Royal Institution but not allowed him to enter and to practise his science. As a highly innovative scientist Faraday was not simply articulating his metaphysical beliefs but in the course of his work was developing his own ideas, interacting with the ideas of others and the results which he (and other scientists) obtained in their laboratories. Moreover, as several historians have emphasised, Faraday worked painstakingly and unremittingly at the laboratory bench and any account of him which ignores this aspect is necessarily incomplete.

In this chapter I want to locate Faraday within the context of his research work, move him back to his laboratory and concentrate on the development of certain research projects. But to enter Faraday's laboratory in the basement of the Royal Institution is to set foot in an area where few people were allowed to go. This was a very private space for Faraday and we should tread carefully, not only to avoid breaking the delicate glassware that lines the shelves but also because this is where Faraday communed with God's Creation. A reverent silence is as appropriate here as in the Sandemanian meeting house. The silence that fills this research area contrasts sharply with the chatter that permeates the public rooms of the Royal Institution. Similarly we must contrast Faraday's private research activity (as recorded in his *Diary*) with the public presentation of his discoveries in the Institution's lecture theatre or before the assembled Fellows of the Royal Society. As we will see, much of his research never left the private domain, and that which did was reinterpreted for public consumption.

To do full justice to the development of Faraday's very extensive researches would require a very different book from the present study: one closer to L. Pearce Williams' biography of 1965 or any of the many papers which trace in detail particular aspects of Faraday's research.[1] My intention in this chapter is far more modest, since I will be using but two examples in order to make these general points and to suggest how the issues discussed in the previous two chapters intersected with Faraday's scientific investigations. In discussing these examples I shall therefore be drawing attention to the metaphysical and methodological themes already in play. In particular, I want to argue that Faraday's conception of the economy of nature was a dominant feature of his scientific research. This is not to say that other aspects of his thoughts and skills were unimportant but rather that the economy of nature underpinned his scientific practice and directed his research effort.

The two examples I have chosen are Faraday's first and last major scientific investigations. The first of these is his early work on electromagnetism and electromagnetic induction, without which no biography of Faraday would be complete. The other topic under discussion is Faraday's attempt to discover the relation between the electrical and the gravitational force. There are, as we shall see, many themes common to these two investigations but there are also two significant differences. One is that while the electromagnetic researches resulted in success and were largely responsible for Faraday's international reputation, the gravitational work ended, in a sense, in failure. Yet this difference should not prevent us from recognising that Faraday was committed to the existence of both these phenomena. Indeed, at the time of performing these two investigations he was more confident about the electric–gravitation relation, which he failed to observe, than the electromagnetic one, which he detected. This is the second difference. In the 1820s Faraday, a lowly member of the British scientific community, was dogged by insecurity which only began to disperse shortly before his 1831 investigation of electromagnetic induction. When he first turned seriously to the gravitational problem about 1837, and when he returned to it in 1849 and 1859, he was the outwardly self-assured, successful and widely respected scientist.

9.1 ELECTROMAGNETISM

9.1.1 *Historical sketch*

Faraday's earliest scientific papers were almost entirely concerned with chemistry, including physical chemistry. The first visible sign of a new direction to Faraday's research is contained in a *Diary* entry for 21 May 1821, in which he reported a set of experiments, which he performed with Davy at the London Institution, to determine the effect of a horseshoe magnet on the 2½ inch arc produced by a powerful battery of 2,000 plates.[2] By the end of that year Faraday was immersed in the study of electromagnetism, had made several discoveries and become recognised as a significant contributor to that field. Faraday's interest in the area probably predates May 1821 by several months and like many other scientists he was first alerted to the topic by the publication in the October 1820 issue of *Annals of Philosophy* of an article by Hans Christian Oersted, who had reported that a small magnet was affected when placed either above or below a parallel wire carrying an electric current. Investigating the nature of this action, Oersted found that, when below the wire, the magnet moved so that the end adjacent to the negative connection of the battery turned to the west and the other end to the east. With the magnet above the wire, the magnet turned in the opposite direction. With his apparatus Oersted obtained displacements of up to 45°.

During the next few months a number of scientists, including Davy, Faraday and William Hyde Wollaston, repeated and extended Oersted's work. A burgeoning literature, including many forays into the theory of electromagnetism, sprang from this celebrated experiment.[3] Over the summer of 1821 Faraday began writing his 'Historical sketch of electro-magnetism', which shows him engaging this literature and sifting through the experiments performed and hypotheses constructed, in order to gain 'a clear idea of what had been done, and by whom'. His survey of these innovations proved an excellent introduction to the subject.[4]

Faraday was just 30 years old when this paper was published. For the previous 8 years he had served as Davy's assistant. From this lowly position in British science Faraday, the author of a few minor publications, felt insecure in launching himself into this new field of research without Davy's guidance. Moreover, the paper appeared anonymously with numerous apologia and humble expressions of respect for the established authors on whom he was commenting.[5] Faraday's diffidence was also evinced by his concern not to make any mistake nor to misrepresent any author. More importantly, in this first literary excursion into electromagnetism Faraday sought to draw a clear distinction between experimental results and hypotheses. At this early stage in his career Faraday was concerned to found science firmly on facts but he was not prepared to allow much scope to the imagination. Thus he asserted that the facts which Oersted and others had discovered were secure, incorrigible and of great importance to science, whereas the plethora of theories had to be weighed judiciously; indeed most of them were patently false. To Oersted's discoveries Faraday added a series of remarkable ones by Ampère, most particularly his observation that two parallel current-carrying wires affect one another: they mutually attract when they are similarly connected to the battery, but repel when connected in the opposite sense. It was to Ampère that he also attributed the invention of the galvanometer and to Ampère and Arago the first use of helical wires to form electromagnets. In recording these and other developments Faraday achieved his limited aim of placing in chronological order the discoveries of a dozen authors dating from the latter part of 1820 to the early months of 1821.

The 'Historical sketch' concluded with a discussion of the various hypotheses that had been erected to account for the phenomena. Even in this early paper we see Faraday employing criteria that were aimed at curbing the excesses of the imagination. He was clearly unimpressed by several published hypotheses which lacked clarity and were unable to explain the phenomena adequately. By contrast, Ampère's views 'are the most extensive and precise, and have been tested by the application of facts and calculation very far beyond any of the rest'.[6] Yet Ampère's theory was by no means free from the problems that plagued its competitors. In particular, when using the term electricity Ampère was sometimes referring to a body's electrified state and sometimes to the hypothetical electric fluid flowing in a wire – Faraday, it should be remembered, remained sceptical about the existence of such fluids and currents. Moreover, Ampère had postulated not one but two mechanisms involving the contrary flow of two opposing electric fluids, strategies which both failed to explain the phenomena and merely gave vent to his unrestrained imagination.[7]

Faraday's response to Oersted is equally interesting. Oersted was an adherent of *Naturphilosophie*, a metaphysical system based on Kantian idealism and developed principally by Friedrich Schelling. This system, which had an extensive following on the continent and few adherents in Britain, postulated a universe filled with powers, or, more precisely, by pairs of antagonistic polar powers. Commentators, Faraday included, have generally found Oersted's philosophical excursions both abstruse and difficult to follow. However, in his early works Oersted offered a number of metascientific principles which could potentially be put to experimental test. In particular, he had asserted the identity between electricity and magnetism, but in a form which one recent commentator describes as 'a poorly articulated general discussion of these similarities'.[8] At one point, however, Oersted expounded this relation specifically in terms of electricity and magnetism, the former being, he claimed, a more latent form of the latter. Although Faraday cited this view of Oersted's, he also admitted that 'I have very little to say on M. Oersted's theory, for I must confess I do not quite understand it', referring probably to Oersted's contention that the conflict between polar forms of electricity gives rise to the magnetised state of matter.

Despite Faraday's ambivalence, two points stand out clearly. Firstly, he believed that chance played no role in the discovery. The rejection of serendipity was not only a standard Sandemanian position – all events resulting from God's will – but it also cohered with Faraday's view that scientific discovery required adequate intellectual preparation and training in the practical skills of science. Secondly, he claimed that Oersted's 'theory rather led to the experiments, than the experiments led to the theory'. What impressed Faraday was that Oersted had predicted that electric and magnetic forces were interrelated, and that 'the thoughts were conceived, and the experiments devised, some time before they were made'.[9] Although the centrality of *Naturphilosophie* to Oersted's discovery is now doubted,[10] Faraday evidently perceived that some metascientific principle – presumably the identity of magnetic and electrical forces – provided the crucial theoretical background to Oersted's experimental success.

The early electromagnetic discoveries of Oersted and others doubtless illustrated, confirmed and helped fix Faraday's growing belief that the economy of nature could be expressed through the relation between the forces in nature, although Oersted's *a priori* identity of different forces was too speculative and lacked empirical confirmation. Faraday's interest in this field of research and his subsequent investigations indicate that he conceived it not only as a rapidly advancing subject and one through which he could advance, but also as an area in which he could deploy his insight that definite relationships between the divinely ordained powers of nature existed.

9.1.2 *Electromagnetic rotation*

Faraday's first sustained set of experiments on electromagnetism covered a period of 8 days in early September 1821 and provided him with the material for two papers which were published in the October issue of the *Quarterly Journal of Science*, establishing his name as a major contributor to this area of research.[11] With great care he continued and extended the work of Davy and others who had studied the orientation of magnets in the neighbourhood of a current-carrying conductor. As Figure 9.1 shows, Faraday mounted vertically a wire (*A*) which carried a large current delivered by a battery. He then determined the positions in which a magnetised needle – shown by an arrow – was either strongly attracted or repelled by the wire. Contrary to expectation (in terms of centrally directed forces) he found (para. 3) that each pole of the needle was thus affected in four positions – two attractions and two repulsions. On the next line (para. 4) he redrew the diagram so as to show, as if

Sept. 3RD, 1821.

Electromagnetic expts. with Hare's Calorimotor. To be remembered that this is a single series?

ELECTRO-
MAGNETISM

1. Position of the expt. wire A˙.
2. Positions at first ascertained were as follows

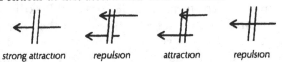

strong attraction   repulsion   attraction   repulsion

3. On examining these more minutely found that each pole had 4 positions, 2 of attraction and 2 of repulsion, thus

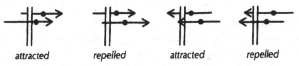

attracted   repelled   attracted   repelled

4. Or looking from above down on to sections of the wire

attracted   repelled   attracted   repelled

5. Or

6. These indicate motions in circles round each pole, thus

Hence the wire moves in opposite circles round each pole and/or the poles moves in opposite circles round the wire.

Figure 9.1   Steps to the discovery of electromagnetic rotation.

Source: Faraday's *Diary*, vol.1, p.49.

looking down from above, the positions in which the needle was either repelled from or attracted towards the current-carrying wire. These relationships were then brilliantly reconceptualised in para. 5 – with the needle stationary the positions in which the wire is attracted (A) and repelled (R) are shown: two positions of attraction and repulsion are associated with both the north and the south poles. In the left-hand part of the final diagram (para. 6) Faraday imposed a new conceptualisation. Instead of two attractive and two repulsive positions in the neighbourhood of the needle's south pole, he conceived the wire possessing a natural rotation round the pole in a clockwise direction. Likewise the attractive and repulsive positions in the neighbourhood of the north pole were transposed into an anticlockwise motion of the wire round that pole.

What this experiment clearly shows is the manner in which Faraday was not merely experimenting but reconceptualising the evidence. Moreover, unlike several of his contemporaries, he was not understanding the observations in terms of traditional Newtonian forces that would have acted by direct attraction and repulsion between the wire and each pole. Instead, he conceived the spatial relationship in terms of circles. Faraday's lack of education in Newtonian mechanics is relevant here, but it is also important to notice that the circle, which he adopted, has been the traditional symbol for the economy of nature. Moreover, he characterised the relation between the motions of the pole and the wire in a symmetrical manner: 'the wire ought to revolve round a magnetic pole and a magnetic pole round the wire'. These predictions were soon confirmed in a number of instances, Faraday rapidly refining his experimental design until he achieved his most ingenious demonstration of electromagnetic rotation with the experiment shown in Figure 9.2, which was briefly described in the first paper but became the sole subject of the second paper.[12]

Throughout this investigation Faraday was continually seeking further ways of conceptualising the relation between the electric and magnetic forces. For example, we find him trying to draw generalisations from cases where, say, two wires or one wire and one pole are involved.[13] In addition, the cautious Faraday remained ambivalent about Ampère's theory that magnetic action is due to electricity; while admitting the existence of much confirmatory evidence, Faraday also drew up a list of apparent differences between the actions of a magnet and an electromagnet.[14] He was cautiously feeling his way on this question, not prepared to go beyond the evidence, at least in a public forum. Yet his caution did not extend far enough in one direction, for in his hurry to publish his results he incurred the wrath of Davy, who charged him with plagiarising the work of Wollaston and not adequately acknowledging his own role. However insubstantial the charge, the bitterness it created affected Faraday deeply and robbed him of the joy of his first significant scientific discovery. That this attack on his 'honour and honesty' erupted but a few weeks after Faraday had made his confession of faith at the Sandemanian meeting house only highlights the depth of the 'distress' he felt.[15] As he subsequently claimed, the affair had taught him 'to be cautious upon points of right and

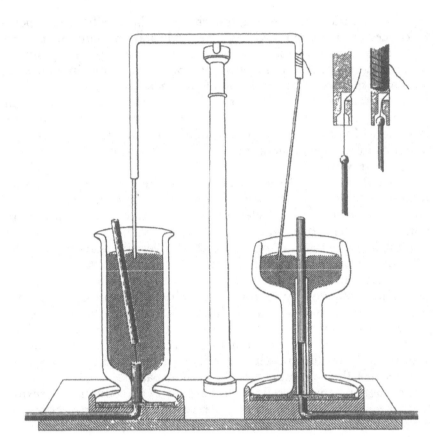

Figure 9.2 Faraday's apparatus for displaying electromagnetic rotation. The two vessels are made of glass and contain mercury. A copper pin is inserted through the base of each glass vessel, while the arms of a brass pillar are positioned so that its ends stand directly above each pin. In the left-hand cup, the pillar is extended so as to touch the surface of the mercury and a cylindrical magnet is fixed at one end to the copper pin by a short thread so that it is almost vertical and its other end protrudes slightly above the surface of the mercury. When a battery is connected between the brass pillar and the copper pin, the magnet rotates about the axis of the glass cup. In the right-hand glass vessel a cylindrical magnet is positioned along its axis and extended a short distance above the surface of the mercury. Suspended from the brass pillar is a rigid wire which reaches to the surface of the mercury. When electricity is again applied between the pillar and the copper pin, the wire rotates about a vertical axis. Notice the symmetry between electricity and magnetism in these two arrangements.

Source: Faraday, 'New electro-magnetic apparatus', *Quarterly Journal of Science*, 12 (1821), 186–7.

priority' and it left a deep scar that was only partially erased with Davy's death in 1829.[16]

9.1.3 *Research during the 1820s*

Only scanty evidence remains to help us plot Faraday's route from the electromagnetic rotation research of 1821 to the discovery of electromagnetic induction 10 years later. For most of that decade he was occupied with many other scientific researches, principally on chemistry, and his multifarious duties at the Royal Institution. While it would be a distortion of the historical record to interpret Faraday's work in the 1820s as preparation for his momentous discovery of electromagnetic induction, that discovery should nevertheless be conceived within the broad context of a number of problems and issues that repeatedly surfaced throughout that decade. One such problem concerns the relation between forces, most particularly the connection between the electric, magnetic and other forces. Thus, in an unsuccessful investigation of 10 September 1822 Faraday tested his expectation that electricity should alter the polarisation of a beam of light, and 6 years later he experimented, again unsuccessfully, on the presumed influence of light on electricity.[17] A variety of possible relations between electricity and magnetism were subjected to experimental test in January 1823, December 1824 and April 1828.[18] As will be discussed below, a set of experiments performed on 28 November 1825 enabled Faraday to examine whether a current in one wire induced a current into a neighbouring wire. A few days later he was again examining the suspected analogy between magnetism and electricity.[19] A final example of Faraday's search for the relation between different powers is provided by an unsuccessful attempt to produce electricity by heat, an experiment which predates his successful series on electromagnetic induction by a mere 3 months.[20]

Thus throughout the 1820s we find Faraday performing a number of experiments to determine the relation between one power and another. At this stage he did not articulate any metascientific view about the conservation of the natural powers but rather he worked implicitly with a notion that powers were mutually related to one another: electricity and magnetism, electricity and light, and electricity and heat. That many of these early experiments proved unsuccessful indicates that Faraday did not have any clear conception of how to evoke these relations in practice. Instead, he was pursuing a trial and error policy in the hope – indeed, the expectation – of making a discovery of similar importance to Oersted's. His general commitment to an economic relation between powers was, by itself, insufficient; what he lacked was a clear theoretical insight into how the powers were propagated across space and how an experimental arrangement could be constructed to detect the relation between these powers. We now turn to Faraday's views on this subject and their development through the 1820s. Although the evidence provides no coherent story, some relevant strands can be detected.

As L. Pearce Williams has emphasised, Faraday's ambivalence to Ampère's theory was one crucial factor. From the early 1820s Ampère had firmly established himself as the leading theoretician of electromagnetism and had produced a number of mathematically sophisticated papers on the subject. If Faraday could not fully understand the mathematical arguments, he nevertheless perceived the rigid physical assumptions which Ampère employed. Most importantly, Ampère denied that electricity and magnetism are of equal significance and should be treated theoretically on an equal footing. Instead, he roundly asserted the physical reality of electricity, and of electric currents, and reduced magnetism to a mere effect of local electric currents. His impressive discovery of 1820 that two parallel current-carrying wires attract (or repel) one another was strong evidence that forces act between elements of electric current. That magnets exhibit similar attractions and repulsions was no surprise, since these arose from the interaction – through the interjacent forces – of the electric currents flowing in the magnets. Electromagnetism was equally subsumable within this context, for Ampère had reduced it to the force interaction between the current in the wire and the current in the magnet.[21] As we have seen, Faraday expressed considerable reservation about Ampère's theory in the 'Historical sketch' of 1821–2. In contrast to Ampère's professional security and his deft but dogmatic handling of his theory of electromagnetism, Faraday could only respond with criticisms and not with an alternative theory. While Faraday's scepticism of theoretical structures was later to enhance his science, the lack of theoretical grounding in the 1820s left his work in this area rather undirected. Yet it was given some direction in responding to the work of Ampère and others, particularly a series of experiments by Arago, Babbage and Herschel.

Arago reported in 1824 an impressive new phenomenon which invited explanation. He arranged a disc made of copper – a non-magnetic – so that it could rotate in a horizontal plane, and above it he positioned a freely suspended magnet. When the disc was rotated, the magnet was first caused to move from its initial position, and it was subsequently dragged round by the disc. However, this force was not apparent when the disc was stationary. A year later Charles

Babbage and John Herschel presented a paper to the Royal Society in which they varied the arrangement, for example, by rotating the copper disc between the poles of a powerful horseshoe magnet. They also determined the magnetic 'susceptibility' – their term to describe the observed effect – of different substances. Another important observation was that the magnetic effect produced by the disc was largely destroyed when it was punctured by radial slits. The observed phenomena were explained on the assumption that the interaction between, say, the rotating disc and the sympathetically moving magnet, was due to induction and, moreover, the particles comprising the magnet were affected by an inductive process. The other important point to note about this paper is the authors' insistence that the inductive process does not occur instantaneously but that *'time enters as an essential element'*.[22]

Arago's disc was the subject of Faraday's *Diary* entry for 2 December 1825, when he attempted an analogous version in which the copper disc was subjected to an electric, rather than a magnetic, field. That he should attempt such an experiment is further evidence for his belief in the mutual relation between electricity and magnetism. Presumably he hoped that electrical induction would occur in the same way that, on Herschel and Babbage's account, magnetic induction was supposed to account for the result observed by Arago. In the electrical analogue the particles of the disc should have suffered electrical induction.

A few days earlier Faraday had conducted three experiments aimed at detecting electrical induction and probably informed by Ampère's theory. Each of these experiments comprised a primary circuit connected to a moderately powerful battery and a secondary circuit attached to a galvanometer. If induction occurred, a current would have been induced into the secondary circuit and detected on the galvanometer. Three arrangements were tried. In the first, each of the two circuits comprised a straight parallel wire 4 feet in length and separated by two 'thicknesses of paper'. Secondly, the primary was composed of a helix of covered wire through the centre of which was a straight wire, being the secondary circuit. Finally, the straight wire was made the primary and over it was suspended the helix connected to the galvanometer. No current was detected in any of these trials.[23]

One can but speculate what theoretical considerations preceded these experiments. One possibility, as Williams has argued, is that Faraday may have followed Ampère in conceiving the electricity in the primary wire as comprising the opposite flow of two positive and negative electric fluids. At any instant a positive electric charge would have induced a negative charge at the corresponding point on the other wire, and a negative charge in the primary would have given rise to a positive charge at its neighbouring point on the secondary wire. Opposing electricities would have flowed in the secondary, giving rise to a galvanometer reading. In the latter two experiments some kind of electomagnetic inductive effect might have taken place, so that the secondary circuit would have been influenced, via magnetism, by the electricity in the primary.[24] In the light of Faraday's ambivalence towards Ampère's theory we should be careful not to read too much of Ampère's theoretical baggage into Faraday's very brief account of these experiments. That he used the term induction on three occasions in this short discussion is, however, significant, since it indicates the kind of process he envisaged linking the primary and secondary circuits, but it is difficult to determine whether he had a more elaborate theory of inductive action than the assumption that the passage of electricity in one wire should have an equal and opposite effect in the other – an assumption that encapsulated the economy of nature.

In his papers of the early 1820s Faraday was uncertain which electromagnetic theory to adopt. Some of his comments show that he was thinking in traditional terms of centrally directed forces, – e.g. forces that act between a current-carrying wire and the poles of a magnet. Although he was less than satisfied with this type of account, he continued to use it, along with other very different concepts. For example, his discussion of electromagnetic rotations shows that he conceived the powers in bodies as relating in a manner that cannot readily be reduced to centrally directed forces. Another important relation was that of mutual reactivity. Thus if a magnetic pole moves round a current-carrying wire, such a wire should move round a magnetic pole.[25] At one level Faraday worked with a conceptual symmetry between magnetism and electricity (and, indeed, the other powers). An action in A invites an equal and opposite reaction in B; while an action in B results in an equal and opposite reaction in A: thus the back e.m.f. produced in a coil could, ideally, negate the electrical potential applied to it. Despite appearances, this is not Newton's third law of motion, which pertains specifically to dynamics, but a far more general metascientific principle, which is a formulation of Faraday's commitment to the divine economy of nature. It applies to all of nature's powers, including magnetism and electricity. Thus in a *Diary* entry written at the close of 1824 Faraday stated that he '[e]xpected that an electro magnetic current passing through a wire would be affected by the approach of a strong magnetic pole to the wire so as to indicate *some effect of reaction* in other parts of the wire'.[26] All physical phenomena were caught up in an infinite chain of actions and reactions.

Induction was a key term in early nineteenth-century science.[27] During the latter half of the eighteenth century and particularly following the invention by Volta of the electrophorus, it was a commonplace to think of one type of static electricity inducing the opposite kind on a neighbouring body; thus, for some writers, a positive electric charge induced a negative one, or, for others, resinous electricity induced a charge of vitrious electricity.[28] Similarly with magnetism, the positive pole of a magnet in the vicinity of a piece of soft iron would induce in it a neighbouring negative pole. Perhaps owing to his apprenticeship in chemistry and the influence which Ampère exerted on his early electromagnetic work, Faraday did not use that crucial term in the early 1820s but instead employed, although not consistently, the vocabulary of forces.

Concern with the subject of induction was the impressive novel element in Faraday's notes of late 1825, which were followed almost immediately by a further reference to Arago's disc experiment.[29] The published paper that was read before the Royal Society on 24 November 1831 opened with a definition of *induction*, the term which he selfconsciously extended from its conventional electrostatic context to include the induction of electric currents; thus in its new signification induction expressed 'the power which electrical currents may possess of inducing any particular state upon matter in their immediate neighbourhood, otherwise indifferent'.[30] What is also significant is that his several references to induction in that paper concentrated on the magnetic induction displayed by Arago's disc rather than the electrostatic case. It therefore seems likely that Arago's experiment and its attendant literature induced Faraday to consider inductive action seriously.

However, there are two ways in which inductive action can be treated. Firstly, it can be taken as a fact that, say, a (detectable) positive electrostatic charge on one body is considered as producing a (detectable) negative charge on its nearest neighbour. Secondly, induction can be understood to occur at the level of the particles comprising either a substance or the electric and magnetic fluids flowing in that substance. Thus in their explanation of the Arago disc experiment Babbage and Herschel appealed to the inductive action on the 'molecules' comprising the magnet. From Faraday's *Diary* entries of late 1825 it would appear that he had by that date accepted magnetic induction as a fact and was searching for an electromagnetic analogue. These entries give no indication whether he conceived induction in terms of the interaction of particles. However, throughout the 1820s Faraday rarely speculated about matter or force either in his *Diary* or in his early published papers. It appears that his theoretical insights into these topics only began to

mature a year or two before his 1831 investigations into electromagnetism. The timing is significant and, as I shall argue below,
Faraday's willingness to speculate should be interpreted as part of
his new-found feeling of self-assurance at that period.

Many earlier writers had articulated the view that electricity is
propagated like a wave, sometimes even as the conflict between two
waves travelling along a wire in opposite directions. Faraday was
certainly familiar with this latter theory which he attributed to
Oersted in his 'Historical sketch', although at that time he evinced
little sympathy for the theory.[31] As Williams has pointed out,
another, related part of the jigsaw is provided by the wave theory
of light, which offered a further way of accounting how one body (or
particle) affected another. Faraday's work on optical glass, which
began in 1824, doubtless increased his interest in light and he was
clearly impressed by Fresnel's version of the wave theory, which
appeared in a simplified exposition in the *Quarterly Journal of Science*,
edited by Brande, in 1827–9.[32] In an experiment of 26 September
1828 Faraday attempted to produce electricity in a plate of copper
on which a solar spectrum was projected.[33] Presumably the particles
of copper should be affected – perhaps being vibrated or strained –
so as to produce an electric current. Moreover, he considered
electrical induction to be like a vibration, which, as with light and
sound, propagates in time. The connection with sound is further
suggested by a number of papers and lectures dealing with the
production and transmission of musical sounds and by a series of
experiments on vibrating plates that Faraday carried out in the
spring and summer of 1831. In pursuing his investigation into what
are called Chladni's figures, Faraday became skilled at displaying
musical notes visually.[34]

9.1.4   *Electromagnetic induction*

Turning to the discovery of electromagnetic induction, we should
note that Faraday's *Diary* entry for 29 August 1831 is remarkably
succinct. (There are no earlier notes for this investigation, so it is
impossible to know with any certainty the preparation leading up to
this entry.) He states simply that he wound two helices of copper
wire round a torus of soft iron (see Figure 9.3). The circuit,
including one of these helices, was completed by a length of copper
wire passing over a parallel, magnetised needle; this would detect
any electricity flowing in that helix. The other helix was connected
to a battery. When this connection was made, the needle moved
sensibly in one direction before settling back to its original position;
when the circuit was broken, the needle moved similarly but in the

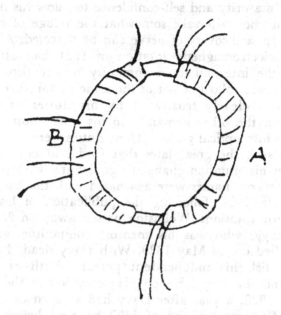

Figure 9.3 Faraday's induction ring. On 29 August 1831 Faraday wound two coils of insulated wire round this soft iron ring. The circuit, including coil *B*, was completed by a long wire which passed over a parallel magnetic needle. When a battery was either connected or disconnected to coil *A*, 'immediately a sensible effect on needle' resulted. 'It oscillated and settled at last in [its] original position.'

Source: Faraday's *Diary*, vol.1, p.367.

opposite direction. By the end of the third paragraph of his *Diary* entry of the experiment, Faraday reported his success.[35] His report makes the experiment appear simple and his steps assured. He displays no hesitancy and makes no false moves. It is as if he had been meditating on this experiment for a long time and then performed it with clearsighted accuracy. The reader may easily – too easily – be swayed by Faraday's confidence into thinking that he had a full premonition of the result before entering his laboratory on that day.

While the 1831 paper, in both the *Diary* and published forms, contains a brilliant experimental investigation, it also displays the remarkable self-assurance that Faraday achieved both in his experimental discourse, in which he seems to move effortlessly from one experiment to another, but also in his willingness to speculate. This self-confidence was not present in his papers of the 1820s and only emerged at the very end of that decade.[36] By about 1830 he possessed the maturity and self-confidence to allow his imagination more freedom and to forsake somewhat the refuge of experiment. More generally, a recovery of nerve can be detected. Although his discovery of electromagnetic rotation in 1821 had attracted the attention of the international community of scientists, his subsequent papers were solid but not of the same importance.

Faraday was clearly frustrated by his duties at the Royal Institution and the other demands on his time, most particularly his researches into optical glass.[37] These duties were in conflict with his desire to seek the great laws that God had employed in the Creation. Yet his situation changed significantly with the death of the two men whose names were associated with the humiliation he had suffered in 1821, following the publication of his paper on electromagnetic rotations. Wollaston passed away on 22 December 1828 and Davy, who was his mentor, competitor, accuser and taskmaster, died on 29 May 1829. With Davy dead, Faraday was free to establish his independent persona both at the Royal Institution and the Royal Society. He only joined the Council of the latter in 1828, a year after Davy had stepped down from the Presidency. Towards the end of 1829 he read before the Royal Society the results of his extensive research on the manufacture of optical glass. This paper formed the Bakerian lecture and thus represented a major honour, his first from England's leading scientific society, as well as recognition of the considerable time and energy he had expended on the subject. Another substantial paper, on acoustical figures, followed in the summer of 1831.[38] Both these publications show Faraday as the mature scientist who could handle with ease major scientific problems. He was now his own master and was thus psychologically and technically prepared to engage the subject of electromagnetic induction when he entered the laboratory on 29 August 1831.

In a strong sense the new experiment drew extensively on the earlier series of unsuccessful experiments, especially the three performed on 28 November 1825, in which he attempted to interrelate different natural powers – the central problem arising from his conception of the divine economy of nature. All these experiments were arranged with a battery in the primary circuit and a galvanometer in the secondary. They were all founded on the supposition that *induction* would take place between the primary and secondary circuits. Like the parallel wires experiment of 1825 (and unlike the other two) the new experiment was conceived in terms of a symmetry between the primary and secondary circuits. Like the last two experiments of the 1825 series (but not the first) the new experiment utilised helices, but the magnetic effects of the helices were concentrated by a soft iron core. This difference is probably attributable to Faraday's familiarity with experiments in which soft iron cores could be used to enhance the power of electromagnets. From his reactivity principle he may have envisaged that a strong electromagnetic field produced in the soft iron would, in turn, give rise to a strong current in the secondary. As he subsequently made clear in his published account, the well-known production of magnetic action at right angles to a current-carrying wire strongly suggests that the magnetic action thus produced should induce an equally strong current into a wire.[39]

Throughout these investigations Faraday sought to effect in practice a mutual relation between electricity and magnetism. His very first *Diary* entry reads, 'Expts. on the production of Electricity from Magnetism, etc. etc.', while in a subsequent entry he noted that an experiment manifested a 'distinct [case of] conversion of Magnetism into Electricity'. Again, in the published paper, he hailed the experiment in which a disc rotated in a strong magnetic field as demonstrating 'the production of a permanent current of electricity by ordinary magnets'.[40] As yet he held no strong thesis about force conservation but rather conceived the relation he was investigating in causal terms. Closely connected with this were a series of other causal relations which were effected when electricity flowed in the secondary circuit. Thus the battery produced electricity from a chemical reaction and the galvanometer was not merely a detector but was also dependent on the production of magnetism from the electric current. Faraday, moreover, connected to the secondary winding a number of other standard pieces of apparatus, encapsulating other conversion processes which he expected to produce observable effects when the battery in the primary was connected and disconnected. However, he could not obtain the expected chemical or heating effects, although after introducing slight modifications to this experiment he managed to make a frog's leg convulse and he also obtained a spark.[41]

While Faraday sought an inductive interaction between a primary and a secondary circuit, what he could not have known, although he may have suspected, was that the particular arrangement he tried on 29 August 1831 would produce the desired result. But there is a further aspect of the result which he probably did not expect, and this is that the galvanometer in the secondary circuit showed a *transient* flow of electricity, rather than a continuous current. Like other scientists working on this problem Faraday had, throughout the 1820s, assumed a continuous effect.[42] The *Diary* entry betrays no marked surprise at the transient result obtained in 1831, although, if asked before the experiment, he would probably have predicted a continuous current on the grounds of a continuously operating cause. However, he was already familiar with several transient electrical phenomena, such as the Leyden jar (what we would call a condenser) which could produce an electric shock or spark. As he noted, the electricity in the secondary 'partook more of the nature of the electrical wave passed through from the shock of a common Leyden jar[,] than of the [constant] current from a voltaic battery'. Arago's experiment provided a second analogy, since it, like the new electromagnetic induction effect, involved a transient phenomenon. As he asked rhetorically in a *Diary* entry towards the end of his first day's research, 'May not these transient effects [obtained with the helices wound round the soft iron torus] be connected with causes of difference between power of metals in rest and in motion in Arago's exp[erimen]ts'.[43] Just as Arago's disc had to be in motion in order to produce a current, so electricity in the secondary helix was only produced by a change – the connection or disconnection of the battery. This action produced a sudden inductive effect which resulted in the response of the galvanometer magnet, as in the analogous Arago's disc situation.

A more problematic part of the analogy concerns the role of time. As Babbage and Herschel noted, '*time enters as an essential element*' into Arago's experiment. At several places in the published account Faraday wrestled with the question whether electromagnetic induction likewise involved time. Although the induced current in the secondary circuit appeared 'to continue for an instant only', Faraday was nevertheless convinced that electromagnetic induction required 'sensible time' to operate and that the galvanometer's response was activated by a short, sharp electric current but one which nevertheless operated in time.[44] Thus although the action appeared to be instantaneous on connecting the battery, Faraday conceived the event operating in strictly temporal terms. This conclusion is not implied by the empirical evidence and it is possible that he was influenced by the paper of Babbage and Herschel. However, in numerous later writings he was so insistent that physical processes, with the possible exception of gravitation, occupy time that some other factors may be relevant. If we return to the economy of nature as a basic and recurrent theme in Faraday's science, his emphasis on the continuity of time in his 1831 paper and elsewhere can be interpreted as reflecting his concern that physical processes must operate within this economy and cannot occur instantaneously. It is a commonplace among theological writers that the processes in human history, including the past events and prophecies related in the Bible, occur in time. God did not create the world instantaneously, but in time, and it is also in time that the providential plan, including human history, is acted out. Moreover, just as Faraday rejected the spatial discontinuities implied by atomism, so he viewed time as continuous and all physical processes as performed in real time.

While Babbage and Herschel blandly attributed Arago's result to the induction of molecules, Faraday offered some deeper insights into what occurred when the battery was connected (or disconnected) in his experiments. He likened the effect in the secondary circuit, as indicated by the sudden movement of the galvanometer, to the 'nature of the electrical wave'. In his *Diary* Faraday likewise mentioned the 'apparently very short and sudden' 'wave of electricity' produced in the secondary circuit when the battery was connected or disconnected.[45] If induction could produce such a wave, Faraday may have speculated on the possibility of other sources. In another experiment he plunged a magnet into a helix connected to a galvanometer, which likewise moved but then returned to rest. This effect was so close to the torus experiment that Faraday noted that 'a wave of electricity was so produced from *mere approximation of a magnet* and not from its formation in *situ*'.[46]

How the primary circuit could have affected the secondary is far from clear in Faraday's early notes; however, in the published paper he introduced what he called a 'new electrical state or condition of matter' and coined the term 'the *electro-tonic* state' to describe it. Although he was soon to reject this notion, its brief appearance in the November 1831 paper tells us a great deal about Faraday's views at that time. This concept was introduced in an apparently *ad hoc* fashion to resolve the problem why the current in the secondary circuit appeared to be instantaneous and not continuous. In a divinely ordained universe cause has to balance effect; hence a continuously operating cause, such as the current in the primary circuit, has to give rise to a continuously operating effect, and not the short, sharp induced current as indicated by the galvanometer. A further effect of a continuous nature must therefore also operate on the secondary circuit and one which 'resists the formation of an electric current in it'. When a current flows in the primary, an induced current is produced in the secondary, but this is soon blocked by the resistive condition – the electro-tonic state – into which it is thrown. Faraday described the electro-tonic state as an 'electrical condition of matter' but admitted that this 'peculiar condition shows no known electrical effects whilst it continues; nor have I yet been able to discover any peculiar powers exerted, or properties possessed, by matter whilst retained in that state'.[47] The brief currents observed in the secondary when the primary circuit is either closed or opened were produced when the electro-tonic state of the secondary circuit was either established or destroyed.

In this paper Faraday was not only prepared to speculate on the existence of a new state of matter for which he possessed no independent evidence, but he was also willing to characterise that state in terms of the action of particles of matter. As he insisted, 'the electro-tonic state relates to the *particles*, and not to the mass, of the wire'. Moreover, when in this state 'the homogeneous particles of matter [comprising the wire] appear to have assumed a regular but forced electrical arrangement in the direction of the current, which if the matter be undecomposed, produces, when relieved, a return current'.[48] The use of the term 'undecomposed' reminds us of Faraday's background in chemistry and, in particular, of the contemporary theories of electrochemical action to which he was soon to contribute. His reference to the alignment of these particles is also reminiscent of the theory of magnetism and, more specifically, Faraday's *Diary* entry for 24 September in which he argued that 'magnetic action tends to arrange particles longitudinally in the direction of its own axis and is itself powerfully arranged by previous arrangement of iron particles – important influence thus exerted'.[49]

From the available evidence it is far from clear how Faraday arrived at his 'discoveries' of electromagnetic rotation in September 1821 and of electromagnetic induction in August 1831. (I have put the word discoveries in quotation marks because discovery is now recognised as a highly problematic concept.)[50] What I have been emphasising in this section are several of the factors that appear to have been involved – Faraday's debts to Oersted, Ampère and Arago, his concern with electrical inductive processes, his familiarity with transient electrical phenomena and with waves. A number of practical skills were also needed in constructing the apparatus and modifying it to enhance and vary the phenomena. But an important factor in 1831, missing in 1821, was his self-confidence, which enabled him not only to make a practical 'discovery' but also to express a profound theoretical understanding of electromagnetic induction.

Finally, and most importantly, we see throughout the history of these discoveries that Faraday was impelled by the belief that nature constituted a divinely ordained economy. From this standpoint electricity must be related to magnetism and we have seen that over a period of a decade Faraday discovered two ways in which this interrelation occurred. This was the problem underlying both 'discoveries' and therefore it must play a central role in historical accounts of both of them. Yet, underpinning and directing Faraday's work in this area was the conviction that nature consists of powers that operate according to laws in a divinely sanctioned economy, as discussed in sections 8.2 and 8.3.

9.2 THE PROBLEM OF GRAVITATION

The second case study engages an apparently similar problem which so greatly puzzled Faraday that he tried to resolve it on several occasions over a quarter of a century. It is enmeshed in the issues raised in the previous chapter and in particular relates to Faraday's insistence that electricity was the principal example of God's power as manifested in the Creation. Electricity seemed to be closely related to both magnetism and chemical affinity, and to possess at least some connection with both light and heat. By contrast, gravity, which played a major role in natural phenomena, such as the fall of bodies close to the earth and the orbits of planets, seemed significantly different from electricity in its mode of operation. If electricity stood at the top of Faraday's scale of forces, gravitation occupied the lowest rung; thus in his 1859–60 juvenile lectures on the forces of matter he began with gravitation and progressed to cohesion, chemical affinity, heat, magnetism and finally electricity.[51] Yet despite that apparent distance between the two, Faraday was clearly convinced that the gravitational force must be related to other forces, particularly the electrical force; indeed his belief in the divine economy of nature required that the gravitational force could not stand alone but had to bear a specific relation to other forces. The precise nature of that relation proved difficult to specify, and we find that over three decades Faraday returned to the problem, bringing to it a number of different intellectual tools and a wealth of experimental research.

We first encounter the problematic state of the gravitational force in a Friday evening discourse in June 1836, when Faraday speculated on the possible existence of powers in nature of a higher order than those already discovered. Electricity was the highest known power, and it could be conceived as the 'universal source of all the other powers except perhaps gravitation'.[52] Since gravitation was singled out as the (possible) exception, he was confronted by the apparently insurmountable problem of how it was related to electricity or any of the other powers which were closely connected with electricity. A few weeks later the problem reappeared when Faraday was speculating on the theory of electrochemistry and suggested that the way in which a particle was polarised by induction offered 'a distant, but a direct, relation between the attraction of Gravity and chemical attraction'.[53]

A far more developed and acceptable solution to the problem, and one derived from a very different source, attracted Faraday's attention before the end of the year. As he wrote excitedly to Whewell on 13 December, he had just received a paper from the

Italian scientist Ottaviano Mossotti, which 'jumps in with my notion which I think I mentioned to you that Universal Gravitation is a mere residual phenomenon of Electrical attraction & repulsion' – a thesis that makes electricity the main source of power in the universe and demotes the role of gravitation. Although unable to understand the details of Mossotti's mathematical reasoning, on which he deferred to Whewell, he clearly grasped the main points of the argument.[54] In his paper Mossotti had considered the claims by Franklin, Æpinus, Coulomb, Poisson and others concerning the explanation of statical electrical phenomena in terms of one or two fluids. While rejecting Poisson's two-fluid theory, he sought to show that all Poisson's results could be obtained from a much simpler set of assumptions which were, moreover, directly related to known physical laws. Mossotti postulated spherical molecules which repel each other by an inverse square law, a repulsive force being required to prevent the molecules being attracted and collapsing into one another. These molecules swim in a sea of electrical ether whose particles mutually repel one another, and also the molecules, with a force varying as the inverse square of the distance. After deriving the conditions for equilibrium, Mossotti showed that, for distances greater than a certain amount, the inverse square law of attraction, usually attributed to gravity, was accounted for by the postulated electrical forces, a conclusion which he hailed as 'mark[ing] in a striking manner the admirable œconomy of nature'. Faraday was so impressed by Mossotti's reduction of gravitation to electrical forces that he devoted a Friday evening discourse to the subject on 20 January 1837, and was also responsible for an English translation of the paper appearing in the first (1837) volume of Richard Taylor's *Scientific Memoirs*.[55]

Faraday's interest in Mossotti's theory was short-lived, and I can find no further references to the paper. There are several dissimilarities between gravitation and electricity which help to explain Faraday's rejection of Mossotti's neat but sterile solution to the problem. These dissimilarities received repeated emphasis in a number of Faraday's later papers. As already noted, he had, by the end of 1837, become convinced that induction operated only on contiguous particles, and this was manifestly different from the gravitational force which acted at all distances, on both contiguous and non-contiguous particles.[56] Secondly, by 1837 Faraday had clearly acknowledged that while gravitation acts in straight lines, the electrical lines of force are curved.[57] Closely related to this is a third, persuasive, argument. In the case of magnetism and electricity the lines of force bend and are thus modified by the substances through which they pass, so that the magnetic and electrical characteristics of the intervening medium form part of the situation which the scientist describes. By contrast Faraday considered the possibility that material bodies carried around their own gravitational fields which were not modified by the conditions of the surrounding space. As he expressed the matter in 1852, gravity appears 'to have no relation to any physical process by which the power of its particles is carried on between them, [but] seems to be a pure case of attraction or action at a distance'.[58]

Two further differences between gravity and the other forces were clearly stated in a course of lectures in 1846. One was that the gravitational force of a particular quantity of matter is unchanged (at a location on the earth's surface) if the scientist operates on it physically or chemically, whereas the body's state of heat or electrical action could readily be changed by external agencies. Thus, unlike heat or electricity, a body's gravity could not be transferred to or from another body.[59] Fourthly, and more significantly, he conceived that electricity and magnetism were dual, polar forces, in the sense that they were capable of both attraction and repulsion, whereas gravitation operated only by attraction. This last point was given further emphasis in lectures delivered in the following year when Faraday noted that while the 'force of *gravity* is very simple in its nature and very sublime in its simplicity . . . [the other forces] are *dual and antithetical* in their nature'.[60] The polar nature of induction and electrolytic action, which were central concepts in Faraday's research, had no counterpart in gravitation. A sixth important difference which emerged in several of Faraday's writings, but perhaps most clearly in a lecture of 1853, is that while all other forces operate in time, gravitation appears to act instantaneously at all distances.[61]

Despite the strength of this evidence, which placed gravitation firmly outside the network of other forces, Faraday remained convinced that although it appeared such an incorrigible, anomalous force, it must form part of the economy of nature and therefore be related to the other forces. As he speculated in his *Diary* on 19 March 1849, '*Gravity*. Surely this force must be capable of an experimental relation to Electricity, Magnetism and the other forces, so as to bind it up with them in reciprocal action and equivalent effect. Consider for a moment how to set about touching this matter by facts and trial'.[62] Over the next few months he speculated and experimented on this tantalising problem and subsequently presented his results before the Royal Society in November 1850.[63] It is instructive to trace in outline Faraday's progress in these investigations, which, had they established a positive relation between gravity and electricity, would have been of the greatest scientific value.

Faraday's *Diary* entry for 19 March begins not with experiments but with a meditation. He postulated that because electricity and magnetism possess a 'dual and antithetic nature', so must gravity. But how can this duality be conceptualised in a force that is always attractive and never repulsive? Faraday interpreted duality in this instance in terms of the motion of a test body along the line in which gravitation acts towards the body under consideration. Thus, taking this second body to be the earth, the state of a test body would be, say, positive in falling towards the earth and negative in moving away from the earth along the same line.[64] Pursuing the analogy with electromagnetism, Faraday assumed that opposite electrical states would be produced in these two instances. Furthermore, as in his earlier electromagnetic experiments, he envisaged the electrical detector to be a helix of wire connected to a galvanometer while that helix was undergoing a change in gravitational force, for example, in falling towards the earth. The effect would be very small but, if it could be detected, it might prove a significant process affecting even the planetary bodies. 'ALL THIS IS A DREAM', he asserted in the final entry for the day, but he quickly checked himself, recognising that his speculations had to be subjected to careful experimental investigation.[65]

Faraday continued dreaming over the next couple of weeks, but he also began to consider how the interrelation between the gravitational and electrical forces might be ascertained in practice. By 6 April he had made preliminary experiments, using a hollow helix of wire connected to a galvanometer. He reasoned that the gravitational power in the vicinity of the helix should be affected by dropping a copper bar through its centre (an arrangement reminiscent of the production of electricity by a magnet plunged into a helix). Contrary to expectation, or hope, the galvanometer remained unmoved. He next tried dropping the helix and also inserting cores of other materials. Still no positive result. Perhaps, he surmised, the helix's fall and its rapid deceleration on stopping produced contrary electrical effects which cancelled each other?[66]

When he returned to the subject on 25 August, he had constructed a more sophisticated piece of apparatus which enabled the helix to fall a distance of 36 feet (11 m, from high in the Royal Institution's lecture theatre), thus allowing adequate time for the production of the current before the helix reached the ground (or, more precisely, his lecture table). The helix was attached to a cord passing over a pulley; this enabled the helix to be raised and also ensured that its axis remained vertical throughout its descent (see Figures 9.4A and 9.4B).

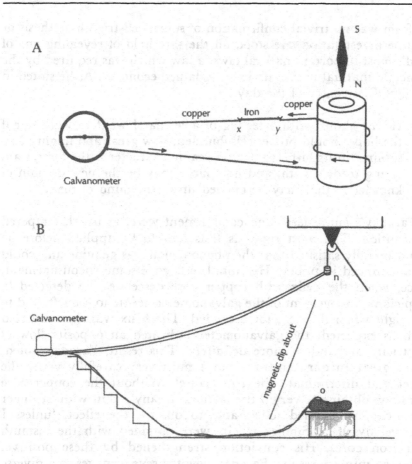

Figure 9.4 Two of Faraday's diagrams from his gravelectric research of 25–30 August 1849. The upper diagram (A) shows a coil, wound on a non-magnetic core (such as copper or bismuth), and connected to a galvanometer. This could be activated either by introduction of a heated iron wire (*xy*) in the circuit or by the approach of a magnetic pole (N). The lower figure (B) shows Faraday's arrangement for the fall of the coil from n to o while keeping the axis of the coil vertical.

Source: Faraday's *Diary*, vol.5, pp.156 and 167.

Cores of various materials could be inserted into the helix and rigidly fixed, so as to concentrate the local field and also ascertain how the effect, if any, depended on the substance used. Copper was selected for its density and good electrical conductivity, bismuth for its diamagnetic qualities and poor conductivity and iron for its eminent magnetic qualities. The experiment he was about to perform was no trivial confirmation of some well-tried hypothesis or routine investigation. He stood on the threshold of revealing one of God's most important natural laws, a law which was required by the principle that nature is a divinely-ordained economy. As he stated in his first *Diary* entry of the day,

> It was almost with a feeling of awe that I went to work, for if the hope should prove well founded, how great and mighty and sublime in its hitherto unchangeable character is the force I am trying to deal with, and how large may be the new domain of knowledge that may be opened up to the mind of man.[67]

Faraday's enthusiasm and commitment were, as usual, tempered by caution. The most rigorous tests had to be applied before he could be fully satisfied that the phenomenon was genuine and could be announced in public. His initial tests gave some encouragement, since, when the helix with copper core descended he detected 'a suspicion' of movement of the galvanometer needle to the left, and to the right when the detector ascended. The helix was then inverted and, as expected, the galvanometer indicated an opposite flow of electricity and did so more definitely. This result, Faraday noted, 'gives great encouragement – but I must very carefully verify the effect and discriminate the true cause'. Without the copper core Faraday obtained 'very little action, if any', but with stronger spectacles he claimed to be able to observe the effect, 'unless I deceive myself'.[68] Similar results were obtained with the bismuth and iron cores. His conviction strengthened by these positive, although minute, results, Faraday devoted several pages to a diverse range of insightful speculations. The analogy with his earlier work on electromagnetism was evident throughout, and he even offered a simple means of illustrating the direction of the electric current for an object falling towards the earth and also for one moving away from the earth. Moreover, he introduced a new term for this source of electricity – the 'Gravelectric current' – and outlined a programme to demonstrate that it produced the same effects as other kinds of electricity.[69]

Returning to these experiments 5 days later, Faraday was in a more sober mood. In one set of observations with the helix ascending, the galvanometer needle moved in the same direction both before and after the helix had been inverted. This was contrary to expectation and Faraday suspected that his observation might be due to a loop in the galvanometer wire which would induce an electromagnetic current as it passed through the earth's magnetic field. After twisting the wires together and repositioning them, he found no current when the helix was moved either up or down. 'At all events', he noted, 'we are purifying the inquiry from interfering causes'.[70] Faced with this null result and the manifest problem of moving the helix through a considerable distance, Faraday set about designing a machine which would allow him to investigate the gravelectric current in a fixed helix or one subjected to a small regular movement. The machine, which was constructed by the instrument maker John Newman, is illustrated in Figure 9.5. By turning the handle, *h*, the wooden lever, *ddd*, is made to oscillate about its fixed axis, *e*. Held between the ends of the lever, *dd*, the test cylinder moves back and forwards through the centre of the helix, *u*, as the handle is cranked. By the use of a commutator, *r*, any small effects produced by individual motions of the core through the helix could be added together so that their sum would register on the sensitive galvanometer, *g*. (The pedigree of this machine can be traced back to September 1831, when an apparatus of similar design was used to generate electricity from magnetism.)[71] After many adjustments so as to eliminate unwanted effects, Faraday found no evidence of the gravelectric current. A similar null result was obtained when the helix was disconnected from its fixed support and allowed to oscillate with the metal core. His *Diary* entry for 21 September 1849 ends with the words, '*Not a sign of any thing resulting from gravitating action*'.[72] Although he briefly returned to the subject on three or four occasions over the next year, no new evidence caused him to alter this conclusion.

However, when Faraday presented the null result of these experiments to the Royal Society on 28 November 1850, he ended on a sanguine note. Although the results were negative, he admitted, they '*do not shake my strong feeling of the existence of a relation between gravity and electricity*, though they give no proof that such a relation exists'.[73] Given his depth of commitment, it is not suprising that Faraday soon returned to the subject, although in a conceptual rather than an experimental context. In his twenty-eighth series of electrical researches, dating from late 1851, he strove to articulate his theory of lines of magnetic force. Such lines in, say, the neighbourhood of a bar magnet are illustrated by the orientations of a small compass needle or, more easily, by the directions of iron filings sprinked on a sheet of paper. In principle these lines run between the two magnetic poles and they define the magnetic 'field' in space. The concept of lines of force could readily be extended to the case of gravitation, with the lines being radial to a gravitating body. Yet, as he noted in a lecture in June 1852, while gravity could be represented in this manner, gravitational lines of force lacked the physical reality that magnetic or electric lines seemed to possess. The

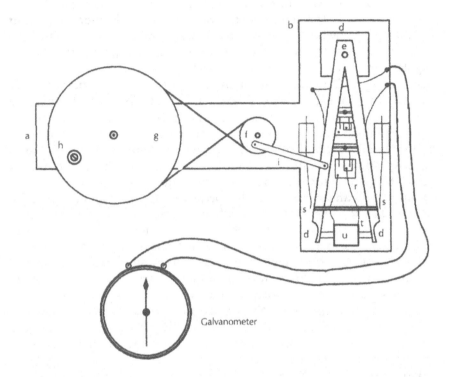

Galvanometer

Figure 9.5   Apparatus devised by Faraday in September 1849 to determine whether changes in the local gravitational field would produce electrical effects in a coil, *u*. Thus, with *u* stationary the handle (*h*) was turned and the cylinder (made of copper, bismuth, glass, etc.) moved briskly through the centre of *u*.

Source: Faraday, *Experimental researches in electricity*, vol.3 (1855), p.165.

gravitational lines were purely abstract in that they were not disturbed by the presence of other bodies (so he claimed), nor were they modified by the intervention of 'screening' substances. The mutual character of the gravitational force and the necessity of more than one body also troubled him, as did gravity's instantaneous action.[74]

Despite these difficulties, Faraday received encouragement from William Whewell, who implied that Newton had held a view similar to Faraday's.[75] In a later paper Faraday quoted Newton's third letter to Richard Bentley as showing that 'he was an unhesitating believer in physical lines of gravitating force'. The quoted passage offered less than convincing support, since Newton merely denied that gravity is innate and also that it can act at a distance – a point on which Faraday would have agreed. Instead, Newton argued that some material or immaterial agent was required. Yet in 1852–3 it is clear that Faraday was able to bring gravitation at least partially within his framework of lines of force and thus partially bridge the apparent divide separating the phenomena of gravity from those of electricity.[76]

Although he had failed in 1849 to discover the relation between gravity and electricity empirically, and possessed no adequate physical theory to relate the two, the problem continued to worry him. In a Friday evening discourse on 19 January 1855 he expressed his dissatisfaction with what he understood to be the received view about gravitation. Here he gave particular emphasis to the argument that the standard theory of gravitation required force to be created or annihilated. We can illustrate his argument with a simple example. Consider, firstly, the sun as the only body in space and therefore not affected by any gravitational force. Next introduce another body, say the earth, which is moved towards it. Both bodies now experience forces impelling them towards each other. However, he argued, these forces, which did not previously exist, appear to have been created out of nothing. Again, if the earth is now moved away from the sun, the mutual gravitation will decrease and therefore force will be annihilated.[77] Just over 2 years later an almost identical argument became the thrust of Faraday's lecture on the conservation of force. He began, however, by erecting the principle of force conservation as a necessary truth with which all physical theories must be compatible. The truth of this principle rested on a firm theological foundation, since only God could create or destroy force. Turning to gravitation, he argued that in moving a body across a gravitational field force would be created out of nothing. Gravitational theory is therefore incompatible with force conservation, which, as we have seen, was an expression of the divine economy of nature. Gravitational theory must therefore be altered, so that the apparent creation of the gravitational force is balanced by the decrease in another form of force. In this lecture, as in 1855, he suggested that this other force – the 'cause of gravitation' – may be suffused throughout all of space.[78]

Faraday's criticisms of the theory of gravitation provoked a number of responses. In an encouraging letter of 9 November 1857 James Clerk Maxwell pointed out that Faraday was incorrect to think that the lines were straight and were directed radially from each massy body. Instead, the presence of, say, a planet in the sun's gravitational field would cause the lines from the sun to deviate considerably away from the planet, and those emanating from the planet would likewise be turned well away from the sun. Thus, according to Maxwell's theory, the lines of gravitational force *are* affected by the presence of objects and as such the analogy with magnetic lines of force is preserved. Maxwell politely indicated that the case of gravitating bodies was perfectly in accord with the principle of energy conservation and, indeed, the conservation of lines of force. He ended by pointing out that he did 'not think gravitation a dangerous subject to apply your methods to'.[79]

Faraday's 1857 paper was presumably subjected to further criticism, since, in the March 1859 issue of the *Philosophical Magazine*, he published a short response to his critics in which he acknowledged that he used the word force in a different way from them. For Faraday force was stripped of its usual connotation in Newtonian mechanics and instead was a cause or source of power in bodies; as such he was articulating his belief in the divine economy of nature. Far from changing his view about forces, Faraday was adamant that his argument in the conservation of force paper remained untouched and that his critics still needed to resolve the apparent conflict between their erroneous account of gravitation and conservation theory.[80] In the April issue William Rankine, the Professor of Civil Engineering at Glasgow, published a curt note criticising Faraday's paper and arguing that energy, not force, is conserved. Moreover, he defined the conserved quantity in terms of the kinetic and potential energies and provided general mathematical expressions for the potential function. In a second paper, which appeared in the following issue, Rankine softened his criticism of Faraday but pointed out that an adequate analysis of the gravitational case must take cognisance of the kinetic energy (which Faraday had ignored) measured as $\frac{1}{2}mv^2$, as well as the potential energy. Rankine thus treated Faraday to a double lecture on basic mechanics, a subject which Faraday had consistently failed to comprehend.[81]

These exchanges in the late 1850s provide the context for Faraday's last major scientific investigation. On 10 February 1859 he once more confronted the problem of relating gravity to the other forces in nature. His method of attack was similar to the one he had adopted 10 years earlier. Turning to his *Diary*, we find a series of entries which begin with brilliant flights of imagination, showing that his imaginative faculty had not been diminished by advancing age. Three resolutions arising from his earlier failures informed these new investigations. Firstly, that in order to detect any gravitational effect very large masses must be used, one of which was the earth and the other a large quantity of lead. Secondly, since the amount of force converted would depend on the difference in gravitational field strengths, the distance through which the test body falls must be maximised. We therefore find Faraday peering down the stairwell at the Royal Institution, climbing the clocktower at the Houses of Parliament and finally settling for the tower in Lambeth used by Walker, Parker and Company in the manufacture of lead shot. Thirdly, he queried whether the gravitational field would produce an electromagnetic force; perhaps an electrostatic force or a heat force would instead be produced. He therefore replaced the helix he had used in 1849 by other forms of detector. While attending to these prosaic details, Faraday speculated that all the phenomena in the earth's atmosphere might result from conversions involving the gravitational force. Volcanos, earthquakes, waterfalls and the great forces in the planetary system might all bear on the subject, which opened up an 'entirely new relation of natural forces'.[82]

In the midst of these speculations he proclaimed, 'Let the imagination go, guiding it by judgment and principle, but holding it in and directing it by *experiment*'.[83] Over a period of some 4 months, beginning in early March 1859, Faraday sought experimental support for these informed conjectures. One extensive set of experiments was based on the assumption that a body should collect an opposite electrostatic charge when falling as when rising. This experiment proved exceedingly difficult to execute. Using large quantities of lead, often several hundred pounds in weight, he devised a sling and tackle that allowed the lead to be raised and lowered but without influencing the amount of static electricity. After many tests and trials, Faraday devised a gutta percha sling which would safely carry the 280lb (127kg) pig of lead he used in his final experiment on 9 July and yet insulate it from the winding gear. At the shot tower in Lambeth this mass of lead was given an electrostatic charge and was then raised and lowered 165 feet (50.3m) while Faraday and his assistant employed delicate electrometers to see whether the charge changed during these transits. With some satisfaction Faraday noted that the 'experiments were well made' but, he was forced to admit, 'the results are negative'.[84]

Faraday's other set of experiments was aimed at detecting the heat produced by a body's movement between two places differing in gravitational field strength. His expectation was that in moving downwards a body would lose heat and gain it in moving up. Several instruments were constructed for accurate temperature measurement and were hauled up and down the Royal Institution stairwell and the Lambeth shot tower. On one of the first trials

Faraday obtained the temperature changes in the expected directions. However, he added, 'I can hardly think it possible that this is a Gravity effect, for the height is only 36 feet: but it is encouraging, for the philosopher must hope against hope or he will do nothing great'. This early optimism was soon dashed, and after an extended series of trials Faraday had to admit that he had failed to detect any conversion of the gravitational force into heat.[85]

In the summer of 1860 Faraday submitted to the Royal Society a paper containing the null results of these experiments. In his opinion the results did *not* show that there was no conversion between gravity and the other forces; rather, he was inclined to attribute his failure to the very small differences in gravitational field across which his apparatus operated. The paper was soon, however, withdrawn on the advice of George Gabriel Stokes, the Lucasian Professor of Mathematics at Cambridge University. Stokes not only agreed that the difference in the gravitational field strength between the top and bottom of the tower was far too small to produce a detectable result, but he also clearly doubted whether the supposed phenomenon existed. As he – and, he implied, others – had no prior expectation of a positive result, he felt that Faraday's paper was not worthy of a place in the *Philosophical Transactions of the Royal Society of London*.[86] For Stokes, Faraday's prediction and the assumptions on which it was based had no place in the energy physics of the 1860s.

Faraday was increasingly looking like an isolated figure, a representative of a past era in natural philosophy. He had not mastered either mathematics or mechanics, which formed the basic language and concepts of so much contemporary physics. Faraday was, however, unbowed and he had concluded his withdrawn paper with the statement that, although the results were negative and the phenomenon may not exist, 'yet I cannot accept them [i.e. the results] as conclusive'.[87] Thus, despite the lack of empirical evidence and many counter-arguments, Faraday still firmly believed that the gravitational force (as he understood it) should produce other forces in accordance with the principle of the economy of nature. In this instance, and in many others, religiously based metaphysics triumphed over empiricism.

9.3 CONCLUDING COMMENTS

In the preceding examples we see how strongly Faraday was wedded to the view that nature is an economical system fashioned by God. This was a necessary principle governing the physical world and one which directed his research, firstly, into electromagnetic rotation, secondly, into electromagnetic induction, and, thirdly, into the gravelectric effect. These three research projects were motivated by the problem of finding the laws governing the connections between two different types of force. While Faraday was not alone in seeking such connections, he was particularly strongly impelled to concentrate on problems of this kind. Moreover, he lacked the concern with industrial efficiency which directed some of his contemporaries, such as Sadi Carnot, James Prescott Joule and William Thomson, towards outwardly similar problems. Faraday's motivation was fundamentally derived from his beliefs about how God had created the world and how the scientist must seek God's laws.

In his writings Faraday made repeated attempts to express the divinely fashioned natural economy in terms of metascientific principles, most importantly his principle of the conservation of forces. In his work of the 1820s and early 1830s the economy of nature was more implicit than explicit. By the late 1830s, however, it was beginning to emerge in a variety of different metascientific statements and to take on a more fixed form. Only with his far more explicit formulation of the principle of force conservation in the mid-1850s do we see it proposed as an incorrigible principle and one which must take precedence over all scientific theorising.

In his research into the gravelectric effect we see how strongly religious-based metaphysics directed his thought and action. He felt he was on the brink of an immensely important discovery. He fantasised about the gravelectric effect and its global significance, and tried on several occasions to detect this effect experimentally. That he failed to do so barely lessened his conviction that gravity must be related to electricity. Had he been a good 'empiricist' (as has often been claimed), then he would have forsaken this project at a very early stage. We learn a lot about Faraday (and indeed science) from this example, for it shows how strongly his science was motivated by metaphysical beliefs. Paradoxically this metaphysics directed both his highly successful work on electromagnetism and his unsuccessful search for the gravelectric effect.

As stressed in section 8.2, Faraday conceived this economy as implying that nature was perfectly designed by God with no wastage of physical force. Each and every expenditure of force must therefore possess a purpose. As he asked at the conclusion of his 1850 paper on atmospheric magnetism,

> What is the final purpose of this magnetic condition of the atmosphere . . .? No doubt there is one or more, for nothing is superfluous there. We find no remainders or surplusage of

action in physical forces. The smallest provision is as essential as the greatest. None are deficient, none can be spared.[88]

When discussing the purposes of natural processes, Faraday acknowledged that God had designed nature to be of use to mankind and we therefore have a duty to apply the laws of nature in improving our technology. However, a far more important role for science was to show how God had created His universe. God had created a perfectly economical system. Moreover, the study of nature and the demonstration of how the forces of nature are interconnected provided Faraday with ways to praise and glorify God. As he stated to Prince Albert in a private lecture in 1849, magnetism appears to exert a necessary influence on every particle of matter in the universe. However, 'What its great purpose is, seems to be looming in the distance before us . . . and I cannot doubt that a glorious discovery in natural knowledge, *and of the wisdom and power of God in the creation*, is awaiting our age'.[89] Thus not only was the principle of the economy of nature a necessary truth but the determination of precisely how the different forces were interrelated was a profoundly religious activity. From this standpoint we can see that Faraday's investigations into electromagnetic rotation, electromagnetic induction and the gravelectric effect cohered with his religious worldview.

The two examples discussed in this chapter show how Faraday's theological-centred conception of the world affected the scientific problems he chose to work on and the manner in which he proceeded in solving those problems. In these investigations he operated with a complex mixture of metaphysical and metascientific principles, speculations, hard-won laws and incorrigible facts; and with imagination and the technical skill to interrogate nature in an incisive fashion. No account of Faraday can afford to omit any of these factors, and yet at various levels his science was so marked by his Sandemanianism that it provides a crucial key to the whole.

10 *Faraday's Personality Revisited*

Having examined Faraday's religious, social and scientific experience, we turn to an analysis of his personality, since several aspects of his personality provide us with themes common to these three areas of his life. We start therefore by reviewing the traditional portrayal of his personality and then offer a very different account, drawing on psychological insights.

10.1 IMAGES, BRIGHT AND TARNISHED

There were numerous contemporary accounts of Faraday by people who either knew him or heard him speak in public. These writers were almost unanimous in portraying him as an extraordinary person who possessed a highly attractive personality. For example, John Tyndall, who was not often given to hyperbole in praise of religious sentiments, paraphrased 1 Timothy 3:2–3 in claiming that Faraday was 'blameless, vigilant, sober, of good behaviour, apt to teach, [and] not given to filthy lucre'.[1] The great German physicist Hermann von Helmholtz likewise claimed that the 'perfect simplicity, modesty, and undimmed purity of his character gave him a fascination which I have never experienced in any other man'.[2] John Gladstone even devoted a whole chapter of his *Michael Faraday* to his subject's character, showering on Faraday such epithets as 'kindness, friendship, geniality, . . . benevolence', 'truthfulness', 'childlike simplicity', 'unworldliness', 'wonderful playfulness', 'faithfulness' and 'a reverential attitude towards Nature, Man, and God'.[3] These examples can be multiplied *ad nauseam*. Faraday's many friends, and even those who knew him only from a distance, found him a delightful, unpretentious and humble person.

That a scientist of the first rank should possess such an attractive personality has produced a highly favourable public image of the scientist. In contrast to Mary Shelley's Frankenstein, who is destroyed by his own ugly creation, or the numerous more recent portrayals of the scientist as amoral and threatening, Faraday has come to symbolise the scientist as humane seeker after truth. This positive image has proved very serviceable for science. To give one example among many, a recent article in *The Lady* ends with the

words: 'A simple, religious man, Michael Faraday searched constantly for the truth in everything. Today we reap the rewards of his
search'.[4] However, we should not accept such judgements at face
value but recognise that the currently accepted images of Faraday
are complex amalgams resulting from cultural pressures acting over
a period of more than a century and a quarter, and also from the
careful and protective manipulation of 'Faraday', particularly by his
earliest biographers, in league with close members of his family, such
as Jane Barnard.

If we examine these biographies, which played a major role in the
affirmation (if not creation) of the mid-Victorian 'Faraday', we
encounter a recurrent message – not only did Faraday possess an
extraordinary personality but his personal qualities explain his
success as a scientist. After all, Faraday inherited no advantage
from his background – his father was a poor ironsmith – and thus his
rise as a scientist resulted solely from personal effort. This Smilesian
dimension to his biography is not lost on Margaret Thatcher who
named Faraday as her historical 'hero' in a television interview:

> A remarkable person, typical of so many of our great people.
> Very ordinary background, father ran a smithy, son had
> virtually no education.... But he was brilliant. And he ...
> didn't go to university.... But extraordinary – brilliant. The
> Good Lord's no respecter of backgrounds, never has been, he
> plants genius the world over and *it's up to us to find it*.[5]

Probably unwittingly Mrs Thatcher reiterated a Victorian stereotype of Faraday by pointing to his personality and inner strength as
the key to his success. The early biographers were, however, far
more explicit about the nature and source of this strength. Henry
Bence Jones ends his two volume biography by noting 'the beauty
and the nobleness' of Faraday's character, his energy, kindness and
truth – 'His noble nature showed itself [especially] in his search for
truth'.[6] Here, as elsewhere, we find the implicit claim that the
discovery of scientific truth is only available to a person of the
highest moral calibre. The *Good* and the *True* seemed necessarily
connected to many Victorians.

If Faraday's success could be traced to his sterling personal
qualities, then these qualities were perceived, in turn, as being
derived from his devout Christianity. Overlooking the profound
disagreements that Faraday, like other Sandemanians, had with
both the established Churches and with other denominations, the
example of Faraday was tirelessly used by clergymen to argue that
Christianity provided the right moral grounding for the successful

scientist. For example, Samuel Martin, a leading Congregationalist minister, claimed that Faraday's 'character was moulded and his conduct guided by the best portions of his religious belief', and he explicitly connected Faraday's character with his remarkable scientific career. For Martin and many other theological writers Faraday was the Christian philosopher *par excellence*.[7]

A further personal quality which friends frequently perceived in Faraday was equanimity, for he appeared to possess great inner strength and to be untouched by doubts or insecurity. Not only was he not drawn by worldly lusts but he also radiated inner stability and contentment. This aspect of his character was sometimes referred to as his 'peaceableness',[8] but found its clearest expression in a poetic obituary published in *Punch*:

> A priest of Truth: his office to expound
> Earth's mysteries to all who willed to hear –
> Who in the book of science sought and found,
> With love, that knew all reverence, *but no fear*.[9]

Faraday was, to use William James's term, 'healthy-minded', in that he conveyed to others the optimism, contentment and security founded in his sincere religious beliefs.[10] Faraday emerges as the paradigm of mental health. He was the perfectly balanced individual and his science and his religion were in perfect stasis.

We have here one of the most potent images not only of Faraday but also of the empiricism he came to exemplify. Truth is not merely revealed to the good, but virtuous qualities facilitate the right scientific method. Faraday's equanimity, his 'peaceableness' and his ability to face nature with love – but without fear – enabled him to look at phenomena in an unprejudiced light. The word prejudice is frequently encountered in Faraday's writings, and Joseph Agassi has correctly located Faraday's recurrent concern to minimise, if not eliminate, prejudice in scientific matters.[11] Moreover, in terms of the mid-Victorian image of Faraday there was a clear connection drawn between his inner stability, his lack of prejudice, his pursuit of an inductive method and his success as a discoverer. The image is both familiar and powerful, and it seemed natural and unquestionable to his biographers.

Faraday's Victorian biographers took great care to ensure not only that this image looked plausible but also that any counter-vailing evidence was suppressed. Thus Bence Jones, Faraday's physician and biographer, passed quickly over the breakdown which Faraday experienced in the early 1840s and instead devoted over thirty pages of *The life and letters of Faraday* to the rather tedious journal that Faraday kept while recuperating in Switzerland. Bence Jones mentioned but failed to analyse his breakdown, while most other biographers do not even allow it a place in their narratives.[12] Mental disturbance was eschewed in the standard hagiology. On this and a number of related issues the early biographies and biographical sketches inhibit the historian, since they consistently cast a veil over certain aspects of Faraday, particularly those facets of his personality that he did not project in public. It is not that the congenial public image was manifestly false, but rather that it was incomplete, and incomplete in ways that consistently obscure those crucial psychological drives and processes operating behind the veil.

The problem facing the historian is how to by-pass the Victorian stereotype and discover other aspects of Faraday's psyche. Published sources are of very limited assistance. However, there is a remarkable, although unpublished, passage by Tyndall which casts doubt on one aspect of the conventional image of Faraday. Tyndall, who worked closely with Faraday for many years, asserted that his character nearly 'approached what might, without extravagance, be called perfect'. Yet Tyndall was not convinced that 'dogmatic religion' is necessarily connected with 'moral purity and grace' and cited Charles Darwin – 'the Abraham of scientific men' – as a counter-example. He could not, he concluded, 'as so many desire, look upon Faraday's religious belief as the exclusive source of qualities shared so conspicuously by one [Darwin] uninfluenced by that belief'.[13] Tyndall has often been portrayed as a materialist who mercilessly attacked religion in his infamous 'Belfast Address' before the British Association in 1874. A more coherent interpretation of that address takes account of his Irish Protestant background and his opposition, not to religion as such, but to the dogmatic theology and the opposition to science that he had witnessed in Ireland, particularly in the Catholic school system. Moreover, Tyndall encompassed a pantheistic philosophy. From this standpoint he became fascinated by Faraday's deep religious sensibility towards nature which he partially shared, though in an agnostic form. At the same time his agnosticism allowed him to criticise this one facet in the public's perception of his mentor.[14]

In his *Faraday as a discoverer* (1868) Tyndall offered a more complex analysis of Faraday than is contained in most Victorian biographies, for although he pointed out that his mentor was indeed tender, gentle and equable, he also enumerated other countervailing tendencies:

Underneath his sweetness and gentleness was the heat of a volcano. He was a man of excitable and fiery nature; but through his high self-discipline [repression?] he converted the fire into a central glow and motive force of life, instead of permitting it to waste itself in useless passion.[15]

This passage is almost unique in delineating a vein in Faraday's character which starkly opposes the composure for which he is generally renowned. Yet, if one looks carefully, there are a few other accounts of Faraday where minor cracks in the image are evident. One instance occurs in Bence Jones's biography where he allows the voice of one of Faraday's more neglected nieces to be heard. Margery Ann Reid, who lived with the Faradays from 1830 to 1840, reported that 'When dull and dispirited, as he sometimes was to *an extreme degree*, my aunt used to carry him off to Brighton, or somewhere, for a few days, and they generally came back refreshed and invigorated'.[16] A second example is provided by Jane Barnard, the niece who was companion to the Faradays during her uncle's last decade. Jane, who usually emphasised Faraday's manifestly positive qualities, nevertheless portrayed him in Walter Jerrold's biography, *Michael Faraday: man of science* (1891), as possessing repressed passions; however, she could recollect only two occasions when 'Faraday, even for a moment, let his passion get the better of him'.[17]

Not only do these comments of Tyndall and the passing remarks of the two nieces who lived with him give us reason to pause but in this post-Freudian era we tend to be suspicious of 'nice guys' and we are prone, perhaps all too prone, to question whether a person's apparent virtuous qualities might not be manifestations of darker subconscious drives. Yet few historians of science, and even fewer historians of physics, have been prepared to utilise psychology in trying to understand their subjects. However, in two pioneering papers Elspeth Crawford has sought to understand Faraday's creativity by studying his learning processes. As she rightly emphasises, learning about nature was for Faraday a form of self-knowledge and one that led to a particular emotional state by which he could, through humility, free himself from prejudice and give rein to his curiosity and creative powers. While Crawford draws on Wilfrid Bion's affective theory of learning, Ryan Tweney has turned to cognitive psychology in order to analyse Faraday's laboratory procedures. Despite their differences in approach, both Crawford and Tweney show that Faraday's laboratory practice cannot be subsumed under any simplistic methodological category – such as 'induction' or 'hypothetico-deduction' – but that psychological categories need to be used in understanding his research work.[18]

The only other published psychoanalysis of Faraday that I have encountered appeared in an American medical magazine in 1967. In his very short article the author, Dr Lyle Eddar, notes that Faraday would sometimes attempt 'to reconcile two diametrically opposed attitudes of thought', and cites in support of his thesis Faraday's combination of science with his Christianity. 'This marked ambivalence in his psychological make-up', writes Eddar, 'must have caused intolerable stresses within his psyche. The result was a schizophrenic episode which lasted three years' – presumably referring to Faraday's breakdown in the early 1840s. While Eddar's all-too-brief analysis leaves many questions unanswered and is predicated on the false assumption that science and religion are necessarily incompatible, he nevertheless recognises that Faraday possessed a psyche.[19]

10.2   FAMILY RELATIONS

The starting point of any psycho-biography must be the subject's relation to its parents. There is little extant information about Faraday's childhood or his early relation with his parents or siblings, but what there is provides some insight into his family situation. The family's poverty has already been stressed but to this must be added James Faraday's ill-health, which significantly reduced his earning ability. Writing in 1807 to one of his brothers in Yorkshire, he complained that although he was 'very seldom off work for a whole day together, yet I am under the necessity (through pain) of being from work part of almost every day'. James turned to his religious faith to comprehend his suffering, adding, 'But we, perhaps, ought to leave these matters to the overruling hand of Him who has a sovereign right to do what seemeth good to Him, both in the armies of heaven and amongst the inhabitants of the earth'.[20] James's entry in the roll book of the London Sandemanian meeting house provides two further points of interest. First, he was at some point excluded from the sect, although this is not unusual and occurred to a high proportion of its members, including approximately half of those listed in Appendix B. Second, to this undated addendum is added the word 'Reconcil[e]d', whereas the term usually used was 'Restored', suggesting that James subsequently made a second confession of faith, possibly on his deathbed, and died in the faith.[21]

Although Faraday's extant correspondence does not cover the period of his father's life, there are no references to him in his later letters. Doubtless James's physical illness prevented him from fully pursuing the role of father and there are many aspects of Faraday's character which seem to have been shaped by his father's difficulties and the dominant role adopted by his mother. While the smithy continued to evoke pleasant childhood memories throughout Faraday's life, his antipathy towards trade may perhaps be seen as a reaction to his father's disability, which was exacerbated by heavy work at the forge.[22] There is also one reminiscence of his father which sheds a little light on their relationship. This story concerns a visit Faraday made to the forge where his father had worked many years earlier. There he related how, during his childhood, he had almost been killed while playing in an upstairs room. He stepped back from his game and fell through a hole in the rafters into the room below where his father was working at an anvil. Had it not been for his father, he recounted, he would have fallen on the anvil with serious if not fatal results. 'As it was, my father's back just saved mine.'[23] James emerges from this story as a hero and a saviour.

Further light is shed on family relationships by the fact that Michael Faraday and two of his three siblings encompassed the religion which helped sustain James, a religion which made the world stable and comprehensible. That world was under the dominion of God the Father, who showed through the Bible and through Christ's example how Christians should live. As the Sandemanians frequently emphasised, their aim was to lead their lives in imitation of Christ's. In imitating the Son, they too sought to act as perfect sons. From this standpoint we may perceive God the Father as fulfilling the role of a surrogate father in Faraday's life. A partial analogy may be drawn with Newton, whose father died before his birth. Frank Manuel has argued that Newton conceived God to fulfil the role of a father figure while he took on the mantle of God's elected son, with the special duty of rediscovering the laws that God had chosen to regulate the physical world.[24] Faraday likewise devoted his life to determining the God-given laws that physical objects obey. As a Sandemanian, he also committed himself to obeying the moral laws that God had written in the Bible.

Within 3 years of his father's death (and 8 years before he made his confession of faith) Faraday found a potential father figure in Humphry Davy. Although a mere 13 years his elder, Davy had achieved an impressive reputation as a chemist of eminence and as a luminary of London science. In many respects the two men were strikingly similar. Like Faraday, Davy came from a poor background, his father being a woodcarver who died when Davy was 15 years old; but he soon attracted the attention of a number of patrons who helped launch his scientific career, first in Cornwall, then in Bristol and finally in London where he became installed at the Royal Institution in 1801. Uninhibited by Faraday's strict religious principles, Davy rapidly moved up the social ladder – he married a wealthy widowed bluestocking named Jane Apreece, gained a knighthood in 1812, a baronetcy in 1818 and established himself as the President of the Royal Society in 1820. From that seat of power the woodcarver's son lorded over British science. As David Knight has pointed out, Davy's progeny were not confined to his scientific discoveries but also included two younger men who served as his apprentices. The first was his brother John, who proceeded to a medical education and became an army surgeon, spending much of his career abroad. Yet John became the guardian of his more eminent brother's reputation by publicly supporting his claims about muriatic acid against other chemists, writing his biography and by editing the nine-volume edition of his *Collected works* (1839–40).[25]

If John Davy outwardly fulfilled the role of the virtuous son, Faraday's relationship to Humphry Davy was more complex and ambivalent. Failing to attract the attention of Sir Joseph Banks, the Royal Society's President, the young Faraday sought a position with Davy, who accepted him as his assistant in 1813 but soon extended his duties to looking after the apparatus and, from 1821, superintending the Royal Institution and its laboratories. Faraday thus received a scientific position and served his apprenticeship under Davy; for that early support he remained ever in Davy's debt. Their relationship was largely confined to their mutual interest in science, and while initially on friendly terms they were never very intimate. For example, Faraday always referred to Davy in formal terms and expressed dutiful feelings towards him. Moreover, their relationship soon became strained. In 1813 Faraday joined Davy and his wife on a continental tour that lasted a year and a half. Since Davy's valet let him down shortly before sailing, Faraday also adopted the role of servant both to his master and to his master's wife. Faraday had no objection to serving Davy, who had not (yet) been 'corrupted by high life', but he soon came to despise Jane Davy who, in his opinion, 'endeavour[ed] to thwart me in all my views & to debase me in my occupations'.[26] The role of a servant was by no means foreign to Faraday's character, but he could only adopt this role if he could uphold it with dignity. Moreover, there was evident competition between him and Jane for the attentions of Sir Humphry. His relationship with John Davy was similarly marred; for example, the two men collided in 1836 when John charged Faraday with not adequately acknowledging his brother's work on both the condensation of gases and electromagnetic rotation.[27]

In casting a suspicion of dishonesty on Faraday John Davy raised a spectre dating from the early 1820s, when Humphry made a final bid to dominate Faraday. Faraday's first major contribution to science was the discovery of electromagnetic rotation in early

September 1821. He rushed into print with a paper in the October issue of the *Quarterly Journal of Science* in which he related the astonishing new phenomenon.[28] In a letter of 8 October he complained to a more experienced friend that various scientists were questioning his 'honour and honesty', although nobody had yet accused him directly. He understood, however, that he was being charged with failing to acknowledge information he had received from Davy and with plagiarising the views of William Hyde Wollaston, who was accorded priority in this area of research. On the first of these issues Faraday then confronted Davy, who 'I hope and believe . . . is satisfied'.[29] Although he thought that he had not wronged Wollaston, he was deeply hurt by the mere possibility that he had behaved dishonourably and he therefore sought an interview with Wollaston in order to set matters right. A letter of Wollaston's of 1 November indicates that he had taken no offence.[30] There the matter rested or would have rested had not a report appeared claiming that Davy had repeated the charge of plagiarism before the Royal Society on 13 March 1823. Davy subsequently denied the charge and asserted that he had been misrepresented.

About that time Faraday was researching not only electromagnetism but also the liquefaction of gases, another subject all too close to Davy's work. With one major discovery to his credit and a decade working in Davy's shadow, Faraday doubtless felt slighted and betrayed by this latest attack. For his part Davy, resenting the success of his earnest and hardworking assistant, was trying to keep him down. Master and servant were in direct competition and their rivalry may have motivated Davy's attempt to block Faraday's membership of the Royal Society.[31] As a result of Davy's autocratic behaviour, Faraday's friends helped move his membership application, which was read at a meeting on 1 May. Davy, the President, demanded that the certificate be taken down but Faraday refused, since it had not been posted by himself but by his sponsors. By the time of Faraday's election on 9 January the certificate had attracted twenty nine signatures, including that of Wollaston.[32] Although Davy was unsuccessful in preventing Faraday's membership, this incident proved a turning point in their relationship. More than a decade later Faraday admitted, 'I was by no means in the same relation as to scientific communication with Sir Humphry Davy after I became a Fellow of the Royal Society as before that period'. The incident made such an impression that Sarah recounted in 1870, over half a century after the event, that Faraday 'was almost painfully anxious to be correct in all moral responsibilities & I never heard him doubt himself in that case, but he felt very grievously that he was some how or other not fairly treated'.[33]

When Faraday became Davy's assistant in 1813, Davy appeared an admirable mentor, since he was a successful, respected chemist who had risen rapidly in the ranks of science through his successes as a researcher and lecturer. In the early years Faraday worked closely with Davy and learnt a great deal from him. However, particularly after the searing events of 1821–3, he found his subservient role oppressive and progressively sought to develop his independent persona. This ambivalence can be traced at a number of different levels. For example, while his early views on matter and electrochemistry bore many similarities to those of Davy, his later views often stand in marked contrast to Davy's. Again, while Davy played a prominent role in fashionable society and in the politics of science, Faraday the Sandemanian distanced himself considerably from both of these areas. Through his Christianity Faraday would not only have recognised the moral bankruptcy of Davy's social values but also reaffirmed his own role as the dutiful, loyal son of God.

In contrast to his ambivalence towards Davy, Faraday was on close and constant terms with members of his immediate family. After James's death in 1810 the major burden fell on Faraday's elder brother Robert, who worked initially at a smithy and later achieved financial security as a brass manufacturer and installer of gas fittings. There appears to have been the closest friendship between the two brothers, and Robert frequently supported Michael as he entered an uncertain career in science. Thus, for example, Robert gave his brother the money that enabled him to attend John Tatum's science lectures in the early 1810s. Later Faraday helped Robert and his son James obtain contracts for their gaslight-fitting business. Both brothers were members of the London Sandemanians and were appointed to the office of deacons in 1832. Robert's sudden death in 1846 from a carriage accident came as a great shock to Faraday.[34]

Evidence about his mother, who died in 1838, is very sparse. However, one biographer, William Jerrold, speaks of the 'tenderest affection' between Faraday and his mother, and of his 'beautiful chivalrous kindness and deference' towards her.[35] The truth of these assertions, made more than fifty years after her death, is difficult to judge, but in several letters to his mother he repeatedly expressed not only his deep affection for her but also his concern to behave correctly towards her as would a dutiful son. These letters convey a sense of an over-rigid and slightly distant relationship, as if Faraday had assumed the attributes of gentleness, affection and social duty towards her in order to live up to her expectations.[36]

Very little information remains about Sarah, who, at least in public, appeared to subordinate herself to her famous husband. Her surviving letters contain little about herself but often concern her relation to Faraday. Thus in 1871 she wrote to Bence Jones that 'it was delightful to be his companion & wife, & the remembrances left to me are a great pleasure & consolation in my widowed state'.[37] Judging by her pose in Figure 10.1 she could be both doting and

Figure 10.1 Sarah and Michael Faraday.

Source: Royal Institution of Great Britain.

protective. Cornelia Crosse recalled her visit to the Faradays' apartment in 1850. During the conversation Sarah made only two interjections, both expressing her concern that her husband's mind should not be overtaxed.[38] One recent commentator calls her a 'hausfrau', contrasting her plainness and lack of social status with Davy's fashionable wife.[39] Although she came from an affluent and successful Sandemanian family, she may have looked rather out of place in the public rooms of the Royal Institution.

If this is an unflattering portrait of Sarah, her letters show her to have been a person of considerable spiritual strength and she must have exerted a great influence on Faraday, supporting him during his periods of illness and darkness. She and Faraday were deeply committed to one another and, except in the realm of science, they appear to have lived in great intimacy. His letters to her overflow with love and with unsophisticated expressions of affection, as if he were asserting that their close personal relationship was unaffected by his success in public.[40] Most importantly, they shared their faith in God and their connection with the Sandemanians.

Faraday's relationship with his younger sister, Sarah and the nieces who tended them, indicates his dependence on these women. Being largely self-taught, he was not supported by any of the (totally male) institutions that have traditionally provided for scholars and scientists, such as the Cambridge colleges. Moreover, by the early 1820s he could no longer count on the support and patronage of Humphry Davy. The women in his life and the Sandemanian community, to which most of them belonged, thus offered him the practical and emotional support he needed to pursue a life in science with all its attendant uncertainties. Throughout his life the female members of his family offered constant support to Faraday. If this constituted an exploitative relationship, Sarah and her nieces appear to have been willing accomplices, who found direction in their lives through their connection with Faraday and his high-profile career in science.

Many women clearly found Faraday an attractive, empathetic person and they often doted on him; not only his younger sister, his wife and the nieces who helped look after him in later life but also those, such as Mary Somerville, Cornelia Crosse and Ada Lovelace, who heard him lecture in the Royal Institution.[41] In his private life he was regularly in the company of women, and women far outnumbered men in the London Sandemanian community. He invariably treated women with great kindness and respect and regarded them as equals, and in return he gained their respect and admiration. He enjoyed a similar relation with some men, but he recoiled from those who manifested power and egotism through aggressive, selfish and bombastic behaviour. Although a person of immense inner strength, Faraday hardly ever displayed stereotypical forms of masculine aggression.

He was inordinately fond of children, whereas Victorian males conventionally left the rearing of children to women. Not having any children of his own – his was the only non-prolific marriage among the Faradays and Barnards of his generation – he delighted to join in childrens' games. His nieces and other children with whom he had contact found him a charming and lively companion and mentor who dissolved all barriers of status and age. For example, when John Gladstone's family visited him at the Royal Institution Faraday played hide-and-seek with the children 'in the lecture theatre, and afterwards amused them upstairs with tuning-forks and resounding glasses'.[42] But perhaps his most enduring connection with children was his founding of the juvenile lecture series for their edification and education. Many years later Sir Alfred Yarrow related that at the end of the lecture Faraday would 'stop sometimes for quite half an hour talking to a lot of us boys, and sometimes making us go through some of his own experiments with our own hands'.[43] At the beginning of one series of lectures Faraday noted that although old and infirm he would 'return to second childhood and become, as it were, young again amongst the youth'.[44] As one biographer perceptively commented, 'he was still a child himself'.[45] The ease with which Faraday could identify with both women and children indicates that he experienced difficulty in relating to men, particularly in those social situations where conventional masculine values of power, aggression, property and money were paraded. This trait further suggests that Faraday's childhood was dominated by his mother.

Evelyn Fox Keller's insightful studies of the relation between science and gender shed further light on the connection between Faraday's lack of strongly masculine characteristics and the science he pursued. Keller argues that science possesses a marked gender bias, such that nature is portrayed as female and the scientist as male. In pursuing science the scientist seeks to dominate and subdue nature. Moreover, science has traditionally been described as a rational and objective activity, and these adjectives are likewise equated with masculinity in our culture. By possessing such associations science stands in opposition to subjectivity and the emotions, which are instead linked to the female. Hence, 'our image of science' is 'antithetical to Eros'. Keller also argues that the supposed objectivity of science imposes a great distance between the scientist and nature. Objectivity requires that the scientist examines nature from afar and refuses to allow emotions or subjectivity to intervene.[46]

It is interesting to note that Faraday does not cohere with Keller's characterisation of science but departs from it in a consistent manner. As noted in section 7.6 Faraday portrayed nature as female but he did not consider that he, the scientist, dominated nature. Instead he saw himself collaborating with nature. Again, far from being the distant objective observer, Faraday expressed a close emotional bond with nature and stressed the importance of engaging directly in intimate experimental investigations. Outside the laboratory he also acknowledged the importance of affective relationships, since, while he was no devotee of Eros, he was instead committed to another form of love – Christian love, which he, as a member of the Sandemanian community, practised to a very high degree. In these respects Faraday departed significantly from the male stereotype. The picture that emerges is that Faraday not only fails to fit the masculine image of science identified by Keller, but he brought to his science a variety of values including those traditionally associated with the female. Faraday thus emerges as atypical among scientists in displaying these female traits. Moreover, Keller's thesis provides a connection between Faraday's gender characteristics, his percepion of nature and the intimate way he pursued his science.

10.3   FEARS AND CRISES

Having been brought up in a Sandemanian environment, married into another Sandemanian family, made his confession of faith, been appointed to the offices of deacon and elder, Faraday was immersed in the sect, in its practices, its doctrines and social norms. His actions were conditioned by this constant exposure to the Church of Christ. In it he lived, moved and had his being. It affected his choices and responses to the outside world and to the practice of science. Yet the primary psychological significance of Sandemanianism for Faraday was that it provided him with security, with an assurance that if he followed its precepts, he would be acting in accordance with God's laws and wishes and would thereby gain acceptance with God the Father. Faraday's Christianity (like his father's) also assured him that life is not chaos and suffering but that there is a meaningful plan behind events. The kingdom of heaven is eternal, sure and unchanging. The Bible is the key to this knowledge and in the tightly disciplined world of the Sandemanians there was almost no role for uncertainty or for human fallibility. This assurance we find in many of Faraday's letters to his co-religionists. For example, referring to a devout lady who was close to death he wrote, 'she is very patient and comforted by the scriptures in the great & glorious hope of relief not only from these things but from all sorrow & sighing through Jesus Christ'.[47] In this letter Faraday was concerned not only with Mary Straker's dire situation but also with his own need for patience, comfort and relief. Again, he found in the recent death of another pious Sandemanian 'evidences that God both can and will keep his people who put their trust in him'.[48] Finally, in a letter of 1859, in which he noted the progressive failure of his 'worldly faculties', he asserted that 'they leave us as little children trusting in the Father of mercies and accepting His unspeakable gift'.[49]

To the support offered by the beliefs and doctrines of Sandemanianism must be added the social cohesion of the group. Here was a community of love, fellowship and consensus in which Faraday could feel safe and not threatened. His physical as well as the spiritual needs were adequately supplied. When he fell sick, in body or in mind, his fellow Sandemanians rallied to his support. A stranger on the crowded streets of London, he was welcomed as a brother into his co-religionists' homes. When travelling outside London, he would, where possible, merge into the local Sandemanian community. He was a member of this invisible church.

Although the Sandemanian church created a haven for Faraday, there were dangers continually lurking beyond the sect's clearly demarcated boundary. Yet even here Sandemanianism generally offered ways of controlling these perils and removing external threats. Thus he could expatiate on the laws of physics and chemistry before his audience at the Royal Institution without becoming pulled into their world of antithetical values. As long as he could meet others on his own terms, the boundary held firm. However, there were occasions when that boundary was in danger of being ruptured and we see Faraday suddenly confronting a potentially dangerous situation. From letters on a variety of different subjects we thus gain brief glimpses of another Michael Faraday – not the eminent lecturer of that name who seemed to have under his firm control both nature and his polished public discourse, but a Michael Faraday who hovered on the edge of the chasm of fear.

Faraday manifested a number of closely related fears, all of which concerned his need to maintain control of himself within a safe, ordered environment. One such fear was his concern that he might be a hypocrite, for the hypocrite is a split personality who is neither true to himself nor to his Father but is consumed by his own lies.[50] A related fear was that he would fail to differentiate clearly between reality and imagination: for example, in one of his letters he expressed the hope that talking with friends would 'remove a great heap of imaginary matter from my mind'.[51] Another of his fears, which has both religious and political overtones, concerns the breakdown of order. The word 'confusion' appears on a number of occasions; for example, he readily endorsed Bacon's aphorism that truth is more likely to emerge from error than from confusion and he also recoiled from the confusion he perceived in many aspects of this world, especially in politics.[52] Another revealing passage occurred in a letter to Whewell of 1853 which begins 'You frighten me'. The cause of this fright was Whewell's failure to understand some of his papers dealing with lines of force and Faraday was disturbed by the possibility that his work appeared *'obscure and confused'* to Whewell. Communication had broken down and Faraday found himself isolated and misunderstood, thus manifesting a fear similar to that of being lost.[53] This catalogue can be extended but it is sufficient to show that Faraday was not at peace with himself as the Victorian biographers generally maintained. Moreover, these fears appear to be interconnected in ways to be explored in the next two sections.

Such fears were never far from Faraday's consciousness, and they could be triggered by fairly trivial events, such as someone failing to respect his self-imposed boundaries to his action and discourse. Yet there were a few occasions when Faraday encountered situations which severely challenged his normal equable state and threatened to destroy his world and his place in it. These 'crises' will be discussed in some detail, since they provide important clues to understanding Faraday's psychology. In referring to these events as crises I am identifying them as noticeable departures from his normal behaviour.

The first crisis occurred, significantly, in October 1840, when he agreed to become an elder in the London Sandemanian meeting house. This important event was the source of inner conflict since, as he described to William Buchanan, an elder in Edinburgh, his acceptance of the office was an act of 'folly' which caused him 'distress of mind'. He explained that he 'doubted not that it was of the Lord's doing' and recounted Romans 9:17, where God tells the Pharaoh, 'Even for this same purpose have I raised thee up, that I might shew my power in thee'. However Faraday was sorely aware of his 'utter unworthyness' and 'unfitness' for the task. Pulled in these opposite directions, he declared to Buchanan, 'Dear Brother I am afraid to trust myself at present for I feel greatly like a *hypocrite* in all I say or do'.

This is not the composed Faraday sustained by his Victorian biographers. Nor was this the only occasion on which Faraday labelled himself a 'hypocrite', a term that weighed heavily with him and expressed one of his deepest concerns – his perceived failure to be true to himself and to God. In Matthew 23:2–33 the Pharisees are called hypocrites for their deceit and commitment to false ideals, and if Faraday were found to be a hypocrite not only would God see through his pretence but his very persona would be destroyed. For Faraday the call to the elder's office was the source of inner conflict expressed through the following ambivalent feelings: 'I cannot but fear and tremble – and then again when I feel comforted and happy...' This tension continues throughout the letter, even though 'the love of the brethren & their consolations and above all I hope the power of God has removed much of this trouble'.[54] Faraday was concerned that in accepting the elder's office he might not prove worthy of this morally demanding position and any failure would not only expose his unworthiness before God but also lead to severance from the church and its sanctuary. However, he did accept the office and served as elder for a period of 3½ years without apparently feeling like a hypocrite, unless the illness he suffered in the early 1840s is to be interpreted as a symptom of the strain he was under at that time.

One of his greatest fears was to lose his religious faith, its support and the support of the Sandemanian community. Exclusions were frequent events among the Sandemanians and when he, an elder, was excluded in 1844 (for reasons discussed in section 4.3), it 'brought me very low in health and spirits'.[55] His exclusion was also the cause of the 'shame' he felt, 'and I may well indeed be ashamed seeing how I have helped with the wicked to bring great reproach & contempt on the Church of God and deepest condemnation on myself'.[56] His feeling of shame was doubtless accentuated by the trust, respect and high moral standards that the Sandemanians expect from their elders. Although he was restored a few weeks later, the fear of a further exclusion weighed heavily, since, according to the church's disciplinary code, a second exclusion could not be condoned but led necessarily to severance from the church and irrevocable exclusion from its sanctuary.

After 1844 Faraday thus felt particularly vulnerable. On the evidence of a series of letters written in the autumn of 1850, it is clear that he came very close to the second and final exclusion and that this precipitated a mental crisis.[57] The first cry of pain and anguish is to be found in a letter written on Thursday, 31 October, at *3 o'clock in the morning* as Sarah reached out desperately for help from Faraday's close friend William Buchanan:

> My beloved husband['s] mind has been much disturbed in the
> view the Church takes of not receiving an excommunicant more
> than once – & he has had some conversation with our Elders

which so far does not seem very satisfactory.... May I ask you
to write me by return of post if possible – your view of the
subject from the scriptures alone. He does not know of my
writing, & I hope I am not wrong in so doing.[58]

Sarah's second letter was written 3 days later and in it she thanked
Buchanan for writing to her husband and for sending him an
unspecified enclosure. She was greatly relieved by Buchanan's
communications and by 'the voice of the Church today [Sunday 3
Nov. 1850]'. Faraday had obviously responded positively and yet
the matter had still not been resolved but was due to come before
the elders on the following Wednesday evening. Sarah's comments
on her husband's state of mind offer some insight into his conflict:

I have hopes that my dear husband will be allowed to see that
we have been reasoning beyond what the Scriptures allow. ...
My fear now is that his mind is quite over Taxed & he seems
almost as if reason would fail. This subject has pressed with such
intense weight upon a brain already worn with much study &
he again & again says 'I may not be a hypocrite'. He seems to
me as if he could hardly take any arguments but I hope he will
get sleep & his mind be restored.[59]

Here, once again, we find Faraday confronting the self-accusation of
hypocrisy but this time trying to convince himself that the charge
will not stick. Yet the chasm was plainly in view.
    Further insight into his crisis is to be gained from a scribbled letter
which Faraday wrote to Buchanan but declined to send; it was,
however, enclosed with Sarah's second communication with
Edinburgh. It now becomes clear what was at issue when Sarah
stated, in the letter quoted above, that he had been 'reasoning
beyond what the Scriptures allow'. Faraday was specifically concerned with 2 Corinthians 2:6–11, which offers the consoling message
that the offender who genuinely repents should not be inflicted with
further punishment, 'contrariwise ye ought rather to forgive him,
and comfort him'. This passage offered Faraday scope to challenge
the Sandemanian doctrine of exclusion, but in so doing he had
pressed his own personal interpretation so as to conflict with the
sect's disciplinary code and the biblical passages on which it was
based. Through his arrogance and pride he had pursued his own line
of reasoning and had failed to heed Matthew 18:17, 'hear the
church'! Thus Faraday's perception of the situation was that his
'deep distress' lay in his presumption and pride which prevented him
from listening to the church, 'the body of Christ[,] the pillar &

ground of the truth' (1 Timothy 3:15). Paradoxically, the sect's strict discipline over exclusion and Faraday's worry that he might be excluded from the church threatened to precipitate a second exclusion. Thus in the autumn of 1850 Faraday almost placed himself outside the church.[60]

The final letter in the sequence was written on the evening of Wednesday 6 November, following Faraday's interview with the elders – Benjamin Vincent, Thomas Boosey and Stephen Leighton. He was clearly relieved to write:

> God has wrought with me when I was against him & broken down my pride & false reasoning & has this evening shewn me what love there is in the Church to an erring brother for Christ['s] sake[.] I believe all my difficulties are taken away. I hope so but fear to say much for though I rejoice it is with great trembling[,] remembering how ready I was to give up & to cast him from behind me. It is of the Lord's mercies that we are not consumed but how great is the occasion I have to apply that truth to myself.[61]

Here, then, is clear evidence of Faraday hovering on the brink, reaching a position where he was close to being excluded a second and final time from the sect and then being pulled back to safety. It was for him a miracle that the 'Almighty power could deliver us'; that God could save him. The fear and tension created by this incident was immense, as Faraday poured out his troubles and (later) his relief to Buchanan. But Buchanan did not merely listen to Faraday's outpourings but also offered brotherly love and assistance. Faraday thus acknowledged that 'Your love has drawn me out and I am very thankful for the tender words in your letter which has given me this courage'.[62]

A third crisis occurred in the summer of 1864, when Faraday was invited to fill the Presidency of the Royal Institution following the protracted illness and impending death of the current President, the Duke of Northumberland. It seems strange that the incident received only passing mention in Henry Bence Jones's *The life and letters of Faraday*, especially since Bence Jones, the Secretary of the Royal Institution, was the person who tried to persuade Faraday to accept the office.[63] A more direct but, as we shall see, totally misleading interpretation was given by Bence Jones's son Archibald in a note attached to one of Sarah Faraday's letters: 'It is one of the incidents in his life which brought Faraday into conflict with the particular branch of the Sandemanian body to which he belonged'.[64] While this note fails to explain the source of the conflict, other evidence suggests that Faraday, far from being in conflict with the church, was in conflict with Henry Bence Jones. Of the three letters surviving from that period, the first is from Sarah to Bence Jones and is dated (Tuesday) 31 May:

> My poor husband has been so troubled ever since Friday [25 May] with the thoughts of the *Presidency*, that it has quite affected his health – just as the thoughts of giving a friday evening [discourse] did formerly – and now he has returned from his visit to you quite distressed & I am sure you would be quite grieved if you heard his expressing. he says 'the Dr [Bence Jones] says it is the only thing that he ever asked me, & I *cannot* do it, it would make my brain turn soft & I should lose my mind. I am deeply grieved, but I should feel degraded in my own eyes if I allowed it to go on, it is quite inconsistent with all my life and views.' &c. &c. &c.
>
> We have in vain tried to comfort & sooth[e] him & it is my full opinion that you must give it up. We have had our accounts & considered how we could live upon a reduced income, if we had to leave the Institution – it is no use to reason so we go into all his views, & it is by his wish, (as I must say I feel with him) that I write.[65]

This remarkable letter offers some insight into Faraday's inability on this occasion to mediate between his value system and the social demands made on him by his public position. In pressing the Presidency on him Bence Jones had failed to recognise the boundary that Faraday had continually drawn between his private (Sandemanian) world and the outside turmoil. Faraday was the servant of the Royal Institution and he could not be translated into the role of its master. The Presidency, Faraday insisted, 'is quite inconsistent with all my life and views'. He could not bridge the chasm that had opened up between himself and the Royal Institution, and he even doubted whether he could henceforth continue in its employment. Moreover, if he had undertaken the Presidency, he would have run the risk of compromising his Sandemanian faith, since he felt that public office was inconsistent with his Christian values. Thus the risk of a second exclusion weighed heavily.

In this letter we also see Faraday responding as he did to Lord Melbourne in 1835 over the question of honours for scientists,[66] dogmatically rejecting the incursion of the world and its uncongenial values. In 1864 it was not a career politician but a friend, indeed, his physician and one of his closest non-Sandemanian friends, who was pressing honours and worldly responsibilities on him. In his perturbed state, doubtless exacerbated by ill-health and old age, he was beyond being reasoned with by either Bence Jones or Sarah. It is also interesting to note from this and the following letter that she considered the Presidency of the Royal Institution a suitable honour for her husband, although not under the prevailing circumstances, whereas (if Tyndall is to be believed) she helped to dissuade him from accepting in 1858 the Presidency of the Royal Society, the first scientific society in the land. Perhaps her intimate connection with the Institution and her recognition of her husband's long-term association with it made her appreciate that he deserved the honour that Bence Jones wished to confer.

Faraday continued in this distraught state for a further week, directing his anguish against Bence Jones, to whom Sarah wrote again on (Monday) 6 June.

> I feel from the warm interest you have always shown in my husband that it is due to you to say a little more than I could when in anxiety and grief I wrote last Tuesday [see previous letter] – We had then passed a painful night, such as you, our medical friend & one well acquainted with his nature and feelings may better understand than many of our other friends – and the consequence was that I sat down & wrote to the other Elders in our Church, of what I feared would be the result if something were not done to relieve an over anxious mind from pressure, and now I am happy to say, though our friends do not see his deficiencies as we do, & tho' they regret much to lose his assistance & counsel which they value highly & most affectionately – he has been encouraged to lay down the office entirely; he is not very well & I would fain have had him consult you a little but he said 'No while the Dr feels any coldness towards me I cannot', and tho' this may be a misconception on his part & I have no doubt is so, you know his sensitive nature and [you should] make allowance.
>
> I know all your wish was to do him honour & I must confess that it would have been a great gratification both to myself & Jane to have seen him in the President's chair, but it was not to be, and we must resign ourselves, for his peace and comfort is beyond every thing.[67]

As this letter suggests, Faraday resigned the office of elder in the Sandemanian meeting house on the following Sabbath (5 June). Contrary to Archibald Bence Jones's claim, cited earlier, there is no evidence of any conflict between Faraday and the church. Instead, the offer of the Presidency raised the spectre of the much-feared second exclusion. Far from being excluded, he refused the Presidency and, owing to the crisis the incident produced, he was persuaded to lay down his eldership. Moreover, there appears to have been full accord between Faraday and other members of the church, who gave him their emotional support. This is evident in a short letter that Faraday wrote to Benjamin Vincent on Wednesday, 8 June, which shows that he had largely regained his composure. The letter ends, 'Wonderful has been the love shown to me'.[68]

Previous biographers have failed to notice that Faraday hovered on the brink in 1840, 1850 and 1864 (and perhaps other occasions) but some of them have recognised, *en passant*, another but perhaps less dramatic crisis in Faraday's life. By the early 1830s he reached maturity as an experimental scientist, reporting his impressive electrical discoveries principally in a sequence of papers read before the Royal Society. Beginning in 1831 this sequence reached the seventeenth series in early 1840, after which there was an hiatus lasting some 3 years. Further evidence that his research suffered severe interruption in the early 1840s is provided by his *Diary*, which contains no entries for 1841 and very few for 1840 and 1843.[69] He also relinquished some of his duties at the Royal Institution: for example, he was relieved of organising the Friday evening discourses in 1842.

Both Bence Jones and Tyndall attributed this inability to continue research to a long-term problem which reached its culmination at that time. 'He was now feeling the effects of the mental strain to which he had been subjected for so many years', wrote Tyndall. 'During these years he repeatedly broke down.' Bence Jones also noted that loss of 'memory and giddiness had long, occasionally, troubled Faraday and obliged him to stop his work'.[70] Both biographers, then, acknowledge the long-term nature of this problem, one which, like his father's illness, frequently interfered with his work.

As Faraday's letters of the period, particularly his correspondence with Schoenbein, make clear, loss of memory was the most marked symptom. For example, in May 1843 he noted that his memory was 'so treacherous, that I cannot remember the beginning of a sentence to the end'.[71] Edward Hare, a psychiatrist who has examined not only Faraday's statements of that period but also the casebook of Peter Mere Latham, the physician who attended him, concludes that in December 1839 Faraday probably suffered an episode of brain damage owing to part of the brain being deprived of blood for a time. Such an attack might explain other symptoms, such as headache, giddiness and confusion. Yet Hare is less certain whether Faraday's long-term problem with his memory can be attributed simply to brain damage unless the memory centres of the brain sustained damage.[72] Faraday certainly complained of a bad memory on many occasions before 1839; indeed, in his earliest letter to Benjamin Abbott, dated 12 July 1812, he complained that he kept forgetting the appropriate words to express his ideas.[73] That Faraday experienced particularly distressing lapses of memory in the early 1840s there can be no doubt. However, he certainly worried a great deal about his memory and believed it to be deficient. Thus he confided to Schoenbein in March 1841 that, although his physicians considered that he was suffering from mental fatigue, he believed 'I am permanently worse'.[74] Although memory ability seems to differ widely between individuals, Faraday's vast scientific output and the facility with which he could quote passages from the Bible suggest that his memory was not manifestly deficient, except during the early 1840s and his final years.

Faraday thus seems to have convinced himself that his memory was substandard and his worry on this score would have made matters worse. He was particularly worried that his memory would let him down in public. For example, in 1858 he wanted to deliver a Friday evening discourse at the Royal Institution on Schoenbein's research on ozone. However, he confided in Schoenbein that he was 'terrified at the thoughts of trying to do so, from the difficulty of remembering . . . I have never before felt so seriously the evil of loss of memory and of clearness in the head'.[75] His conviction that his memory had failed was oppressive. Moreover, having to rest for long intervals, Faraday was forced to make the most of his more lucid periods. Although he previously allowed himself little time for diversions, his illness in the early 1840s left him with 'scarcely a moment to spare for any thing, but work'.[76]

Loss of memory was not the only long-term symptom of which he complained. As early as 1828 he complained of 'nervous headaches', while in November 1840 he referred to a headache 'which as some people say I have *enjoyed* for the last four months'.[77] A rather more complete diagnosis is offered in 1847, when he wrote that his doctors attributed his headaches to 'mental occupation. The result is loss of memory, *confusion* and giddiness; the sole remedy, cessation from such occupation and head-rest'.[78] Faraday's frequent visits to the seaside, especially Brighton and Folkestone, and occasionally to other places – he spent 3 months in Switzerland in the summer of 1841 – were often connected with his need to recuperate from these stress-related headaches.

It is, however, doubtful whether either the putative cause – 'mental occupation' – or the prescribed cure – rest – is adequate. An alternative hypothesis has been suggested by Dorothy Stein in her biography of Ada, Countess of Lovelace. Here she suggests that Faraday was suffering from chronic chemical poisoning produced by the chemicals released in his experiments. He frequently handled nitrogen tetrachloride, hydrochloric acid, hydrofluoric acid, nitrous oxide and ether, and there were a few explosions while he was operating with these substances. Moreover, his early electric motors used basins of mercury and he also prepared benzene based anilines. 'Dizziness, confusion, difficulty in concentration, and memory loss are among the symptoms of aniline poisoning', writes Stein.[79] Certainly these were some of the symptoms displayed by Faraday, but against Stein's argument must be set his vast scientific output not only in the 1830s but also in the decade and a half following his breakdown. During the latter period he produced twenty three series of investigations for his *Experimental researches in electricity* and a number of other important papers, while his *Diary* covered over 1,400 pages when printed. Moreover, the periods of recuperation when Faraday was away from the laboratory were sometimes successful but also sometimes unsuccessful in removing his symptoms. One of the strongest arguments against Stein's thesis is that Faraday's lifespan of nearly 76 years seems difficult to reconcile at least with severe heavy metal poisoning. A further problem raised by her thesis is that it is impossible to know the quantities of various chemicals that entered his system.[80]

Historians have been debating for some years whether, and to what extent, Newton's breakdown in 1693 was of physical, chemical or psychological origin.[81] The case of Faraday appears similarly difficult to decide, and while there is considerable virtue in Stein's hypothesis, it is far from convincing and also fails to relate Faraday's breakdown of the early 1840s to the crises, both large and small, discussed above. The approach of the present chapter provides an alternative way of understanding this protracted illness and it also possesses the advantage of relating it to the three crises mentioned earlier. On this account Faraday's headaches, etc., were the result of stresses arising from the usually suppressed tensions and fears, which, as we have seen, occasionally exploded into the open. Unresolved psychological tensions often manifest themselves as muscle contraction (or 'tension') headaches. Moreover, sufferers from this type of headache also suffer depression and lack concentration – two symptoms that Faraday manifested. Faraday's 4-month headache may have been of this type. It is also relevant that, as neurologists have noted, people who suffer from tension headaches often possess a 'meticulous, energetic personality' – an apt description of Faraday.[82]

10.4 ORDERING EXPERIENCE

In this section I shall employ both theories of personality and theories of mental development to help us understand Faraday

better and, in particular, to provide a common context for understanding both his science and his religion. In short, I shall argue that
both science and Sandemanianism were strikingly similar responses
to his environment, since they both offered ways of ordering his
experience and rendering the world safe.

As regards his personality, one descriptive term that is strongly
implied by the material in the preceding section is *anxiety*. Although
psychologists usually trace the source of anxiety to infancy, lack of
early biographical information prevents us from entering this part of
the topic. However, it is clear that later in life he manifested a high
degree of anxiety on a number of occasions, such as the three 'crises'
discussed above. In each of these instances the immediate source of
anxiety was a specific external event that posed a threat to his safety
and was the source of emotional conflict – Faraday's election to the
elder's office in 1840, the likelihood of a second exclusion from the
Sandemanian church in 1850 and Bence Jones's proposal that he
accept the Presidency of the Royal Institution in 1864. He
experienced anxieties of a lesser degree on many other occasions:
for example, prior to delivering a Friday evening discourse in his
latter years or when he found himself isolated from Whewell who
had failed to understand some of his papers.[83] While it is clear that
he manifested anxiety to various degrees, it is also important to
notice that over extensive periods of time he was able to displace
these anxious feelings and to present himself in public as a very
equable person. Although anxiety often hovered in the wings and
sometimes engulfed him, Faraday could equally create a haven of
contentment that was guaranteed by his religious and scientific
understanding of the world as a safe and ordered place. Anxiety
and contentment are apparently opposing mental characteristics
that he seemed capable of manifesting to a high degree.

A number of the symptoms of which he complained may have
been connected with anxiety, e.g. the headaches that he frequently
experienced. He also often complained of feeling weak and tired: in a
letter to Edward Magrath, for instance he wrote that 'My strength
fails[,] my nerves grow feeble and I am become a patient that does
no credit to anyone'. Other mental symptoms that probably cohere
with this pattern are 'loss of memory, confusion and giddiness'.[84]
Finally, he sometimes suffered feelings of depression, although these
periods were usually of short duration.

Faraday possessed to a high degree several character traits, such
as orderliness, meticulousness, overconscientiousness and perfection-
ism, that are associated with an *obsessional personality*.[85] The first of
these traits will be discussed further below, but let two examples
suffice for the present. Between 1831 and 1855 Faraday read before

the Royal Society a series of thirty papers that were also published as his three-volume *Experimental researches in electricity*. Each paragraph in this series was numbered consecutively from 1 to 3361. Moreover, between 1832 and 1860 he likewise numbered in a single sequence the paragraphs in his *Diary*, the final entry being 16041. I know of no other scientist who so consistently and forcefully imposed orderliness on his work through this method of numbering paragraphs; these two sequences, one public and the other private, each covered a period of nearly 30 years. It is as if Faraday had to create in his work the same order that he believed existed in the universe. Science thus became the mirror of nature.[86] Secondly, his attraction to Sandemanianism can be traced to similar roots, since Sandemanians interpret the Bible as prescribing a clear code of conduct for the Christian to follow accurately. God's laws had to be followed in their most exacting details. As psychologists have noted, people with obsessional personalities often pay considerable attention to religion, 'especially its form, ritual, and . . . detail'.[87]

The psychological drives that manifest themselves in obsessional character traits render some people totally incapable of creative work, while in other cases the causes of these traits may be the source of great creativity. Some manifestations of Faraday's psychological drives are readily identifiable. He was a workaholic and experienced no difficulty cutting himself off from society for long periods of time during his intense work binges. His advice to an aspiring young scientist – 'Work, finish, publish' – is a succinct statement of this personal philosophy. Again, in 1845, while speculating and experimenting on the relation between light, magnetism and electricity, he claimed, 'I have scarcely a moment to spare for any thing, except work'.[88] There is much more to creativity than hard work, and one further aspect has been identified by Elspeth Crawford, who rightly insists that in his scientific work Faraday did not follow a set method or algorithm. Instead, in his creative thinking he put aside prejudices, allowed novel thoughts to enter his mind and then freely played with them, selecting those that allowed him to reorder his experience.[89] From this perspective what is interesting is the interplay between the obsessional, rigid aspects of Faraday's character and his ability to freewheel in clearly defined mental spaces created within those constraints. These two apparent opposites may be closely interrelated, since Faraday's creativity operated within a limited space that was shielded from danger by the assurance that the world is stable under God's dominion. Likewise, as we saw in sections 7.1 and 7.3, Faraday effectively synthesised the apparent opposites of hard-nosed empiricism and wild flights of fancy. Yet not all opposing tendencies could be synthesised.

Incompatible oppositions feature prominently in Faraday's writings, and he often used pairs of terms with opposing emotional charges, one member of the pair expressing an anxiety, the other the alleviation of anxiety – for example, danger and safety or confusion and order. The use of such opposites forms a central strand in the theory of personal constructs, a theory of psychological development, which George Kelly proposed in 1955. Since I have found Kelly's theory helpful in articulating some of Faraday's mental processes, I shall briefly outline it.

Kelly postulated a developmental view of mind based on a dynamic network of polar constructs. Thus a person might organise his/her view of himself/herself and of the world in terms of such opposites as danger:safety, confusion:order or theist:atheist. Kelly emphasised that we use these constructs to anticipate what the world is like and to organise our experience. For most people, most of the time, experience confirms the adequacy of their construct systems, but sometimes a novel experience challenges those constructs. Any one of a number of scenarios may follow. We might meet the challenge by changing our constructs – this occurs when, say, an adolescent is maturing. If, on the other hand, our construct system is too rigid to accommodate the new situation, then it poses a threat. 'We are threatened when our major beliefs about the nature of our personal, social and practical situation are invalidated and the world about us appears to become chaotic.' In extremis this can lead to attempted suicide.[90]

Kelly's theory is useful for analysing Faraday and how he responded to the world in which he lived, since the theory draws attention to the way a person creates pairs of opposites in his/her construct system and how those oppositions are maintained (or dissolved) through that person's development. Thus, as we have just seen with Faraday, a constructive synthesis of opposites can sometimes be obtained. Often, however, oppositions produce rigid structures so that experience is pigeonholed in one of two incompatible ways. Thus some people categorise others as either right or wrong, virtuous or evil, allowing no room for intermediate gradations or alternative ways of interpreting the situation. It is very noticeable that on a number of topics in both science and religion Faraday frequently employed pairs of opposing terms. In using these terms, he was drawing a sharp distinction between what was acceptable to him and what was not. In other words, he conceived many situations in terms of stark contrasts between black and white, rather than as slowly varying shades of grey. Kelly's theory can be linked with the foregoing discussion of Faraday's personality by noting that one defensive strategy often employed by people with obsessive personalities is to make such rigid polar judgements.[91]

Turning first to Faraday's religion, we noted in section 1.2 that Sandemanians drew a sharp divide between this world and the kingdom of heaven. Areas of experience could then be pigeonholed in one of these antithetical domains. Politics, ambition and the quest for worldly wealth were clearly mundane matters and were therefore of no positive relevance to the true Christian. By contrast, to follow the ordinances laid down in the Bible, to practise brotherly love and to lead spiritually pure lives enabled entry into the kingdom of heaven. The sect's disciplinary code helped maintain this divide and to make clear what was and what was not acceptable conduct. Those who forsook the kingdom of heaven and became too attached to the mundane world had to be excluded from the sect. Moreover, this contrast between the two realms was apparent in the high boundary that the Sandemanians drew between the sect and the rest of society.

While much of the Sandemanians' understanding of their experience was controlled by this dichotomy, they did not possess any strong prescription for categorising science. The Bible certainly contained a number of truths about the physical world, especially its creation, but apart from accepting these a Sandemanian was free to pursue science. What is particularly revealing is that Faraday did not classify science (except for the politics of science) as a mundane activity. Instead he conceived science to be closely akin to true Christianity, since order and peace were to be found in both. Moreover, in engaging in science he was pursuing a quasi-religious activity, since he was discovering the way God had constructed and ordered the physical universe. As noted in section 6.1, another similarity was that the scientific community should behave like 'a band of brothers' – in other words, like members of the Sandemanian fellowship.[92] However, whenever the politics of science entered, science was debased and it became a thoroughly mundane activity.

As we saw in section 5.5, Faraday employed opposing constructs in his conception of social space. Thus the Royal Institution was clearly demarcated between private and public spaces, the latter being part of the mundane realm. In his private living quarters and in the laboratory Faraday created an ordered existence, practised his religion and pursued his science. His social relationships were likewise marked by dichotomies, for not only did he draw a sharp divide between the circle of Sandemanians and fashionable London society, but he also cast himself in the role of a loyal servant of the Royal Institution. In his early scientific career he had been cast in a subservient role to Davy and only asserted his independence when Davy tried to block him and forsook scientific research for the pleasures of high society. What he experienced difficulty in accepting was the role of master either to other scientists or to the Royal Institution. The offer of the Presidency in 1864 precipitated a crisis.

Strong dichotomies also appeared in his writings. In his 1854 mental education lecture, in particular he drew a sharp contrast between prejudice and judgement, and he insisted that in all matters, but especially in science, the correct use of judgement was essential. By contrast, we are prone to many forms of prejudice, and, instead of allowing it to engulf us, we must work tirelessly to extirpate it.[93] This pair of terms was likewise used by Faraday and other Sandemanians in the context of religion. Moral issues were also usually portrayed in terms of oppositions. Thus, for example, honesty and truth were contrasted with dishonesty, falsehood and hypocrisy.

A further pair of terms is particularly important for understanding Faraday, since it features strongly and frequently in his religion, his science and his comments on society. This is the contrast between order and confusion, which in turn map on to the contrast between safety and danger. For Faraday Sandemanianism provided security, the assurance that if he followed its precepts accurately he would be acting in accordance with God's laws and commands. He would thereby gain acceptance with God the Father. Faraday's Christianity also assured him that life is not chaos and suffering but that there is a meaningful plan behind events. The kingdom of heaven is eternal, sure and unchanging. The Bible is the key to this knowledge, and in the tightly disciplined world of the Sandemanians there was almost no room for uncertainty or human fallibility. His religion was an oasis of peace, and on the few occasions when his peace of mind did break down, his fellow Sandemanians helped him regain his composure.

Science was also an island of security on which he could take refuge from the storms of the outside world. In practical terms he could retire to his laboratory, where he and the ever-faithful assistant, Charles Anderson, pursued their honest toil in privacy with the door firmly closed. In his science Faraday was contemplating nature, applying his considerable practical and intellectual skills to discover the laws that God had employed in the Creation. But danger lurked in science, since, as discussed in section 7.3, the free rein of fantasy and hypothesis threatened to cast the scientist on the rocks of prejudice. Empirical facts were necessary to curb excessive speculation. Nature could be beautiful, even startling, but Faraday had to hold fast on to facts while seeking the underlying laws. He was firmly committed to the view that all phenomena were produced by unbreakable laws framed by God at the Creation. However, as we have seen, there was some tension between his avowed empiricism and his metaphysical commitment to the economy of nature. Yet this commitment expressed his belief that nature is an ordered, stable system.

In contrast to the highly ordered, calm domains of religion and science, the social world exemplified confusion. The Sandemanian social philosophy laid down how to deal with that world and prevent its incursion into the kingdom of heaven. In maintaining order and control, and thus keeping confusion at bay, Faraday and other Sandemanians adopted various strategies, such as refusing to engage in party politics. For much of his life the line held firm and he could create his own environment of order. However, on the three occasions discussed in the preceding section his construct system was severely threatened and (to paraphrase a statement from Kelly's theory) 'the world about him appeared to become chaotic'.[94] The chaos was, however, not long-lasting, and he managed to draw back from it after a few days. Two other events discussed above severely challenged his construct system. The first was Davy's attack on his honesty following his discovery of electromagnetic rotation in 1821, a charge which, although apparently rebutted at the time, continued to trouble him for the rest of his life. The other was his exclusion from the Sandemanians in March 1844, which brought him 'low in health and spirits', so that even 6 months later he was still afflicted by the shame it had brought him and the church.[95]

The rigid pairs of dichotomous constructs discussed above enable us to understand how Faraday organised his experience. They are also very helpful in analysing the fears and crises that were produced by threatening situations. Their main use, however, is to provide a set of terms that underpinned both his science and his religion. In this way both science and religion are brought into a common context.

11 *Epilogue: Science and Religion*

Writing to a fellow Sandemanian in 1792, James Morison of Perth related the parable of the 'celebrated Dr. [Andrew] Wilson, who many years since, was I believe six weeks in the Church'. The doctor's short connection was terminated because he had been 'engaged through philosophy [i.e. science] & vain deceit [Colossians 2:8] in remarkable opposition to the truth'. On his deathbed he had seen the error of his ways, repented, and was reported as 'rejoicing in the abundance of Divine Grace'.[1] If at the end of his life Wilson gained acceptance before God, he signally failed to do so earlier. Wilson's life had been eventful: he had studied medicine at Edinburgh, later practising as a physician in Newcastle and London, where he also held a position in a lunatic asylum. He published several works on medicine and natural philosophy which show that he was attracted to the writings of John Hutchinson, and he likewise based his natural philosophy on Hutchinson's peculiar interpretation of Scripture, which, as we have seen, was rejected by John Glas. According to Morison, Wilson's exclusion from the Sandemanian fellowship was owing to his preoccupation with science, including Hutchinsonian natural philosophy. Morison's account also encapsulates a conventional view of the science–religion relationship, for it portrays the two as opposed and in conflict. Wilson carried, as it were, the serpent within his breast.

Science and religion have frequently been perceived in this way, and one of the major theses about their interrelation is often called the 'conflict thesis'.[2] However, there is no consensus over the meaning of this term, and several different conflict theses have been propounded. For some writers science is necessarily opposed to religion, because these two activities employ different and antagonistic methods: for example, science is considered to be empirical and rational, whereas religion is presumed to trade on irrational belief. Such stark contrasts do not withstand analysis but some writers, positivists included, have read the historical record from this standpoint. Indeed, the black-and-white contrast between religion and science has proved a highly effective but a manifestly self-serving strategy. For other writers, however, the conflict has not been between science and religion *per se*, since they conceive certain specific aspects of religion commensurable with science. For example, advocates of liberal, latitudinarian theology have often conceived science as a natural ally, but have portrayed both activities as necessarily opposed to Catholicism and fundamentalism.

A further version of the conflict thesis centres on the logical incompatibility of certain statements derived from the two systems of thought. Thus, to take the standard example, Darwin's theory is opposed to 'creation science', since the former requires a vastly longer timespan for the earth's history. Either the earth is some 6,000 years old, or it is not. One side must be right and the opposition must be wrong. The battle lines are clearly drawn. A final interpretation of the conflict thesis does not interpret the warfare as being over either ideas or conceptions of method; instead, the opponents fight for power. On this account the Church has traditionally been vested with power and authority. Science enters the ring to wrest this authority from the Church and thus to become the dominant force in our culture. There is therefore a necessary head-on collision between the two parties.

The various forms of the conflict thesis have attracted much support, but they are not adequate as general claims about how science and religion have been interrelated in history. To extend the military metaphor, the conflict thesis is like a great blunderbuss which obliterates the fine texture of history and sets science and religion in necessary and irrevocable opposition. Much historical research has invalidated the conflict thesis. Thus Galileo can no longer be portrayed as the harbinger of truth and enlightenment who was pitted against reactionary priests who refused to look through the telescope. Instead he counted many Jesuits among his supporters, but his censure resulted partly from his mishandling of a sensitive diplomatic situation.[3] The other paradigmatic conflict concerns the Darwinian theory of evolution and centres on the Huxley–Wilberforce confrontation in 1860. These opponents are now viewed as trading minor insults in the heat of debate and not as exemplifying the necessary conflict between science and religion. Moreover, in the latter nineteenth century Darwinian theory was not generally seen as antagonistic to scriptural religion.[4]

Not only is the conflict thesis a general thesis about the relation between science and religion, it is also a subject for historical analysis. Thus the two best-known Victorian versions of the thesis – John Draper's *The warfare between science and Christianity* (1873) and Andrew White's *A history of the warfare of science with theology in Christendom* (1896) need themselves to be set in historical context. Draper's virulent anti-Catholicism and White's more measured defence of liberal theology against fundamentalism can best be understood in biographical terms. Thus White's response arose from attempts by fundamentalists to inhibit the liberal regime he instigated at Cornell University where he was President.[5] Such specific responses to local situations provide inadequate starting points for a balanced understanding of the subject. Yet science and religion are two of the great strands of our culture and neither can be characterised adequately in the simplistic manner which the conflict thesis demands. Indeed, the work of a number of historians during the last few decades has not only demonstrated the untenability of the conflict thesis but has also highlighted the need for detailed case studies.[6] It is only through such case studies that we can better appreciate the complex and subtle interrelation between science and religion. The present study of Faraday is intended as a contribution to this burgeoning literature.

Each version of the conflict thesis implies a preferential relationship in which science and religion should exist. For example, there has been a long tradition in which science has been viewed as a natural ally of liberal theology and as necessarily opposed to fundamentalism. In place of conflict the correct theology must then support and be supported by science. An extreme version of this argument is sometimes referred to as the revisionist thesis, and its proponents (usually Protestants) argue that Christianity and Christianity alone provided the essential soil from which modern science grew. There are many arguments against this strong version of the revisionist thesis: for example, it depreciates Greek, Islamic and Chinese science, and it also fails to explain adequately why modern science had to wait in the wings during some 16 centuries of Christianity. At a more textual level revisonists are hard put to find passages in the Bible that are manifestly conducive to modern science. On close examination revisionism turns out to be as untenable as the conflict thesis.[7]

Advocates of the conflict and revisionist theses offer rigid definitions of both science and religion. They thereby tend to portray both activities, and their interaction, in anachronistic ways. However, there is an alternative to this essentialist approach which instead focuses on the *individual* and examines the ways in which that person's science and religion were interrelated. The two activities might turn out to work in constructive harmony, or there might be areas of tension between the two, or any of a number of other possibilities. Such an approach, which has been adopted in this study of Faraday, portrays the scientist as an active agent and not as the passive recipient of prepackaged views about both science and religion.

Many previous commentators have cited Faraday as a paradigm example of an individualised version of the revisionist thesis. Indeed, Faraday has often been used by religious apologists to show that a scientist of the first rank can also be a biblical Christian.[8] Moreover, as the revisionist requires and as we have seen in previous chapters, there was no outward conflict between Faraday's science and his Christianity. The great appeal of Faraday to revisionists is that he not only refutes the conflict thesis but he also exemplifies a strong positive interaction between his science and his religion.

Several *caveats* must, however, be entered against this revisionist interpretation of Faraday and the tendency to extend this example into a general revisionist argument. His scientific work was confined to physics, chemistry and electricity, and he carefully avoided several other sciences – especially the history of the earth and of species – which might have impinged uncomfortably on the literal interpretation of the Bible. Even in the sciences he researched, he adopted highly idiosyncratic views on matter theory, electricity, space, scientific method, etc. Thus, when compared with the scientific community, Faraday's work does not exemplify contemporary scientific practice. Indeed, he pursued his science in a highly individualistic manner and in a way that made it cohere fairly unproblematically with his religious beliefs. Faraday's Sandemanianism likewise placed him far from the contemporary religious norm. Thus both his science and his religion were peculiar, and although he achieved a highly personal synthesis between the two, we cannot use him as a legitimate example of the general revisionist thesis, since revisionism looks to unchangeing, atemporal aspects of both science and religion and not distinctly local and idiosyncratic versions of the two activities. Moreover, even if we were to grant that Faraday exemplified the individualist version of the revisionist thesis, the above *caveats* prevent us from using him as an example to confirm the general revisionist position.

A more serious historical thesis about the relation between science and religion was canvassed by Susan Cannon in the 1960s and 1970s and subsequently developed by other historians.[9] Cannon conceived the scientific leaders in the second quarter of the nineteenth century as a group of 'broad churchmen' closely associated with Cambridge University. Their non-dogmatic theology and their commitment to rational, liberal principles allowed them to play a full role in science and its institutions, especially the British Association and the Royal Society. Natural theology provided a bridge between the two areas and enabled these scientists to dwell, in a publicly acceptable way, on the theological significance of recent innovations in science. These scientists were, however, explicitly opposed to the scriptural geologists and Hutchinsonians, who sought to ground their science explicitly on the Bible. Clearly Faraday cannot be considered a 'broad churchman', although as Cannon has noted he was greatly admired by the Cambridge 'broad churchmen'.[10] Nevertheless, it is important to notice that in public Faraday occasionally proffered comments about, say, God's wisdom as manifested in the physical world, which would have been acceptable to a broad spectrum of theological opinion.

In reaction to Cannon and other historians who have stressed the natural alliance between science and latitudinarian Christianity, several authors have recently argued that many scientists were far more conservative in their religious views than the Cannon thesis appears to allow. For example, William Whewell, one of Cannon's 'broad churchmen', now emerges in the work of John Hedley Brooke as a far more conservative religious thinker, as does Baden Powell, who has been studied by Pietro Corsi. Likewise James Moore, concentrating on reactions to Darwin, has found that early American fundamentalists readily accepted most aspects of evolutionary theory, rejecting only Darwin's apparent denial of divine design.[11] This biography of Faraday provides further evidence that the connection with science is not only to be found among latitudinarians. Faraday was, after all, one of the most successful and highly acclaimed scientists of the nineteenth century, and he was also a sectarian who accepted and lived by a highly literal interpretation of the Bible.

In the preceding chapters I have described and analysed Faraday's highly personal synthesis between science and religion. However, I want to avoid portraying this relationship in strict causal terms, since his understanding of both science and religion was an interrelated part of his social, intellectual and spiritual development. Moreover, as argued in the preceding chapter, both his science and his religion were chosen responses to his psychological needs. Since they fulfilled very similar roles, no strong causal arrow can be drawn from his Sandemanianism to his science (or *vice versa*). Instead, his religion and the kind of science he pursued were very much of a piece.

It is not possible from the available evidence to determine how his science affected his religious views and practices. That he refrained from following the author (possibly Robert Sandeman) of *The philosophy of the Creation* (1835), who sought the geological theory contained in the opening verses of Genesis, and from following John Glas, who analysed the numbers mentioned in the books of Daniel and Revelation, suggests that these forms of biblical exegesis were unattractive to Faraday the scientist.[12]

The influence of his Sandemanianism on his science is more readily documented. Belonging to a sect with clearly prescribed norms, he extended the Sandemanian social philosophy into his science, as discussed in chapters 5 and 6. To act as a true, moral Christian was his overriding concern, and therefore his views and behaviour were natural extensions of his Sandemanianism. To be a scientist did not threaten his religious persona; rather science provided a relatively safe area in which he could practice his Sandemanianism. From this position he engaged the social world around him, though on his own terms. Faraday's scientific work likewise bore the marks of his Sandemanianism. While the forty papers he read before the Royal Society were perfectly acceptable to the eclectic scientific community and did not outwardly bear the sectarian's stamp, I have argued in chapters 7, 8 and 9, that Faraday's science was predicated on strong beliefs about what the physical world is like, how God constructed it, and how he was to understand nature and its laws.

The view proposed here is that Faraday was not free to pursue science in any conceivable manner, but rather that he was constrained by his religious beliefs, which channelled his research in certain general directions. One positive constraint was that his religious conception of the world and of his place in it provided a driving force behind his science. He therefore conceived his vocation as reading the book of nature which God had placed before man – an ideal which motivated much of his thought and action. His theocentric perspective also helped define certain scientific problems: for example, to discover the laws that governed the relations between the various divinely ordained powers. Thus he sought the mutual relation between electricity and magnetism (which proved highly successful) and between gravity and electricity (which did not). Moreover, his views on scientific method were of a piece with his theology.

The theological underpinnings to Faraday's research also help account for his prodigious productivity. As his *Diary* amply testifies, Faraday not only spent long hours in the laboratory but his work was usually firmly directed towards illuminating a particular physical problem or relationship. Particularly after the 1820s much of his research can be seen as directed to discovering a new natural philosophy to replace the Newtonian theory of matter and force. There is also a sense in which Faraday's religion played a role in his success as a discoverer, since he possessed a strong sense of his self-identity and of his intimate relation with God's Nature, whose laws were revealed through his effort. At the same time he understood that his mind was liable to prejudice and thus, while working within prescribed limits, he was sufficiently flexible to entertain and test a variety of ideas. This lightness of touch and approach was an important component in Faraday's scientific style.

While Faraday's religious beliefs helped direct his science, he avoided any outward clash between his science and his religion. The authority of Scripture was supreme, while science, although much purer and more stable than other worldly activities, was necessarily of a subordinate order. Faraday could therefore claim that 'the natural works of God can never by any possibility come into contradiction with the higher things that belong to our future existence, and must with everything concerning Him ever glorify Him'.[13] As a Christian deeply committed to a literal interpretation of the Bible, he was outwardly untroubled by the doubts that assailed many latitudinarians and by the contemporary controversies over the meaning of science for contemporary culture. The absolute authority of the Bible could not be undermined by science, but science, if practised in a truly Christian way, can illuminate God's other book.

The relation between Faraday's science and his religion should also be understood in social terms, since he appreciated the need to deport himself morally in the scientific community. Just as Sandemanianians are expected to behave in a morally upright fashion, so Faraday transferred the Sandemanian social philosophy to science. Scientists, he believed, constituted a fellowship in which there was no place for avarice, parties or personal disputes. Most of all he repudiated the entrepreneurial spirit and the system of rewards and patronage that he saw as undermining the purity of science. Instead, Faraday resisted these trends and sought to define the scientist as a moral agent. Central to Faraday's moral conception of science was his belief that scientific knowledge should not be the sole property of the scientific community. Instead he held that scientific knowledge was morally uplifting and had to be shared with the population at large. Through his lectures at the Royal Institution he sought to spread the scientific spirit.

Since Faraday's day science has become increasingly utilitarian in the service of both industry and the military. It has also become dominated by experts and has lost contact with the wider public. Moreover, science is frequently portrayed as necessarily opposed to religion. In these respects Faraday would find present-day science morally unattractive.

APPENDIX A:
Faraday/Barnard Family Tree

Names of known Sandemanians are in *italics*. Family names well-represented in the Sandemanian community are in CAPITALS.

A. Children and grandchildren of *James FARADAY* (1761–1810) and Margaret Hastwell (1764–1838).

1. *Elizabeth* (1787–1847) = *Adam Greenhow Gray* (b1774)
    Mary (b1812)   = Jeremiah SPENCE (b1810)
    Eliza[beth] (1814–86)   = Edward METCALF
    Sarah (1816–82)   = James MACOMIE
    Jane (1817–55)   = *Albert GILES* (d1855)
    James (b1819)   ?
    Ellen (1821–82)   = William GILES (d1861)
    John (b1823)   ?
    Maria (b1829)   = Barnard Simpson PROCTOR (1829–1903)
    William (b1832)   ?

2. *Robert* (1788–1846) = *Margaret LEIGHTON* (d1865?)
    Abigail (1815–85)   = *George SANDEMAN* (1818–80)
    James (1817–75)   = (a) *Margery Ann REID* (1820–50)
                                = (b) Elizabeth LEIGHTON (1829–62)
    Margaret (b1821)   = *George Cargill LEIGHTON* (1826–95)
    Elizabeth (1822–91)   unmarried

3. *Michael* (1791–1867) = *Sarah BARNARD* (1800–79)

4. Margaret (1802–62) = John BARNARD (1797–1880)
    Anna (1827–98)   = *Edward Ker REID* (1821?-86)
    Frank (1828–95)   ?
    Margaret (b1830)   = *Alexander BLAIKLEY*
    Jane (1832–1911)   unmarried
    Mary (b1833)   = William DEACON
    John (1835–1915)   = (a) Fanny PROCTOR (b1834)
                                = (b) *Ruth BLAIKLEY* (b1844)
    Charlotte (b1840)   = R. H. Davis
    Alfred (b1842)   ?
    Catherine (b1844)   ?
    Rachel (1845-1929)   = *David James BLAIKLEY* (1846–1936)

B. Children and grandchildren of *Edward BARNARD* (1767–1855) and *Mary BOOSEY* (1769–1847)

5. *Mary* (1792–1845) = William Ker REID (1787–1868)
       Edward Ker (1821–86)   = *Anna BARNARD* (1827–98) [=4.1]
       William (1832–55)   unmarried
       Mary (1813–53)   = *Alexander BOYD*
       Margery Ann (1815–88)
       Caroline (1816–90)   = *Thomas DEACON*
       Margaret Elizabeth (1828–58) unmarried

6. *Elizabeth* (1794–1870) = *David REID* (1792–1868)
       Christian John (1816–91 = Mary BOOSEY (1818–1903)
       Mary Barnard (1818–61) unmarried
       Margery Ann (1820–50)   = *James FARADAY* (1817–75) [=2.2]
       Andrew (1823–96)   = *Ellen BOOSEY* (1832–1925)
       Sarah (1825–1908)   unmarried
       Margaret (1827–1905)   = Edward BARNARD (1826–76) [=7.3]
       Elizabeth (*b*1830)   ?
       David (1832–1914)   = Frances Laura DEACON (*b*1831)
       Edward (1834–87)   = Helena Elizabeth BELL (1839–68)

7. Edward (1796–1867) = Caroline CHATER
       Ellen (1823–99)   = *Benjamin VINCENT* (1819–99)
       Eliza   ?
       Edward (1826–76)   = Margaret REID (1827–1905) [=6.4]
       Vernor (1827–1916)   = (a) *Edith LEIGHTON* (*d*1865)
                                 = (b) *Anne VINCENT* (1832–1929)
       Caroline   ?
       James   ?
       Walter (1833–1922)   = Ellen Rutt
       Mary Chater   ?
       Henry   ?
       Charles (*b*1839?)   = Laura Webb
       Louisa   ?
       Frederick (1846–96)   = Alice FARADAY (1847–1910?)

8. John (1797–1880) = Margaret FARADAY (1802–62) [=4]

9. *Sarah* (1800–79) = *Michael FARADAY* (1791–1867) [=3]

10. *William* (1801–48) = *Martha LYON* (*b*1799)
        William (*d*1900)   ?
        daughter   ?
        George Edward (1836–1907) = Jane MORISON

11. *Jane* (1803–42)

Continued on next page

12. *Charlotte* = *George BUCHANAN* (1790–1852)
    Mary
    George (1834–1907) = *Emily BOOSEY* (1834–82)
    Jeanie
    Edward
    Nathaniel

13. *George* (1807–90) = Emma HILLHOUSE
    Jane Florence   ?
    Annie   ?
    others?

APPENDIX B:
Male Members of the London Sandemanian Meeting House, 1821–67

NAME	BORN	ADMIT.	LEFT/DIED	TRADE	MARRIED
BARKER George		1853	1858	Artist?	*Hannah*
BARKER Thomas (D?)	1825	1844	1866(d)	Merchant's clerk	*Sarah W.*
BARNARD Edward (E)	1767	1801	1855(d)	Silversmith	*Mary BOOSEY*
BARNARD John (DE)	1835	1857	1915(d)	Silversmith	*Fanny PROCTOR*
					Ruth BLAIKLEY
BARNARD Vernor	1827	1852	1916(d)	Draughtsman/Decorator	*Edith LEIGHTON*
					Ann L Vincent
BARNARD William (DE)	1801	1826	1848(d)	Silversmith	*Martha LYON*
BASSETT Thomas		1844	1855		
BAYNES James	1766	1792	1837(d)	Artist	*Jane BELL?*
BAYNES John		1854			
BELL Edward R.	1798	1829	1836		*Mary CRICHTON*
BELL Henry A.	1809	1841	1844		*Ann P WILSON*
BLAIKLEY Alexander	1816	1835	1903(d)	Artist	*Jane PRATT*
					Jane Shaw
BOOSEY Thomas (DE)	1795	1827	1871(d)	Bookseller	*Elizabeth CHATER*

NAME	BORN	ADMIT.	LEFT/DIED	TRADE	MARRIED
BOYD Alexander		1816	1831	Smith/Ironmonger	*Mary REID*
BOYD James (D)		1775	1834	Smith/Ironmonger	
CHATER Eliezar (DE)	1767	1783	1835(d)	Coal Merchant	
CHATER Joseph		1833	1838(d)	Stationer/Gentleman	*Hannah?*
CLARKE William (D)		1804			
COOLING William		1815		Tailor?	
CROFT Thomas	1770		1845(d)	Coalminer	
DEACON Daniel	1772	1796	1846(d)	Carrier	*Esther*
DEACON Edward		1856			
DEACON Francis		1801			
DEACON Henry		1838			
DEACON Samuel	1790	1840	1861(d)	Newsagent	
DEACON Thomas (D)		1845			
DOUGLAS William		1857	1871		*Isabella WHITELAW*
FARADAY James	1817	1848	1863/75(d)	Brassfounder/Gasfitter	*Margery Ann REID* *Margaret LEIGHTON*
FARADAY Michael (DE)	1791	1821	1867(d)	Scientist	*Sarah BARNARD*
FARADAY Robert (D)	1788	1821	1846(d)	Brassfounder/Gasfitter	*Margaret LEIGHTON*
FISHER David	1815	1837	1844/49(d)	Bootmaker	*Harriet LORIMER*
FOX Dorcas		1823			
FOX Thomas		1857			
GILES Albert		1838			*Jane GRAY*
GRAY Adam	1778	1810	1855(d)	Saddler	*Sarah Steele?* *Elizabeth FARADAY*
GROSVENER William (D)	1745	1765		Stationer	
HIPWELL John	1744	1770	1825(d)		*Ann?*
HUDDLESTONE James		1826	1850(d)?		*Mary?*
LAWTON Nathan		1805		Chair & Cabinet-maker	

NAME	BORN	ADMIT.	LEFT/DIED	TRADE	MARRIED
LEIGHTON C Blair	1823	1841	1855(d)	Artist/Lithographer	Caroline BOOSEY
LEIGHTON George (E)	1762	1782	1839(d)	Bookbinder	Abigail Egelton
LEIGHTON George C	1826	1845	1895(d)	Printer/Lithographer	Margaret FARADAY
LEIGHTON John (D)	1777?	1799	1857(d)	Printer	
LEIGHTON John (D)	1803	1832	1868(d)	Printer/Artist	Tilda DEACON
LEIGHTON Stephen (DE)	1797?	1823	1881(d)	Printer	Helen Blair
					another
LORIMER Anthony (D)		1827		Bookbinder	
MACC(R)OMIE Alex (D)		1816	1857		
MARTIN David W	1798	1821	1884(d)		Selina DEACON
MILLER George (D)		1824	1840(d)		
NEWLING Samuel		1783	1830(d)		
PARADISE George (D)		1838		Stationer	
PARADISE William (DE)	1788	1805	1843/66(d)		
PEAT Alexander		1826	1835		
RAITH George	1804	1835	1860(d)	Counting House Clerk	Elizabeth
REID Edward Ker (D)	1821		1885(d)	Silver Plate Maker	Ann BARNARD
SANDEMAN David	1814	1834	?/1887(d)	Company Secretary	Margaret Watt
SIMS Robert		1826		Writer?	Susan RIGBY
SMITH Godfrey		<1847		Servant (Linen-draper?)	
SPOONER Joshua		<1848			
STOPARD William	1773	<1844	1894	(Ex)Farmer	
THOMAS David		1824	1832		
THOMAS William		1824	1832		
VARLEY Cornelius	1781	1844	1847/73(d)	Artist/Inventor	Elizabeth STRAKER
VINCENT Alfred	1841	1861	1923(d)		Jessie DOUGLAS
					Sarah Jane Hobbs
					Janet Nicoll
VINCENT Benjamin (DE)	1819	1832	1899(d)	Librarian	Ellen BARNARD

NAME	BORN	ADMIT.	LEFT/DIED	TRADE	MARRIED
VINCENT Charles W	1837	1859	1864/1905(d)	Librarian/Chemist	Ann Ross BAXTER
					Florence Barrett
VINCENT Thomas	1810	1832	1883(d)		
VINCENT Thomas I (DE)	1840	1859	1890(d)	Bookbinder foreman	*Selina MARTIN*
VINCENT William R	1798	1821	1867(d)	Piano(?) maker	*Mary Jane (Fondell?)*
WALKER John		1821			
WASS John (E)	1751	1783	1831(d)		*Rebecca*
WHITELAW David	1845	1863	1879(d)	Publisher's Manager	*Hannah BAXTER*
WHITELAW George (DE)		1827	1872(d)	Publisher's Manager	*Ann BAYNES*
WHITELAW George B	1846	1863	1927(d)		*Emma BAXTER*
WHITELAW Scott	1832		1850	Cabinet-maker	Selina BAYNES
WILSON William	1801		1835(d)		Jane BOYD
YOUNG Arthur (D)	1816	<1858	1888(d)	Average Adjuster	*Agnes PEAT*

Notes:

In first column, D = Deacon, E = Elder.

Admit. = Date of admission to the London Meeting House.

Left = Date of exclusion, departure from London, or death (d).

Capitalised surname of wife, if she was a member of a leading Sandemanian family.

Wife's name italicised, if she was a Sandemanian.

APPENDIX C:
Sandemanian Scientists

Faraday was not the only Sandemanian who contributed to science, although the sum contributions by other members of the sect was small, but not insignificant, when compared with Faraday's. Our first two examples predate Faraday. Samuel Pike (1717–73), the Independent minister who joined the London meeting house in 1762, had earlier written an undogmatic account of John Hutchinson's natural philosophy entitled *Philosophia Sacra* (1757). Secondly, Dr Andrew Wilson (1718–92), an eminent physician, was for a short time connected with the London church and he published on both medicine and natural philosophy, the latter from a Hutchinsonian standpoint.

Among Faraday's contemporaries, the most famous was the painter and inventor Cornelius Varley (1781–1873), a member of the Royal Institution, who contributed a number of electrical and optical inventions and was awarded the Isis medal of the Society of Arts. Varley belonged to the sect for a little over 2 years, while the chemist Charles W. Vincent (1837–1905) was a member for nearly 5 years. Through Faraday's encouragement Vincent studied at the Royal College of Chemistry and he subsequently published his process for producing Burton brewing water (1878) and also edited *The yearbook of facts in science and art* (1874, etc.) and *Chemistry, theoretical, practical and analytical* (1877–9).

The Sandemanian whose scientific interests were closest to Faraday's was his brother-in-law George Buchanan (1790–1852), who was a land surveyor and civil engineer and became a Fellow of the Royal Society of Edinburgh. Educated at Edinburgh University, where he studied natural philosophy under John Leslie, Buchanan subsequently lectured on that subject to the Scottish School of Arts and published several papers on scientific and technical subjects. When Joseph Henry from America visited Faraday in 1837, he was introduced to Buchanan and spent some time with him in Edinburgh. Another Scottish Sandemanian, Dr John Crichton (1772–1860) from Dundee, became a respected surgeon and was particularly noted for his lithotomy operations. Finally, David James Blaikley (1846–1936), a younger member of the London church, published four papers on acoustics between 1878 and 1884.

Contributions to science were also made by several members of Sandemanian families who did not join the sect. Archibald Sandeman (1822–93) was the first professor of mathematics at Owens College, Manchester, and also published several introductory texts. A relative of Sandeman's, John James Waterston (1811–83), quietly produced a succession of scientific papers which were ignored by contemporaries. These, and particularly his researches into the kinetic theory of gases, have subsequently become recognised as profound contributions to science. Henry Deacon (1822–76), who was encouraged by Faraday, became a successful chemical manufacturer in Widnes, patented a number of chemical processes and published several papers, including two attacks on atomism that seem to have been inspired by Faraday's ideas. Various members of the Proctor family of Newcastle were likewise chemical manufacturers. Faraday also encouraged (Sir) William Fletcher Barrett (1844–1925), who began his scientific career as a laboratory assistant in the Royal Institution and later became Professor of Physics at the Royal College of Science in Dublin. While his research covered several topics, he became best-known, if not notorious, for his work on psychical research.

One scientist who has sometimes been identified as a Sandemanian is Alexander Tilloch (1759–1825), who filed several inventions, particularly related to printing, and ran the *Philosophical Magazine* and *Mechanic's Oracle* for a number of years. While Tilloch's theological writings were acceptable to some Sandemanians and he was clearly influenced by Glas and Sandeman, he did not belong to the London Sandemanian meeting house but worshipped at a neighbouring independent chapel.

Bibliographical Essay

The two most useful works dealing with the history and practices of the Sandemanians are the unpublished PhD dissertations by John T. Hornsby (*John Glas (1695–1773)*, University of Edinburgh, 1936) and Derek B. Murray (*The social and religious origins of Scottish non-Presbyterian Protestant dissent from 1730–1800*, University of St Andrews, 1976). Articles based on these dissertations were published in the *Records of the Scottish Church History Society* – Hornsby, 'The case of Mr John Glas', *ibid.*, 6 (1937), 115–37 and 'John Glas: His later life and work', *ibid.*, 7 (1940), 94–113; Murray, 'The influence of John Glas', *ibid.*, 22 (1984), 45–56. Lynn McMillon's *Restoration roots* (Dallas, 1983) is concerned with the rise of the Sandemanians and also their impact in America. The sect's history in America and Canada is also well served by J.M. Bailey and S.B. Hill, *History of Danbury, Conn. 1684–1896* (New York, 1896), pp.198–301; C.St.C. Stayner, 'The Sandemanian loyalists', *Collections of the Royal Nova Scotia Historical Society*, 29 (1951), 62–123; Jean F. Hankins, 'A different kind of loyalist: The Sandemanians of New England during the Revolutionary War', *New England Quarterly*, 60 (1987), 223–49.

The writings of John Glas are collected in *The works of Mr. John Glas* (five vols, 2nd ed., Perth, 1782), while most of Robert Sandeman's writings are to be found in his *Letters on Theron and Aspasio. Addressed to the author* (two vols, 4th ed., Edinburgh, 1803) and his *Discourses on passages of Scripture: With essays and letters* (Dundee, 1857). An important source for the rise of the Glasites is Anon., *An account of the life and character of Mr. John Glas, late minister of the Gospel at Tealing, near Dundee* (Edinburgh, 1813). Other early documents were published in *An epistolary correspondence between S.P. and R.S. with several additional letters, never printed before* (London, 1764); D. Mackintosh, ed., *Letters in correspondence by Robert Sandeman, John Glas, and their contemporaries* (Dundee, 1851); J. Morison, ed., *Supplementary volume of letters and other documents by John Glas, Robert Sandeman and their contemporaries* (Perth, 1865). The main collections of unpublished Sandemanian material are in the possession of Dundee University Library (MS 9), Mrs J.M. and Miss J. Ferguson and the Edinburgh meeting house. (Owing to the closure of this meeting house another location is currently being sought.) Two short but crucial works stating the Sandemanian position on a number of key topics are Anon., *The customs of the Churches of Christ as found in the New Testament* (Edinburgh, 1908) and *An account of the Christian practices observed by the church in Barnsbury Grove, Barnsbury, London, and other churches in fellowship with them* (London, n.d.). The text of the latter work, and probably the former, date from the eighteenth century.

Faraday's books comprise a very short list – *Experimental researches in physics and chemistry* (London, 1859); *Experimental researches in electricity* (three vols, London, 1839–55); *Chemical manipulation: Being instructions to students in chemistry*(London, 1827); *The chemical history of the candle* (London, 1861); *A course of six lectures on the various forces of matter and their relations to each other* (London, 1860). The first two items are collections of Faraday's main papers. Four of Faraday's exhortations have been published in J. R[orie], ed., *Selected exhortations delivered to various Churches of Christ by the late Michael Faraday, Wm. Buchanan, John M. Baxter, and Alex. Moir* (Dundee, 1910). Alan E. Jeffreys produced an invaluable bibliography of Faraday's publications – *Michael Faraday: A list of his lectures and published writings* (London, 1960). The reader should consult this list for Faraday's other publications. Faraday's manuscripts are widely scattered but the two main collections are deposited in the Royal Institution and the Institution of Electrical Engineers, London. Over 1,000 letters to and from Faraday appear in L. Pearce Williams, R. FitzGerald and O. Stallybrass, eds, *The selected correspondence of Michael Faraday* (two vols, Cambridge, 1971). A project to publish all known Faraday correspondence is now in progress under the direction of Frank James of the Royal Institution, and the first volume in the series is due to be published in 1991. Thomas Martin, ed., *Faraday's diary. Being the various philosophical notes of experimental investigations made by Michael Faraday, DCL, FRS, during the years 1820–1862 and bequeathed by him to the Royal Institution of Great Britain* (seven vols, London, 1932–6) is a mine of information about his laboratory research.

Faraday has been the subject of numerous biographies and biographical sketches. The major early book-length biographies are Henry Bence Jones, *The life and letters of Faraday* (two vols, London, 1870); John Tyndall *Faraday as a discoverer* (London, 1868); John Hall Gladstone, *Michael Faraday* (London, 1872); Walter Jerrold, *Michael Faraday: Man of science* (London, 1891); Sylvanus P. Thompson, *Michael Faraday, his life and work* (London, 1901). A spate of biographies appeared earlier this century, mainly to coincide with the centenary celebrations of 1931, but these were largely derived from their Victorian predecessors. The first modern scholarly biography of Faraday is L. Pearce Williams's *Michael Faraday: A biography* (London, 1965), which includes a detailed account of his scientific research. While all the preceding biographies discuss Faraday's religion, James F. Riley's *The hammer and the anvil: A background to Michael Faraday* (Clapham, 1954) discusses his family background and the family's connection with the Sandemanians. This book is supplemented by Riley's 'The faith of a scientist', *Scots Magazine*, 87 (1967), 214–7. Of the many other analyses of Faraday's religion, most are highly derivative, but see R.E.D. Clark, 'Michael Faraday on science & religion', *Hibbert Journal*, 65 (1967), 144–7, and Colin A. Russell, *Cross-currents: Interactions between science and faith* (Leicester, 1985), pp.256–65.

Of the many studies of Faraday's science, the following are particularly important. As the result of a conference held in 1984 a collection of eleven papers were published covering many aspects of his biography and scientific investigations – David Gooding and Frank A.J.L. James, eds., *Faraday rediscovered: Essays on the life and work of Michael Faraday, 1791–1867* (Basingstoke, 1985). Joseph Agassi offered a lively and provocative interpretation of Faraday in his *Faraday as a natural philosopher* (Chicago, 1971).

Faraday's work on electromagnetism is analysed in L. Pearce Williams's biography and S. Ross, 'The search for electromagnetic induction, 1820–1831', *Notes and Records of the Royal Society of London*, 20 (1965), 184–219. David

Gooding has written several insightful articles on different aspects of Faraday's natural philosophy, with particular emphasis on his magnetic and electromagnetic investigations – 'Conceptual and experimental bases of Faraday's denial of electrostatic action at a distance', *Studies in History and Philosophy of Science*, 9 (1978), 117–49; 'Metaphysics versus measurement: The conversion and conservation of force in Faraday's physics', *Annals of Science*, 37 (1980), 1–29; 'Faraday, Thomson, and the concept of the magnetic field', *British Journal for the History of Science*, 13 (1980), 91–120; 'Final steps to the field theory: Faraday's study of magnetic phenomena, 1845–1850', *Historical Studies in the Physical Sciences*, 11 (1981), 231–75; 'Empiricism in practice: Teleology, economy, and observation in Faraday's physics', *Isis*, 73 (1982), 46–67; 'A convergence of opinion on the divergence of lines: Faraday and Thomson's discussion of diamagnetism', *Notes and Records of the Royal Society of London*, 36 (1982), 243–59; '"In nature's school": Faraday as an experimentalist', in *Faraday rediscovered, op. cit.*; *The making of meaning* (Dordrecht and Boston, 1990).

On Faraday's electrochemical research, see S.M. Guralnick, 'The contexts of Faraday's electrochemical laws', *Isis*, 70 (1979), 59–75, and Frank A.J.L. James, 'Michael Faraday's first law of electrochemistry: How context develops new knowledge', in J.T. Stock and M.V. Orna, eds, *Electrochemistry, past and present* (Washington, 1989), pp.32–49. Other helpful analyses of Faraday's natural philosophy are offered by P.M. Heimann, 'Faraday's theories of matter and electricity', *British Journal for the History of Science*, 5 (1971), 235–57; Trevor H. Levere, 'Faraday, matter, and natural theology – reflections on an unpublished manuscript', *ibid.*, 4 (1968), 95–107; Levere, *Affinity and matter. Elements of chemical philosophy 1800–1865* (Oxford, 1971), pp.68–107; Nancy J. Nersessian, 'Faraday's field concept', in *Faraday rediscovered, op. cit.*; Frank A.J.L. James, 'Reality or rhetoric? Boscovichianism in Britain: The cases of Davy, Herschel and Faraday' (forthcoming).

There are several histories of the physical sciences in the nineteenth century. Useful overviews of the development of physics are provided by P.M. Harman in *Energy, force and matter. The conceptual development of nineteenth century physics* (Cambridge, 1982) and by Enrico Belloni in *The world on paper. Studies on the second scientific revolution* (Cambridge, Mass., 1980). A more detailed picture, especially of German physics, is provided in C. Jungnickel and R. McCormmach, *The intellectual mastery of nature. Theoretical physics from Ohm to Einstein* (two vols, Chicago, 1986). Although written nearly a century ago J.T. Merz's *A history of European thought in the nineteenth century* (Edinburgh and London, 1896-1914) is still well worth reading. Also recommended is D. Knight, *The age of science: The scientific world-view in the nineteenth century* (Oxford, 1986). There are also histories of various branches of physics such as D.S.L. Cardwell, *From Watt to Clausius. The rise of thermodynamics in the early industrial age* (London, 1971), G.N. Cantor, *Optics after Newton. Theories of light in Britain and Ireland, 1704–1840* (Manchester, 1983), and J.Z. Buchwald, *The rise of the wave theory of light* (Chicago, 1989).

For the institutional context, see Morris Berman's forcefully argued but problematic interpretation in *Social change and scientific organization. The Royal Institution, 1799–1844* (London, 1978), pp.156–86; Sophie Forgan, 'Faraday – From servant to savant: The institutional context', in *Faraday rediscovered, op.*

cit.; W.H. Brock, 'William Bollaert, Faraday and the Royal Institution', *Proceedings of the Royal Institution*, 42 (1968), 75–86. On the Royal Society see Marie Boas Hall, *All scientists now: The Royal Society in the nineteenth century* (Cambridge, 1984), and for the British Association see J. Morrell and A. Thackray, *Gentlemen of science: Early years of the British Association for the Advancement of Science* (Oxford, 1981).

Notes

The following abbreviations have been adopted throughout the Notes to refer to works frequently cited.

Correspondence	L. P. Williams, R. Fitzgerald and O. Stallybrass, eds, *The selected correspondence of Michael Faraday* (two vols., Cambridge, 1971). The reference *Correspondence*, will be followed by the page number.
Diary	T. Martin, ed., *Faraday's diary. Being the various philosophical notes of experimental investigation made by Michael Faraday, DCL, FRS, during the years 1820–1862 and bequeathed by him to the Royal Institution of Great Britain* (seven vols, London, 1932–6). References to the *Diary* will be followed by the volume number, in italics, and page number(s).
Dundee MS	The manuscripts of the Glasite church in Dundee University Library, followed by call number.
EMH MS	Manuscripts in the Edinburgh Glasite meeting house.
ERCP	M. Faraday, *Experimental researches in chemistry and physics* (London, 1859).
ERE	M. Faraday, *Experimental researches in electricity* (three vols, London, 1839–55). References to *ERE* will be followed by the volume number, in italics, and the page number(s).
Ferguson MS	Manuscripts in the possession of Mrs J.M. Ferguson and Miss Jean Ferguson, Edinburgh
Jones	H. Bence Jones, *The life and letters of Faraday* (two vols, 2nd ed., London, 1870). References to *Jones* will be followed by the volume number, in italics, and the page number(s).
RI MS	Royal Institution, Faraday manuscripts, followed by the call number.
Schoenbein	H.W.A. Kahlbaum and F.V. Darbishire, eds, *The letters of Faraday and Schoenbein, 1836–62* (Basle and London, 1899).
Supplementary volume	J. Morison, ed., *Supplementary volume of letters and other documents by John Glas, Robert Sandeman and their contemporaries* (Perth, 1865).

Both *ERCP* and *ERE* contain papers which were originally published in *Phil. Mag.*, *Phil. Trans.* and other locations. In such cases I have adopted the convention of referring only to *ERCP* or *ERE*. The title of the paper will be given and also, when appropriate, the original date of publication.

In citing published letters I refer to the published source rather than the manuscript location, giving preference to *Correspondence* over other publications.

NOTES FOR CHAPTER 1

1   D. Mackintosh, ed., *Letters in correspondence by Robert Sandeman, John Glas and their contemporaries* (Dundee, 1851), p.125; D.B. Murray, *The social and religious origins of Scottish non-Presbyterian Protestant dissent from 1730–1800* (Unpublished PhD dissertation, University of St Andrews, 1976), pp.76 and 109. The discrepancy between the two figures would be due to a large number of non-Sandemanians attending Sabbath services. In 1834 the Dundee membership was 163 (58 men and 105 women), whereas in 1855 the number had dropped to 102 (26 men and 76 women). Source: 'Diary of Nathaniel Bishop' and meeting house rolls in EMH MS.

2   *Dundee Courier and Argus*, 10 and 18 August 1863; *Dundee Advertiser*, 10 August 1863.

3   J. R[orie], ed., *Selected exhortations delivered to various Churches of Christ by the late Michael Faraday, Wm. Buchanan, John M. Baxter, and Alex Moir* (Dundee, 1910). Faraday's exhortation of 9 August 1863 appears on pp.28–37 and begins with the words, 'On this occasion, my dear brethren, when through the mercy of God we are assembled together, enjoying the great privilege of dwelling together in unity ...' Manuscript copies of this exhortation exist among the Dundee MS and EMH MS. The exhortation was transcribed by a lady in the audience who was learning shorthand and who practised with the sabbath sermons.

4   Faraday to Ada, Countess of Lovelace, 24 October 1844, *Jones*, 2, 191.

5   Anon., *The customs of the Churches of Christ as found in the New Testament* (Edinburgh, 1908), pp.18–19.

6   B. Hargrove, *The sociology of religion: Classical and contemporary approaches* (Arlington Heights, 1979).

7   The deacons and deaconesses play different roles. The other major asymmetry in respect to sex is that elders have to be male.

8   J.F. Riley, 'The faith of a scientist', *The Scots Magazine*, 87 (1967), 214–17. Quotation on pp.216–17.

9   L.P. Williams, *Michael Faraday: A biography* (London, 1965), pp.102–6; D. Gooding, 'Metaphysics versus measurement: The conversion and conservation of force in Faraday's physics', *Annals of Science*, 37 (1980), 1–29; id., 'Empiricism in practice: Teleology, economy, and observation in Faraday's physics', *Isis*, 73 (1982), 46–67; M. Berman, *Social change and scientific organization. The Royal Institution, 1799–1844* (London, 1978), pp.156–60.

10 J.F. Riley, *The hammer and the anvil: A background to Michael Faraday* (Clapham, 1954).

11 J.T. Hornsby, *John Glas (1695–1773)* (Unpublished PhD dissertation, University of Edinburgh, 1936); id., 'The case of John Glas', *Records of the Scottish Church History Society*, 6 (1937), 115–137; id., 'John Glas: His later

life and work', *ibid.*, 7 (1940), 94–113; Murray, *op. cit.* (n.1); *id.*, 'The influence of John Glas', *Records of the Scottish Church History Society*, 22 (1984), 45–56; L.A. McMillon, *Restoration roots* (Dallas, 1983); J.F. Hankins, 'A different kind of Loyalist: The Sandemanians of New England during the Revolutionary War', *The New England Quarterly*, 60 (1987), 223–49.

12 Faraday to Lovelace, *op. cit.* (n.4), pp.191–2.

NOTES FOR CHAPTER 2

1 S. Ross, 'Scientist: The story of a word', *Annals of Science*, 18 (1962), 65–85.

2 J. Morrell and A. Thackray, *Gentlemen of science: early years of the British Association for the Advancement of Science* (Oxford, 1981), pp.546–7.

3 E.P. Thompson, *The making of the English working class* (Harmondsworth, 1984), pp.887–915.

4 O. Chadwick, *The Victorian church* (two vols, London, 1966–70), vol.1, pp.24–47.

5 W.L. Mathieson, *English church reform: 1815–1840* (London, 1923), p.62.

6 Chadwick, *op. cit.* (n.4), vol.1, p.137.

7 Faraday to John Tyndall, 19 November 1850, *Correspondence*, 597.

8 EMH MS, 'Thomas J.F. Deacon's book', f.406, See proclamations of 6 and 10 February 1832 in the *London Gazette*.

9 For accounts of schism, see Anon., *An account of the life and character of Mr. John Glas, late minister of the gospel at Tealing, near Dundee* (Edinburgh, 1813); J.T. Hornsby, *John Glas (1695–1773)* (Unpublished PhD dissertation, University of Edinburgh, 1936); *Id.*, 'The case of Mr John Glas', *Records of the Scottish Church History Society*, 6 (1937), 115–137; D.B. Murray, *The social and religious origins of Scottish non-Presbyterian Protestant dissent from 1730–1800* (Unpublished PhD dissertation, University of St Andrews, 1976).

10 Anon., *op. cit.* (n.9), p.xv.

11 *Ibid.*, p.xvi. Emphasis added.

12 *Ibid.*, p.xxxviii.

13 *Ibid.*, p.xl. Emphasis added. Commenting on this incident, a later Sandemanian thus annotated his copy of *An account*: 'Ecclesiastical courts are, more cruel, and less scrupulous of right and wrong, than the civil power'. Copy in Edinburgh Glasite meeting house.

14 *Ibid.*, pp.ix-x. Original italicised. Cf. Genesis 16:12.

15 *Ibid.*, p.x.

16 J.T. Hornsby, 'John Glas: His later life and work', *Records of the Scottish Church History Society*, 7 (1940), 94–113, esp. 94–5; L.A. McMillon, *Restoration roots* (Dallas, 1983), pp.25–6.

17 See, for example, D. Mackintosh, ed., *Letters in correspondence by Robert Sandeman, John Glas, and their contemporaries* (Dundee, 1851) and *Supplementary volume*.

18 H. Penny, *Traditions of Perth* (Perth, 1836), pp.177–8; Hornsby, *op. cit.* (n.16), p.97.

19 See R. Sandeman, *Discourses on passages of Scripture: With essays and letters* (Dundee, 1857); J.M. Bailey and S.B. Hill, *History of Danbury, Conn. 1684–1896* (New York, 1896), pp.198–301; McMillon, *op. cit.* (n.16), pp.39–67.

20 *Ibid.*, pp.40–2.

21 Bailey and Hill, *op. cit.* (n.19), pp.198–301; Manuscripts in the 'Sandemanian' file and Stevens papers, Scott-Fanton Museum and Historical Society, Danbury.

22 EMH MS, 'Diary of Nathaniel Bishop', f.100.

23 Letter 'From the Church of Dundee to those sojourning in London', signed G.H. Baxter, 10 December 1844: EMH MS, 'Small notebook'.

24 J. Glas, *The testimony of the king of martyrs concerning his kingdom*, reprinted in *The works of Mr. John Glas* (five vols, Perth, 1782), vol.1, pp.1–183, quotation on p.75.

25 J. Barnard, *The nature and government of the Church of Christ* [1761], reprinted in *The Church of the living God* (Perth, 1855), pp.1–71, quotation on p.11.

26 *Ibid.*, pp.3–4.

27 Glas, *op. cit.* (n.24), p.65.

28 Barnard, *op. cit.* (n.25), pp.8–9.

29 [R. Sandeman], *Letters on Theron and Aspasio. Addressed to the author* (4th ed., two vols., Edinburgh, 1803), vol.1, p. 187.

30 *Ibid.*, p.32; Robert Sandeman to John Barnard, 29 August 1759: *Supplementary volume*, 33.

31 Barnard, *op. cit.* (n.25), p.23. Cf. 1 Corinthians 14:33.

32 Anon., *The customs of the Churches of Christ as found in the New Testament* (Edinburgh, 1908), p.12.

33 *Ibid.*, p.13.

34 *Ibid.*, pp.13–14.

35 John Alexander to Alexander Hutton, 15 July 1797: Dundee MS, 9/4/2(14).

36 Anon, 'Historical sketch of the rise, progress, and present state of independency in Scotland. No.III. Glassites, or Sandemanians', *London Christian Instructor*, 2 (1819), 144–9. Quotation on p.145.

37 C.A. Webster, 'John Glas and the Sandemanians' (mimeograph, Dundee, n.d.), p.13. This paper was read at the Dundee Reformed Fraternal on 26 August 1983 and contains a dire warning to present-day churchmen not to follow Glas's dangerous path.

38 D. Bogue and J. Bennett, *History of dissenters, from the Revolution in 1688, to the year 1808* (four vols, London, 1808–12), vol.4, pp.107–25.

39 I. Nicholson, *The substance of a sermon, delivered in Pell Street Chapel, in the month of June 1806, intended as an antidote against the virulent poison of the Sandemanian heresy, diffused in London by the Hibernian stranger* (London, 1806). Quotation forms part of subtitle.

40 See, for example, Barnard, *op. cit.* (n.25), p.11; Bogue and Bennett, *op. cit.* (n.38), p.109.

41 Faraday to Ada, Countess of Lovelace, 24 October 1844: *Jones*, 2, 191. First two emphases added.
42 R. Hall, *A history of Galashiels* (Galashiels, 1898), p.247.
43 J.E. Ritchie, *The religious life of London* (London, 1870), p.317.
44 For example, A.D. Gilbert, *Religion and society in industrial England: Church, chapel and social change, 1740–1914* (London and New York, 1976); K.S. Inglis, *Churches and the working class in Victorian England* (London and Toronto, 1983); H.J. Perkin, *The origins of modern English society 1780–1880* (London, 1969); W.R. Ward, *Religion and society in England 1790–1850* (London, 1972).

NOTES FOR CHAPTER 3

1 W. Wilson, *The history and antiquities of dissenting churches and meeting houses in London, Westminster, and Southwark; including the lives of their ministers from the rise of non-conformity to the present time* (four vols, London, 1810). See also the Wilson manuscripts in Dr Williams's Library, London.
2 *Ibid.*, vol.2, pp.85–98.
3 Robert Sandeman to Samuel Pike, 24 March 1759: *An epistolary correspondence between S.P. and R.S. with several additional letters, never printed before* (London, 1764), pp.130–1.
4 See Wilson, *op. cit.* (n.1), vol.3, pp.365–6.
5 Robert Sandeman to John Barnard, 19 July 1759: *Supplementary volume*, 23.
6 John Barnard to Robert Sandeman, 7 January 1760: *ibid.*, 39.
7 Robert Lyon to ?, 6 February 1781: Scott-Fanton Museum and Historical Society, Danbury, 'Sandemanians' file.
8 John Barnard to Robert Sandeman, 28 June 1766 and 20 June 1868: *Supplementary volume*, 63 and 65.
9 Membership information contained in roll book: EMH MS.
10 John Barnard to Robert Sandeman, 14 January 1769: *Supplementary volume*, p.69.
11 Lyon to ?: *op. cit.* (n.7).
12 Wilson, *op. cit.* (n.1), vol.3, pp.365–6.
13 *Ibid.*, p.276.
14 D. Maxwell, 'Some personal reminiscences', *Hull Literary Club Magazine*, 1 (1899), 144–5.
15 Religious census return, 1851: Public Record Office.
16 J. R[orie], *Letter on the differences which have arisen in the churches originally established by Mr. John Glas* (Dundee, 1886); 'Correspondence book': Dundee MS, 9/4/2(59); Various correspondence and rolls among EMH and Ferguson manuscripts.
17 R. Mudie-Smith, ed., *The religious life of London* (London, 1904), p.173, gives numbers of people entering the two meeting houses on an appointed Sunday. His figures make no distinction between communicants and auditors. Moreover, the numbers for morning and evening

worshippers are added, thus the census taker probably counted many of the same people twice.

18   I have not been able to discover the source of this sketch, which was published in W. Jerrold, *Michael Faraday: Man of science* (London, 1891), p.98. If it was produced specifically for this book, it was probably based on the artist's memory, since the Sandemanians left the building three decades earlier. The hall appears rather too large and the man in the foreground looks suspiciously like Faraday.

19   Anon., 'Paul's Alley, Barbican, 1695–1768', *Transactions of the Baptist Historical Society*, 4 (1915), 46–54; W. Wilson, 'Collection on dissenting ministers and churches': Dr Williams's Library, Ms.63.A, f.58.

20   J.E. Ritchie, *The religious life of London* (London, 1870), p.316; 'Memorandum': Islington Public Library, MsYB208.

21   Anon., 'Faraday memorial', *Electrical Review*, 59 (1906), 867; J. Kendall, *Michael Faraday: Man of simplicity* (London, 1955), p.172.

22   Acts 1:15 & 4:4. J. Barnard, *The nature and government of the Church of Christ* [1761], reprinted in *The Church of the living God* (Perth, 1855), p.9.

23   G.L. Sandeman, *The Sandeman genealogy* (Edinburgh, 1950), p.ix. This invaluable work is based on a manuscript by David Peat that was extended and published by John Glas Sandeman in 1895. The Sandeman coat of arms was first registered in 1780.

24   'Diary of Walter Barnard for 1860': In possession of Dr John Barnard, copy deposited with Faraday Correspondence Project at RI.

25   'A list of the London church established at Glovers Hall in the year 1762': EMH MS. Although this has proved an exceedingly useful document, some people who are known from other sources to have been members of the London church are omitted from the list. Moreover, the roll does not include the dates of death of many of the members and it is incomplete in other ways.

26   Faraday to Christian Schoenbein, 14 October 1856: *Schoenbein*, 274–5. On Charlotte Chater see 1871 census return for 175 Offord Rd., home of Vernor and Annie Barnard.

27   M.L. Barnard, 'Edward Barnard and Sons, goldsmiths and silversmiths of the City of London' (unpublished BA dissertation, Sheffield City Polytechnic, 1983), p.21. My account is largely based on this useful dissertation.

28   Cornelius Varley subsequently became a devout Baptist. See A.T. Story, *James Holmes and John Varley* (London, 1894), p.300.

29   Unlike some puritans, the Sandemanians do not proscribe the making of images and there were portraits of John Glas and other Sandemanian worthies in the Edinburgh meeting house.

30   A. Blaikley, 'Reminiscences of Alec Clydesdale', pp.56-7: Typescript deposited in the Royal Institution; R. Ormond, *Early Victorian portraits* (two vols, London, 1973), vol.1, pp.168–70.

31   D. and F. Irwin, *Scottish painters at home and abroad 1700–1900* (London, 1975), pp.203–4. None of Faraday's 1852 papers include in their titles the phrase 'magnetic lines of' as shown in Bell's portrait. It was probably meant to be 'On the lines of magnetic force', which Faraday presented at

the Royal Institution on 23 January and was subsequently published in condensed form in *Proceedings of the Royal Institution*, 1 (1851–4), 105–8 and *ERE*, *3*, 402–6.

32 Sarah Faraday to Edward Magrath, 14 August 1841: *Correspondence*, 393–4; G. Barnard, *Drawing from nature: A series of progressive instructions in sketching, to which are appended lectures on art delivered at Rugby School* (new ed., London, 1877).

33 A. Graves, *The Royal Academy of Arts: A complete dictionary of contributors and their work from its foundation in 1769 to 1904* (four vols, London, 1905). G.C. Williamson, ed., *Bryan's dictionary of painters and engravers* (three vols, London, 1930).

34 Faraday to Mr [John?] Leighton, 28 July 1828: Smithsonian Institution, Dibner Library, Ms Faraday 57; Faraday to John William Parker, 19 August 1831: *Correspondence*, 200. [Gerrit Moll], *On the alleged decline of science in England, by a foreigner* (London, 1831).

35 John Wass to Anon., 1795: EMH MS, 'Thomas J.F. Deacon's book', f.130.

36 S.P. Thompson, *Michael Faraday, his life and work* (London, 1901), p.286.

NOTES FOR CHAPTER 4

1 R.W. Thompson, *Benjamin Ingham the Yorkshire Evangelist* (Kendal, 1958); J. Burgess, *The lakes counties and Christianity: The religious history of Cumbria 1780–1920* (Carlisle, 1984), pp.84–7; D.F. Clarke, 'Benjamin Ingham (1712–1772), with special reference to his relations with the churches (Anglican, Methodist, Moravian and Glassite) of his time' (Unpublished MPhil dissertation, University of Leeds, 1971).

2 *Ibid.*; J.F. Riley, *The hammer and the anvil: A background to Michael Faraday* (Clapham, 1954), pp.11 and 17.

3 See works cited above and J. Alderson, 'James Allen of Gayle', *The Dalesman*, 41 (1979–80), 293–5.

4 Riley, *op. cit.* (n.2), p.17.

5 E.g., *ibid.*, pp.19–20.

6 Robert Lyon to ?, 6 February 1781: Scott-Fanton Museum and Historical Society, Danbury, 'Sandemanians' file.

7 Robert Ferrier to Mr Foster, 8 May 1790: EMH MS, 'Thomas J.F. Deacon's book', f.127. Despite this rejection of Allen, a number of his books were in the possession of the Edinburgh meeting house at the time of its closure.

8 Lyon to ?: *op. cit.* (n.6).

9 And not his eldest son as Riley (*op. cit.* (n.2)) assumes. See *Supplementary volume*, 108–9. A list in the EMH MS dated 1789 does not include either Margaret or Richard Faraday but only Michael's uncle, John Faraday.

10 The last seems unlikely, since Margaret's name does not appear on the 1789 roll cited in n.9.

11 G.M. Martin, *Dundee worthies. Reminiscences, games, amusements* (Dundee, 1934), pp.118–20; J.F. Riley, 'The faith of a scientist', *The Scots Magazine*, 87 (1967), 214–7.

12   'Then Peter said unto them, Repent, and be baptized every one of you in
     the name of Jesus Christ for the remission of sins, and ye shall receive the
     gift of the Holy Ghost. For the promise is unto you, and to your children,
     and to all that are afar off, even as many as the Lord our God shall call.'
     (Acts 2:38–9). See also J. M[orison], *Notes on Scriptural texts; or the
     testimony of two witnesses* (Perth, 1868), pp.33–40.

13   Ezra Stiles' list of the order of a Sandemanian service is reproduced in
     L.A. McMillon, *Restoration roots* (Dallas, 1983), pp.64–5. One visitor to
     the London meeting house c.1870 noted that the morning service
     commenced at 11am and was not over until at least 1.30pm: J.E.
     Ritchie, *The religious life of London* (London, 1870), p.316.

14   G.R. Sims, *My life. Sixty years' recollections of bohemian London* (London,
     1917), p.18.

15   Faraday to B. Abbott, 25 July 1817: *Correspondence*, 106.

16   'A class book for the reception of mental exercises instituted in July
     1818': RI MS. The members of this group were Michael Faraday, G. (or
     C.) Deeble, George Barnard, Thomas Deacon and J. Corder.

17   'London Marriages: St Augustine & St Faith': Genealogical Society,
     London, Ref. MX98R.

18   Anon., *The customs of the Churches of Christ as found in the New Testament*
     (Edinburgh, 1908), p.20.

19   EMH MS, 'Red ledger', f.47.

20   *Jones, 1*, 297.

21   *Customs, op. cit.* (n.18), p.13.

22   EMH MS, 'A list of the London church established at Glovers Hall in
     the year 1762'.

23   John J. Waterston to George Waterston, 11 July 1832: J.S. Haldane,
     'Memoir of J.J. Waterston', in Haldane, ed., *The collected scientific papers
     of John James Waterston* (Edinburgh and London, 1928), p.xxiii.

24   For example, I Timothy 3 and Titus 1.

25   *Customs, op. cit.* (n.18), pp.12–13.

26   J.H. Gladstone, *Michael Faraday* (London, 1872), p.35; see also S.P.
     Thompson, *Michael Faraday, his life and work* (London, 1901), p.297.

27   For a fuller discussion of the issues, see G. Cantor, 'Why was Faraday
     excluded from the Sandemanians in 1844?', *British Journal for the History
     of Science*, 22 (1989), 433–8.

28   Extracts from Queen Victoria's diary kindly provided by Miss P. Clark,
     Assistant Registrar, Royal Archives, Windsor Castle.

29   'List', *op. cit.* (n.22). Seven members were also separated from the Old
     Buckenham meeting house on 19 April 1844. EMH MS, 'Old Bucken-
     ham visitors' book'.

30   J.H. Gladstone, *Michael Faraday* (2nd ed., London, 1873), p.36.

31   Letter 'From the Church of Dundee to those sojourning in London',
     signed G.H. Baxter, 10 December 1844: EMH MS, 'Small notebook'.
     Was there any connection between these troubles in the Glasite church
     and the contemporary Disruption in the Church of Scotland?

32   R. Cree to ?, 4 June 1845: Ferguson MS.

33   Faraday to Christian Schoenbein, 12 April 1844: *Schoenbein*, 122–3.

34 See *The Referee*, 21 June 1891, p.7.

35 For responses, see *Daily Telegraph*, 3 July 1891; Anon., 'Sandemanianism from the sportsman's point of view', *The Clan*, 3 (1898), 24–6. See also *The Referee* for 28 June and an undated newspaper cutting in David Reid's 'Faraday' notebook, f.382: in possession of Mrs R. Brennand, Richmond, Yorkshire.

36 It appears that Thomas Boosey laid down the elder's office owing to some problem relating to his music publishing business. In his diary entry for 29 January 1860 Walter Barnard drily noted a connection with 'Chappel & Co's affair'. 'Diary of Walter Barnard for 1860': in possession of Dr John Barnard, copy deposited with Faraday Correspondence Project at the RI.

37 *An account of the Christian practices observed by the church in Barnsbury Grove, Barnsbury, London, and other churches in fellowship with them, in a letter to a friend* (London, n.d.), p.4. This pamphlet is a modern edition of a work of a similar title by Samuel Pike, first published in 1766.

38 *Ibid.*, p.5.

39 Four exhortations are published in James Rorie's *Selected exhortations delivered to various Churches of Christ by the late Michael Faraday, Wm. Buchanan, John M. Baxter, and Alex. Moir* (Dundee, 1910), and several copies of these exhortations exist, e.g., Dundee, MS 9/1/4 and in EMH MS. Faraday outlined his exhortations on small cards, noting the scriptural passages he intended to use. Some of these cards are in RI MS and one in the collection of Dr Philip Embleton of Aldeburgh, Suffolk. An example is displayed in *Jones*, 2, 101.

40 C.M. Davies, *Unorthodox London* (new ed., London, 1876), p.172.

41 *Jones*, 2, 99–100.

42 *Ibid.*, 100; W. Jerrold, *Michael Faraday: Man of science* (London, 1891), p.100.

43 Faraday to Benjamin Vincent, 13 August 1852: RI MS, C-37.

44 EMH MS, 'Old Buckenham visitors' book', f.74, contains the certificate of baptism of Alice Louise Loveday at Old Buckenham on 31 August 61; see also f.71. I have been unable to locate the London baptismal register.

45 John Boosey to ?, 23 October 1790: EMH MS, 'Thomas J.F. Deacon's book', ff.326–36.

46 EMH MS, 'Old Buckenham rolls'.

47 Caroline Reid's notes: EMH MS, 'Thomas J.F. Deacon's book', ff.340–50.

48 For example, Faraday to Benjamin Vincent, 27 July 1860: RI MS, E-22a; Faraday to William Buchanan, 29 July 1840: Ferguson MS.

49 Faraday to [David?] Greig, 31 August 1847: RI MS, C-21.

50 'Mr Faraday has not been so well the last month he really requires the relaxation he is about to obtain for a short time at Brighton soon'. Thomas Boosey to ?, 1 February 1850: Ferguson MS.

51 Faraday to Buchanan, *op. cit.* (n.48).

52 Faraday to Benjamin Vincent, 30 August 1861: RI MS, E-29.

53 Anon., 'A Faraday memorial in London', newspaper cutting c.1894: In possession of Dr Philip Embleton, Aldeburgh, Suffolk.

54   T. Bassett to the London church, [early 1855]: Ferguson MS.

55   'The Church of Jesus Christ at London to the Church of Jesus Christ sojourning at Dundee', 7 November 1854: copies in EMH MS and Ferguson MS.

56   Alexander Moir to William Fife, 3 and 26 December 1854: Dundee MS, 9/3(36a&b).

57   'The Church of Christ in London to the Church of Christ at Dundee', 3 and 31 January 1855: Ferguson MS.

58   Faraday to Benjamin Vincent, 12 October 1860: *Correspondence*, 974–5.

59   Faraday to Benjamin Vincent, n.d.: RI MS, G-22.

60   Faraday to Benjamin Vincent, 30 August 1861: RI MS, E-29.

61   Faraday to Benjamin Vincent, 7 August 1863: *Correspondence*, 1016–7.

62   Rorie, *op. cit.* (n.39), p.18. Emphasis added. 'Jesus said unto her, I am the resurrection, and the life: he that believeth in me, though he were dead, yet shall he live: And whosoever liveth and believeth in me shall never die. Believeth thou this?': John 11:25–6.

63   Faraday to Benjamin Vincent, 25 August 1863: *Correspondence*, 1018.

64   Faraday to William Buchanan, 2 May 1850: Ferguson MS.

65   Faraday to Benjamin Vincent, 26 May 1846: RI MS, C-14.

66   Faraday to Thomas Winkworth, n.d. [1824]: Wellcome Institute Library, London, MS 76585.

67   A. Blaikley, 'Reminiscences of Alec Clydesdale', p.82: Typescipt deposited in the Royal Institution.

68   F. Greenaway, M. Berman, S. Forgan and D. Chilton, eds, *Archives of the Royal Institution: Minutes of the managers' meetings, 1799–1903* (fifteen vols, London, 1971-6), 10, 95.

69   Faraday to Schoenbein, 8 August 1843: *Schoenbein*, 112. Schoenbein's book was entitled *Mittheilungen aus dem Reisetagebuche eines Deutschen Naturforschers* (Basel, 1842).

70   Obituary notice in *The Times*, 5 May 1899, p.10.

71   Benjamin Vincent to Schoenbein, 3 July 1843: Offentliche Bibliothek der Universitat Basel, Nachlass Schoenbein, IA 839; Faraday to Schoenbein, 9 December 1850: *Schoenbein*, 190; Faraday to Schoenbein, 6 April 1855: *ibid.*, 243–4.

72   *Minutes, op. cit.* (n.68), 10, 144.

73   *Ibid.*, 12, 239. See also fragment from Faraday's will dated 16 January 1855: University of Newcastle upon Tyne, Special Collections.

74   *Ibid.*, 12, 243. Vincent's second wife Ellen was a niece of Faraday's and the daughter of Edward Barnard, jnr.

75   Obituary, *op. cit.* (n.70).

76   B. Vincent, *A new classified catalogue of the library of the Royal Institution of Great Britain* (two vols, London, 1857–82); *Id.*, *A dictionary of biography, past and present, brought down to September, 1877. Containing the lives of eminent persons of all ages and nations* (London, 1877); *Id.*, *Haydn's dictionary of dates and universal information relating to all ages and nations* (12th ed., London, 1892).

77   S. Forgan, 'Faraday – From servant to savant: The institutional context', in D. Gooding and F.A.J.L. James, eds, *Faraday rediscovered. Essays on the*

life and work of Michael Faraday, 1791–1867 (Basingstoke and New York, 1985), pp.51–67.

78   *Minutes, op. cit.* (n.68), 14, 170–1.

79   *Ibid.*, 10, 152.

80   *Ibid.*, 10, 327–9.

81   *Ibid.*, 11, 62 and 97.

82   *Ibid.*, 11, 198 and 203.

83   C.W. Vincent, *Burton brewing water, a process invented by C.W.V.* (Leeds, 1878); *Id.*, ed., *The year-book of facts in science and art* (1874, etc.); *Id.*, ed., *Chemistry, theoretical, practical, and analytical, as applied to the arts and manufacturers. By writers of eminence* (two vols, London 1877–9); *Id.*, *Catalogue of the library of the Reform Club* (London, 1883; enlarged, 1884).

84   H. Woodbridge, *The Reform Club 1836–1978. A history from the Club's records* (London, 1978), p.110.

85   *Minutes, op. cit.* (n.68), 11, 314–6.

86   G. Caroe, *The Royal Institution: An informal history* (London, 1985), p.106.

87   See *Proceedings of the Royal Institution*, 26 (1929–31), 136–7; 31 (1939–41), 12.

88   *Minutes, op. cit.* (n.68), 15, 57, 61 and 357.

89   *Ibid.*, 13, 198; Faraday to Benjamin Vincent, 25 August 1857: RI MS, E-4.

90   Works by Sandemanians and members of Sandemanian families currently in the Library of the Royal Institution include R. Sandeman, *Grand architectural panorama of London: Regent Street to Westminster Abbey* (London, 1849); P. Sandeman *Meteorological observations* (Greenock, 1851); C.W. Vincent, *Chemistry, theoretical, practical, and analytical, as applied to the arts and manufactures* (two vols, London, 1877–9).

91   *Minutes, op. cit.* (n.68), 9, 52.

92   For example, *ibid.*, 10, 202; 11, 86 and 241; 12, 84; 13, 324, 348 and 354; 15, 4.

93   Thomas Deacon to the Plumstead Peculiars, 18 January 1882: EMH MS, 'Thomas J.F. Deacon's book', ff.14-17, Biblical passage cited is James 5:14.

94   Faraday to William Buchanan, 29 July 1840: Ferguson MS.

95   Faraday to [Edward Magrath?], 22 March 1838: *Correspondence*, 317.

96   Faraday to William Buchanan, 26 August 1846: Ferguson MS (cf. Job 1:21); Faraday to [Margery Ann?] Reid, 1 January 1857: *Jones*, 2, 293.

97   Faraday to William Buchanan, 16 August 1848: Ferguson MS.

98   [Robert Sandeman], *Letters on Theron and Aspasio. Addressed to the author* (4th ed., two vols, Edinburgh, 1803), vol.2, p.29.

99   Faraday to Caroline Deacon, 12 August 1859: *Correspondence*, 925–6; See also Faraday to [Harriet?] Moore, 14 August 1861: *Ibid.*, 1000.

100   Faraday to Buchanan, *op. cit.* (n.96).

101   Faraday to Auguste de la Rive, 5 June 1851: *Correspondence*, 634–5.

102   Faraday to Schoenbein, 6 November 1855: *Schoenbein*, 252.

103   Faraday to Auguste de la Rive, 19 September 1861: *Correspondence*, 1001.

104   Faraday to Comte de Paris, 7 February 1865: *Jones*, 2, 471.

105   Jane Barnard to William Barratt, 28 July 1866: Royal Society, Ms Bs.19.

106   *Ibid.*; Jane Barnard to Henry Bence Jones, 25 November 1866: RI MS, FII; Barnard to Jones, 27 December 1866: RI MS, FII; Margery Ann Reid to Henry Bence Jones, 22 April 1867: *Jones*, 2, 473–5.

107   Jane Barnard to Henry Bence Jones, 26 August 1867: *Jones*, 2, 475; Jane Barnard to Miss Crum, 2 October 1867: Strathclyde Regional Archives, Thomson Papers.

108   J.S. Curl, *The Victorian celebration of death* (Newton Abbot, 1972).

109   D. Whitelaw, *A bonfire of leaves* (London, 1936), p.8.

110   *Jones*, 2, 479.

111   Sarah Faraday to Henry Bence Jones, 22 November 1867: RI MS.

112   Will of Michael Faraday: Principal Registry, Family Division of the High Court, Somerset House.

113   Jane Barnard to Henry Bence Jones, 3 September 1867: *Jones*, 2, 476; Barnard to Crum: *op. cit.* (n.107).

114   Barnard to Jones: *op. cit.* (n.113).

115   W.F. Pollock, *Personal remembrances of Sir Frederick Pollock, second Baronet, sometimes Queen's Remembrancer* (two vols, London, 1887), vol. 2, p.207; *The Times*, 22 June 1869, p.5.

116   Royal Institution, Managers' minutes, vol. 20, 2 March and 6 July 1831.

NOTES FOR CHAPTER 5

1   L.P. Williams, *Michael Faraday: A biography* (London, 1965), pp.357–9; M. Berman, *Social change and scientific organization. The Royal Institution, 1799–1844* (London, 1978), pp.173–86.

2   Faraday to William Whewell, 7 November 1848: *Correspondence*, 528–9.

3   J. Tyndall, *Faraday as a discoverer* (5th ed., London, 1894), pp.176–7. As a non-proselytising sect, the Sandemanians had no reason to publicise their religious views. Religion is a matter of the spirit and human rhetoric plays no role. Moreover, to place their views before others would only invite wasteful disputes of the kind that littered the history of Christianity.

4   'Commonplace book', ff.384–5: Institution of Electrical Engineers, Faraday Papers.

5   Faraday to Elizabeth Faraday, 21 December, 1814: *Jones*, 1, 153–4. Emphasis added.

6   'Travel journal', f.43: Institution of Electrical Engineers, Faraday Papers.

7   *Jones*, 2, 426–8.

8   Faraday to Schoenbein, 24 February 1859: *Schoenbein*, 323–4.

9   W.F. Pollock, *Personal remembrances of Sir Frederick Pollock, second Baronet, sometimes Queen's Remembrancer* (two vols, London, 1887), vol.2, p.207; Dean Milman to Faraday, 3 November 1852: Royal Institution, Conybeare MS, f.94.

10   Faraday to W.H.M. Christie, 25 January 1865: *Jones*, 2, 464. See also Jane Barnard to Henry Bence Jones, 3 September 1867: *ibid.*, 476.

11   J. Glas, *The testimony of the king of martyrs concerning his kingdom*, in *The works of Mr. John Glas* (2nd ed., five vols, Perth, 1782), vol.1, pp.1–183. Quotation on p.85.

12 *Ibid.*, pp.92–5. Emphasis added.

13 Williams insightfully opens his discussion of Faraday's political views by referring to this scriptural passage: *op. cit.* (n.1), p.357.

14 James Morison to Patrick Cochran, 19 December 1830: Dundee MS, 9/3(20). See also James Morison to Alex Hutton, 28 May 1794: Ferguson MS.

15 William Buchanan to Nathanial Bishop, [1848]: Scott-Fanton Museum and Historical Society, Stevens Papers. Emphasis added.

16 W. Norrie, *Dundee celebrities of the nineteenth century: Being a series of biographies of distinguished or noted persons connected by birth, residence, official appointment, or otherwise, with the town of Dundee; and who have died during the present century* (Dundee, 1873), pp.182–4.

17 Obituary of William Buchanan, *Edinburgh Courant*, 21 December 1863.

18 *An account of the Christian practices observed by the church in Barnsbury Grove, Barnsbury, London, and other churches in fellowship with them* (London, n.d.), p.8.

19 Glas, *op. cit.* (n.11), p.115.

20 M. Tait and W. Forbes Gray, 'George Square. Annals of an Edinburgh locality, 1766–1926. From authentic records', *The Book of the Old Edinburgh Club*, 26 (1948), 45; E.C. Anderson, *Christian songs and occasional verses* (Edinburgh, 1903).

21 J.F. Hankins, 'A different kind of Loyalist: The Sandemanians of New England during the Revolutionary War', *New England Quarterly*, 60 (1987), 223–49. Quotation on p.237.

22 C. Rossiter, 'Conservatism', in David L. Sills, ed., *International Encyclopedia of the Social Sciences* (eighteen vols, New York and London, 1968–79), vol.3, pp.290–5. Thatcherism is conservative in that it takes Victorian Britain as its model, but the Thatcherite Revolution is not conservative in this traditional sense.

23 E.P. Thompson, *The making of the English working class* (Harmondsworth, 1984), pp.28–58, esp. pp.35 and 55.

24 M. Chase, *'The people's farm': English radical agrarianism, 1775–1840* (Oxford, 1988), pp.39–42; E. Mackenzie, *A descriptive and historical account of the town and county of Newcastle upon Tyne, including the Borough of Gateshead* (two vols, Newcastle-upon-Tyne, 1827), vol.2, pp.399–402. There are various rolls relating to the Newcastle meeting house among EMH MS.

25 P.H. Marshall, *William Godwin* (New Haven, 1984), pp.23–9. Marshall's claim that the Sandemanians practised a form of communism is over-stated. There was no properly constituted Sandemanian church in Norwich and Samuel Newton's name does not appear on any of the rolls I have been able to locate. He was probably one of many clergymen who were influenced by Sandeman's writings but did not submit to the discipline of the Sandemanian church. A far more accurate account of Godwin's relationship with the Sandemanians is given in D. Locke, *A fantasy of reason: The life and thought of William Godwin* (London, 1980).

26 *Jones*, *1*, 197.

27 Faraday to Auguste de la Rive, 9 July 1849: *Correspondence*, 558.

28 Faraday to Jean-Baptiste Dumas, 18 June 1849: *Correspondence*, 555.

29 Faraday to Schoenbein, 15 December 1848: *Schoenbein*, 182–3.

30 Faraday to de la Rive, *op. cit.* (n.27).

31 Faraday to Jean-Baptiste Dumas, 5 June 1849: *Correspondence*, 552. Emphasis added.

32 'Commonplace book', ff.391, 404 and 434: *op. cit.* (n.4).

33 Berman, *op. cit.* (n.1), pp.173-4. Cf. C.A. Russell, *Cross-currents. Interactions between science and faith* (Leicester, 1985), p.262.

34 *Report of H.M. Commissioners appointed to inquire into the revenues and management of certain colleges and schools and the studies pursued and instruction given therein; with appendix and evidence, Parliamentary Papers*, (1864), [3288], vol.4, p.377.

35 T.H. Thornton, *Colonel Sir Robert Sandeman: His life and work on our Indian frontier. A memoir with selections from his correspondence and official writings* (London, 1895).

36 Charles Grey to Faraday, 17 November 1850: Collection of Dr P. Embleton, Aldeburgh, Suffolk; Faraday to Charles Grey, 10 June 1854: Windsor Castle, Royal Archives; Faraday to Prince of Wales, 18 January 1856: *ibid.*; Faraday to Charles Phipps, 1 January 1858; *ibid.*; Faraday to Ernst Becker, 20 April and 5 May 1858: *Jones*, 2, 393-4; Faraday to Prince of Wales, 16 January 1856: *ibid.*, 369; Faraday to Prince of Wales, 5 January 1863: *ibid.*, 456-7. Invitation to a ball at Buckingham Palace, 10 July 1850: 'David Reid's collection', in possession of Mrs R. Brennand, Richmond, Yorkshire.

37 Faraday to Auguste de la Rive, 7 April 1855: *Correspondence*, 790.

38 Faraday to William Vernon Harcourt, 24 October 1840: E.W. Harcourt, ed., *The Harcourt papers* (fourteen vols, Oxford, 1880–1905), vol.14, pp.96-7.

39 F. Greenaway, M. Berman, S. Forgan and D. Chilton, eds, *Archives of the Royal Institution: Minutes of the managers' meetings, 1799–1903* (fifteen vols, London, 1971–6), 10, 184. See also fragment from Faraday's will dated 16 January 1855: University of Newcastle upon Tyne, Special Collections.

40 Faraday to William Francis Cowper, 3 August 1860: *Correspondence*, 958–9. The name of the applicant is given as Davis (?) by L.P. Williams and as Daines in *Jones*, 2, 430.

41 Faraday to Lord Wrottesley, 10 March 1854: *Correspondence*, 724–5. Cf. Esther 6:6.

42 Faraday to William Jordan, 15 August 1837: *ibid.*, 313; Faraday to Charles R. Weld, 8 January 1848: *ibid.*, 513.

43 Faraday to Thomas Andrews, 2 February 1843: *ibid.*, 409–10; D. Knight, 'Davy and Faraday: Fathers and sons', in D. Gooding and F.A.J.L. James, eds, *Faraday rediscovered: Essays on the life and work of Michael Faraday, 1791–1867* (Basingstoke and New York, 1985), pp.32–49.

44 *Jones*, 2, 97–8; Faraday to Robert S. Mackenzie, 5 October 1844: *Correspondence*, 421–2; Faraday to the Emperor of France, 19 January 1856: *ibid.*, 827; 'Commonplace book', ff.432 & 448: *op. cit.* (n.4).

45 RI MS. See also G.M. Prescott, 'Faraday: Image of the man and the collector', in *Faraday rediscovered, op. cit.* (n.43), pp.14–41.

46 Faraday to Thomas Spring-Rice, 1 December 1837: *Jones*, 2, 96.

47 Anon., 'Shewing how the Tories and the Whigs extend their patronage to science and literature', *Fraser's Magazine*, 12 (1835), 703–9. Quotation on p.707.

48 *Jones*, 2, 56–64.

49 [W. Maginn], 'Michael Faraday, F.R.S., Hon.D.L.C. Oxon, etc. etc.', *Fraser's Magazine*, 13 (1836), 224.

50 M. Weber, *The Protestant ethic and the spirit of capitalism* (London, 1948); J. Wesley, 'The use of money', in F. Baker, ed., *The bicentennial edition of the works of John Wesley: Sermons* (four vols, Nashville, 1984–7), vol.2, pp.263–80.

51 Berman, *op. cit.* (n.1), p.156.

52 R. Sandeman, *Discourses on passages of Scripture: With essays and letters* (Dundee, 1857), p.103.

53 [R. Sandeman], *Letters on Theron and Aspasio. Addressed to the author,* (4th ed., two vols, Edinburgh, 1803), vol.2, p.224.

54 Robert Sandeman to John Barnard, 19 July 1759: *Supplementary volume,* 22.

55 'Sandemanians', in [J. Morison], *A new theological dictionary intended to exhibit a clear and satisfactory view of every religious term and denomination, which has prevailed in the world from the birth of Christ to the present* (Edinburgh and Perth, 1807), pp.788–94. Quotation on p.789.

56 *Ibid.; Account, op. cit.* (n.18), p.8.

57 The church at Kirkby Lonsdale to the church at Perth, 6 October 1919: Dundee MS, 9/3/9.

58 John Barnard to William Sandeman, 4 July 1770: *Supplementary volume,* 70–1.

59 Robert Lyon to ?, 6 February 1781: Scott-Fanton Museum and Historical Society, Danbury, 'Sandemanians' file.

60 'Diary of Walter Barnard for 1860', entry for 29 January: in possession of Dr John Barnard, copy deposited with Faraday Correspondence Project at the RI.

61 EMH MS, 'Diary of Nathanial Bishop', f.35:.

62 Faraday to William Smith, 3 January 1859: *Jones, 2,* 417–18.

63 Faraday to John Henry Pelly, 3 February 1836: *ibid.,* 90–1; Faraday to John Tyndall, 15 November 1859: *ibid,* 421.

64 S. Forgan, 'Faraday – From servant to savant: The institutional context', *Faraday rediscovered, op. cit.* (n.43), pp.51–67.

65 Faraday to Percy Drummond, 29 June 1829: *Correspondence,* 175–7; Charles W. Pasley to Percy Drummond, 25 May 1829: cited in F.A.J.L. James, 'Davy and Faraday as military scientists' (unpublished typescript).

66 Tyndall, *op. cit.* (n.3), pp.180–2.

67 S.P. Thompson, *Michael Faraday, his life and work* (London, 1901), p.286.

68 Faraday to Edward Magrath, 6 December 1851: *Correspondence,* 645.

69   C.W. Smith and M.N. Wise, *Energy and empire: A biographical study of Lord Kelvin* (Cambridge, 1989).

70   Faraday to Sarah, 21 July 1822: *Jones*, *1*, 323–6. Quotation on p.324.

71   Faraday to W. Wright, 11 March 1856: *Notes and Queries*, 11 (1873), 73.

72   Faraday to John F.W. Herschel, 10 November 1832: *Correspondence*, 235; Faraday to John Forbes, 29 December 1841: Linnaean Society; See also Berman, *op. cit.* (n.1), p.174.

73   Faraday to André-Marie Ampère, 17 November 1825: *Correspondence*, 153–4; Faraday to Richard Phillips, 29 November 1831: *ibid.*, 209–210.

74   Charles Lyell to Henry Bence Jones, 22 May 1868: RI MS; Faraday to John Rennie, 26 February 1835: *Correspondence*, 290.

75   Faraday to Harriet Moore, 24 August 1850: *Jones*, *2*, 251.

76   J.H. Gladstone, *Michael Faraday* (2nd ed., London, 1873), p.90.

77   Faraday to J.D. Thompson, 3 July 1832: *Correspondence*, 229. Emphasis added.

78   Faraday to Richard Phillips, 29 August 1828: *ibid.*, 172–3.

79   Faraday, 'Observations on the inertia of the mind': *Jones*, *1*, 230–44. Quotation on p.230.

80   [Sandeman], *op. cit.* (n.53), vol.2, p.29.

81   'Observations', *op. cit.* (n.79), p.240.

82   Sandeman, *op. cit.* (n.52), pp.140–6.

83   Anon., *The customs of the Churches of Christ as found in the New Testament* (Edinburgh, 1908), p.10

84   'I am bound to put it on record that, if their worship does not belie them, the Sandemanians must be the most dismal people on earth.' C.M. Davies, *Unorthodox London* (new ed., London, 1876), p.173. See also J.E. Ritchie, *The religious life of London* (London, 1870), pp.313–320.

85   Faraday to Benjamin Vincent, 30 August 1861: RI MS E-29; Tait and Forbes Gray, *op. cit.* (n.20); Anderson, *op. cit.* (n.20); *Christian songs: In two parts; to which is prefixed the evidence and import of Christ's Resurrection, versified* (14th ed., Perth, 1872).

86   *Jones*, *1*, 383; Tyndall, *op. cit.* (n.3), p.89; W.C. Macready to Faraday, 6 October n.y.: Royal Institution, Conybeare MS, f.165; Faraday to Benjamin Abbott, 25 July 1817: *Correspondence*, 106–7.

87   C.A.H. Crosse, 'Science and society in the fifties', *Temple Bar*, 93 (1891), 33–51. Quotation on p.34. Pollock, *op. cit.* (n.9), vol.1, p.245.

88   Tyndall, *op. cit.* (n.3), p.176.

89   [Sandeman], *op. cit.* (n.53), vol.1, pp.87–9.

90   M. L[loyd], *Sunny memories, containing personal recollections of some celebrated characters* (two vols, London, 1879-80), vol.1, p.66.

91   *Customs*, *op. cit.* (n.83), p.17.

92   *Jones*, *2*, 478.

93   L[loyd], *op. cit.* (n.90), p.65.

94   Faraday, *A course of six lectures on the various forces of matter and their relation to each other* (London, 1860), p.2; *Report*, *op. cit.* (n.34), p.377.

95   Sarah to Henry Bence Jones, 31 May 1864: RI MS, F11.

96   Forgan, *op. cit.* (n.64), p.61.

97   *Ibid.*, p.59.

98   Faraday to Dionysius Lardner, 6 October 1827: *Correspondence*, 168–9; See also Faraday to Richard Phillips, 29 August 1828: *ibid.*, 172–3.
99   James, *op. cit.* (n.65).

NOTES FOR CHAPTER 6

1   L.P. Williams, *Michael Faraday: A biography* (London, 1965), p.357.
2   Faraday to John A. Paris, 23 December 1829: *Correspondence*, 177–8. Cf. William R. Grove's claim that 'philosophers are not what I once thought them & that the cultivation of science by no means necessarily tends to improve the moral qualities'. He proceeded to complain that there was more intrigue, baseness and selfishness in science than in the legal profession. Grove to Christian Schoenbein, 8 October 1857: Offentliche Bibliothek der Universitat Basel, Nachlass Schonbein IA, 511.
3   *Jones, 1*, 189.
4   Faraday to Paris: *op. cit.* (n.2), 48.
5   Faraday to Benjamin Abbott, 9 February 1816; 23 September 1816; 27 April 1819: *Correspondence*, 95–8; 98–9; 113.
6   Faraday to Edward Magrath, 30 July 1834: *ibid.*, 278.
7   Faraday to Thomas Byam Martin, 26 August 1854: *ibid.*, 757; Faraday to Arthur Young, 3 January 1855: Wellcome Institute Library, Ms 67430.
8   Faraday to Charles Babbage, 3 June 1851: *Correspondence*, 632.
9   Faraday to Schoenbein, 18 February 1843: *Schoenbein*, 101.
10   Robert Sandeman to John Barnard, 19 July 1759: *Supplementary volume*, 22.
11   Faraday to John G. Macvicar, 29 December 1856: *Correspondence*, 859:
12   F. Bacon, *Novum organum* in J. Spedding, R.L. Ellis and D.D. Heath, eds, *The works of Francis Bacon* (fourteen vols, London, 1872–4), vol.8, p.210; Faraday to Whewell, 7 November 1848: *Correspondence*, 528.
13   Faraday to André-Marie Ampère, 17 November 1825: *ibid.*, 153–4.
14   Faraday to Carlo Matteucci, 3 March 1853: *Jones, 2*, 314-6.
15   *Ibid.*
16   Faraday to ?, August 1861: *Jones, 2*, 442.
17   A.S. Eve and C.H. Creasey, *Life and work of John Tyndall* (London, 1945), pp.32 and 40.
18   Faraday to John Tyndall, 19 November 1850: *Correspondence*, 597.
19   Faraday to Julius Plücker, 8 April, 1856: *ibid.*, 833–5. Emphasis added.
20   Faraday to Auguste de la Rive, 7 April 1855: *ibid.*, 790–1; Faraday to Benjamin Abbott, 11 October 1812: *ibid.*, 37–9. Cf. Faraday to Schoenbein, 6 November 1855, 22 April 1862: *Schoenbein*, 252–5 and 355; Eve and Creasey, *op. cit.* (n.17), p.43.
21   Faraday to Hans C. Oersted, 15 March 1850: *Correspondence*, 581.
22   *Jones, 2*, 86.
23   Faraday, 'Observations on the inertia of the mind': *Jones, 1*, 230–44; Cf. Faraday to Schoenbein: *op. cit.* (n.9).
24   Faraday to Elizabeth Faraday, 1 July 1814: *Jones, 1*, 130–1.

25   D.P. Miller, 'Between hostile camps: Sir Humphry Davy's presidency of the Royal Society of London, 1820–1827', *British Journal for the History of Science*, 16 (1983), 1–47; M.J.S. Rudwick, *The great Devonian controversy: The shaping of scientific knowledge among gentlemanly specialists* (Chicago and London, 1985).

26   J. Morrell and A. Thackray, *Gentlemen of science: Early years of the British Association for the Advancement of Science* (Oxford, 1981), p.546.

27   On London science, see I. Inkster and J. Morrell, eds, *Metropolis and province: Science in British culture, 1780–1850* (London, 1983); M. Berman, *Social change and scientific organization. The Royal Institution, 1799–1844* (London, 1978); M.B. Hall, *All scientists now: The Royal Society in the nineteenth century* (Cambridge, 1984); Rudwick, *op. cit.* (n.25); J.B. Morrell, 'London institutions and Lyell's career: 1820–41', *British Journal for the History of Science*, 9 (1976), 132–46; B.H. Becker, *Scientific London* (London, 1874).

28   Faraday attended the following meetings of the British Association: Oxford, 1832; Liverpool, 1837; Manchester, 1842; York, 1844; Cambridge, 1845; Southampton, 1846; Oxford, 1847; Swansea, 1848; Birmingham, 1849; Ipswich, 1851; (Hull, 1853?); Liverpool, 1854; Leeds, 1858; Aberdeen, 1859; Oxford, 1860.

29   Faraday to William Vernon Harcourt, 3 August 1831: Quoted in Morrell and Thackray, *op. cit.* (n.26), p.74.

30   Faraday to John Barlow, 18 October 1844: *Correspondence*, 423; Faraday to Julius Plücker, 17 September 1863: *ibid.*, 1019; Faraday to Schoenbein, 23 September 1859: *Schoenbein*, 329–30.

31   Faraday to Richard Phillips, 2 September 1848: Oxford University Museum, Phillips Papers.

32   Faraday to Schoenbein, 10 August 1842: *Schoenbein*, 95–7.

33   Faraday to Sarah, 13 September 1837: *Jones, 1*, 95–6; Faraday to Schoenbein, 21 September 1837: *Schoenbein*, 31–3.

34   See *Transactions of the Society of Arts*, 37 (1820), onwards, and *Journal of the Society of Arts*, 14 (1865–6), 542–3; 15 (1866–7), 643.

35   H.B. Woodward, *The history of the Geological Society of London* (London, 1907), pp.48 and 54–5.

36   T.S. Moore and J.C. Philip, *The Chemical Society, 1841–1941: A historical review* (London, 1947), p.16.

37   *Jones, 1*, 339. See also *ibid.*, 299–314; Williams, *op. cit.* (n.1), p.160; D. Knight, 'Davy and Faraday: Fathers and sons', in D. Gooding and F.A.J.L. James, eds, *Faraday rediscovered: Essays on the life and work of Michael Faraday, 1791–1867* (Basingstoke and New York, 1985), pp.32–49, esp. pp.42–6.

38   Certificate book: Royal Society. Faraday to Thomas Spring-Rice, 23 April 1838: *Jones, 2*, 97.

39   Minutes of the Council of the Royal Society of London: Royal Society.

40   Minutes for 8 January 1835: *ibid*.

41   Certificate book: *ibid*.

42 R.M. MacLeod, 'Whigs and savants: Reflections on the reform movement in the Royal Society, 1830–48', in *Metropolis and province, op. cit.* (n.27), pp.55–90.

43 A.C. Todd, *Beyond the blaze: A biography of Davies Gilbert* (Truro, 1967), p.237.

44 Morrell and Thackray, *op. cit.* (n.26), pp.52–7; J. Barrow, *Sketches of the Royal Society and the Royal Society Club* (London, 1849).

45 Faraday to Carlo Matteucci, 18 February 1843: *Jones, 2,* 172–3. See also Faraday to William R. Grove, 21 December 1842: *ibid,* 171–2.

46 John Barnard to Robert Sandeman, 25 July 1760: *Supplementary volume,* 50–3.

47 Anon., *The customs of the Churches of Christ as found in the New Testament* (Edinburgh, 1908), pp.18–19. 1 John 4:1–6 also cited.

48 MacLeod, *op. cit.* (n.42).

49 Faraday to John G. Children, 23 October 1832: *Correspondence,* 234.

50 Faraday to J.W. Lubbock, 6 December 1833: Ms F2, Lubbock Letters, Royal Society.

51 H. Ward, *History of the Athenaeum 1824–1925* (London, 1926); [F.R. Cowell, *et al*], *The Athenaeum. Club and social life in London 1824–1974* (London, 1975).

52 John Gassiot to John F.W. Herschel, 5 March 1847: Royal Society, Herschel Papers, HS.8.55. See also MacLeod, *op. cit.* (n.42).

53 Todd, *op. cit.* (n.43), p.231.

54 Entry in Herschel's Diary, 12 February 1848: MS 583–6, Royal Society.

55 *The journals of Walter White, Assistant Secretary of the Royal Society* (London, 1898), pp.117–18; Hall, *op. cit.* (n.27), pp.38, 100 and frontispiece.

56 J. Tyndall, *Faraday as a discoverer* (5th ed., London, 1894), pp.183–4.

57 Morrell and Thackray, *op. cit.* (n.26), pp.17–29 and 223–56. Faraday also stood outside the group of mathematical practitioners, many of whom were military men, who were also prominent in this period. While his own style of science differed significantly from theirs, he was on friendly terms with several of their number. See D.P. Miller, 'The revival of the physical sciences in Britain, 1815–1840', *Osiris, 2* (1986), 107–34.

58 J. B. Morrell, 'Professionalisation', in R.C. Olby, G.N. Cantor, J.R.R. Christie and M.J.S. Hodge, eds, *Companion to the history of modern science* (London, 1990), pp.980–9.

59 Rudwick, *op. cit.* (n.25), pp.17–41.

60 William Brande to Faraday, 11 October 1839 and 19 May 1855: *Correspondence,* 348–9 and 795. Faraday to Brande, 6 May 1852: *Jones, 2,* 306–9.

61 C. Smith, 'A new chart for British natural philosophy: The development of energy physics in the nineteenth century', *History of Science, 16* (1978), 231–79.

62 W.T. Brande, *Dictionary of science, literature, & art* (London, 1842), p.217. Emphasis added.

63 William Hyde Wollaston to Henry Hastead, 30 August 1801: University College, London, Gilbert Papers, Box 1, File 1, f.20.

64 H.C. Oersted, 'Experiments on the effect of a current of electricity on the magnetic needle', *Annals of Philosophy*, 16 (1820), 273–6.

65 D. Gooding, ' "Magnetic curves" and the magnetic field: Experimentation and representation in the history of a theory', in D. Gooding, T. Pinch and S. Schaffer, eds, *The uses of experiment: Studies in the natural sciences* (Cambridge, 1989), pp.183–223, esp. pp.193–201.

66 W. Sturgeon, 'Remarks on the preceding paper, with experiments', *Annals of Electricity, Magnetism and Chemistry*, 1 (1836–7), 186–91 and 367–76.

67 Sturgeon set himself up as adversary to many of the leading scientists of the period. Joseph Henry, for example, reported that 'Mr Sturgeon first became dissatisfied with the Royal Institution by a harshe remarke of Sir H Davey who when Mr S shewed him some exp. on magnetism said he had better mind his last than be dabling in science': N. Reingold, *et al.*, eds, *The papers of Joseph Henry* (five vols, Washington, 1972–85), vol.3, p.250. For Sturgeon's Manchester background and other controversies, see D.S.L. Cardwell, *James Joule: A biography* (Manchester, 1989), esp. pp.25–8.

68 Faraday to Julius Plücker, 8 April 1856: *Correspondence*, 833–5.

69 *Report of H.M. Commissioners appointed to inquire into the revenues and management of certain colleges and schools and the studies pursued and instruction given therein; with appendix and evidence*, Parliamentary Papers (1864), [3288], vol.4, p.377.

70 E. Stone, *A new mathematical dictionary* ... (London, 1726).

71 W.F. Barrett, review of J.H. Gladstone's *Michael Faraday* in *Nature*, 6 (1872), 410–13, esp. 412; J. Fenwick Allen, *Some founders of the chemical industry* (London and Manchester, 1906), pp.151–98; A.F. Yarrow, 'An incident in connection with Faraday's life', *Proceedings of the Royal Institution*, 25 (1926–8), 480.

72 Faraday and James Stodart, 'Experiments on the alloys of steel, made with a view to its improvement' and 'On the alloys of steel', *ERCP*, 57–68 and 68–81; Faraday and Richard Phillips, 'On a new compound of chlorine and carbon', *ibid.*, 53–7.

73 D. Stein, *Ada. A life and a legacy* (Cambridge, Mass. and London, 1985), pp.136–7. See also *Jones*, 2, 186 and Faraday to Ada, Countess of Lovelace, 24 October 1844: *ibid.*, 188–92.

74 See, for example, Auguste de la Rive's letters to Faraday of 26 July 1855 and 14 May 1856: *Correspondence*, 803 and 839–40.

75 For example, Auguste de la Rive to Faraday, 24 December 1852: *ibid.*, 675–6; Faraday to de la Rive, 16 December 1859 and 19 September 1861: *ibid.*, 943–4 and 1001.

76 Faraday to Schoenbein, 18 December 1946: *Schoenbein*, 162–3. Cf. G.W.A. Kahlbaum, ed., *The letters of Jons Jakob Berzelius and Christian Friedrich Schonbein, 1836–1847* (London, 1900), pp.92–103.

77 Faraday to William F. Cowper, 3 August 1860: *Correspondence*, 958–9. Cf copy of this letter in *Jones*, 2, 430–1.

78 White, *op. cit.* (n.55), p.256.

79 Joseph Henry to Asa Gray, 1 November 1838: *Papers of Joseph Henry, op. cit.* (n.67), vol.4, p.132.

80 See n.66.

81 'Commonplace book', ff.11–66: Institution of Electrical Engineers, Faraday Papers; Poem in Faraday's hand: RI MS, FL8A 101.

82 [R. Sandeman?], *The philosophy of the Creation, as narrated in Moses' principia, Gen. Chap.I. v.1–18* (Edinburgh, 1835), p.vi.

83 A. Ellegard, *Darwin and the general reader: The reception of Darwin's theory of evolution in the British periodical press, 1859–1872* (Goteborg, 1958); T.F. Glick, ed., *The comparative reception of Darwinism* (Austin, 1974); D.L. Hull, ed., *Darwin and his critics: The reception of Darwin's theory of evolution by the scientific community* (Cambridge, Mass., 1973); J.C. Greene, *The death of Adam: Evolution and its impact on Western thought* (Ames, 1959).

84 *Jones*, *1*, 378–81 and *2*, 140.

85 Cited in Faraday, *A course of six lectures on the various forces of matter and their relations to each other* (London and Glasgow, 1860), p.154.

86 *Jones*, *1*, 351.

87 Faraday, 'Observations on mental education', *ERCP*, 463–91.

88 *Report, op. cit.* (n.69), p.376.

89 *Ibid.*, p.377.

90 J. Oppenheim, *The other world: Spiritualism and psychic research in England, 1850–1914* (Cambridge, 1985); L. Barrow, *Independent spirits: Spiritualism and English plebeians, 1850–1910* (London and New York, 1986); A. Gauld, *The founders of psychical research* (London, 1968); I. Grattan-Guinness, ed, *Psychical research: A guide to its history, principles and practices, in celebration of 100 years of the Society for Psychical Research* (Wellingborough, 1982).

91 Faraday, 'On table-turning' and 'Experimental investigation of table-moving', *ERCP*, 382–91.

92 Faraday to Schoenbein, 25 July 1853: *Schoenbein*, 215.

93 For example, R. Sandeman, *The law of nature defended by Scripture* in Sandeman, *Discourses on passages of Scripture: With essays and letters* (Dundee, 1857), pp.276 and 281.

94 'Mental education', *op. cit.* (n.87), p.465. Latter emphasis added.

95 *Report, op. cit.* (n.69), p.378.

96 'Mental education', *op. cit.* (n.87).

97 *Report, op. cit.* (n.69), p.379.

98 Faraday to Benjamin Abbott, 11 June 1813: *Correspondence*, 54–6.

99 M. L[loyd], *Sunny memories, containing personal recollections of some celebrated characters* (two vols, London, 1879–80), vol.1, p.65.

100 Faraday to Abbott: *op. cit.* (n.98).

101 Faraday to Benjamin Abbott, 18 June 1813: *Correspondence*, 57–9.

102 For example, Faraday, *The chemical history of a candle* (London, 1907), pp.14 and 62.

103 L[loyd], *op. cit.* (n.99), p.67; D.M. Knight, 'The scientist as sage', *Studies in Romanticism*, 6 (1967), 65–88.

104 Faraday, *op. cit.* (n.85), p.155.

105 Faraday to Benjamin Abbott, 4 June 1813: *Correspondence*, 52–4.

106 Faraday, *op. cit.* (n.85), pp.iv-v.

107 *Papers of Joseph Henry, op. cit.* (n.67), vol.3, p.255.

108 Faraday to Abbott: *op. cit.* (n.98).

109 R.M. MacLeod, 'The Alkali Acts administration, 1863–84: The emergence of the civil scientist', *Victorian Studies*, 9 (1965), 85–112.

110 Faraday to Auckland, 29 July 1847: *Correspondence*, 508.

111 Williams, *op. cit.* (n.1), pp.116–20; Berman, *op. cit.* (n.27), pp.164–6; F.A.J.L. James, 'Davy and Faraday as military scientists' (unpublished typescript).

112 Faraday, 'On the manufacture of glass for optical purposes', *ERCP*, 231–91.

113 Faraday to Peter M. Roget, 4 July 1831: *Correspondence*, 199.

114 R.M. MacLeod, 'Science and government in Victorian England: Lighthouse illumination and the Board of Trade, 1866–1886', *Isis*, 60 (1969), 5–38.

115 *Report of the Commissioners appointed to inquire into the condition and management of lights, buoys and beacons, Parliamentary Papers,* (1861), [2793], vol.2, pp.591–3.

116 Faraday to Benjamin Vincent, 12 October 1860: *Correspondence*, 974–5.

117 Williams, *op. cit.* (n.1), p.491.

118 *Ibid.*, p.490. See the forty letters between Faraday and James Chance published in *Correspondence*.

119 Charles Lyell to Charles Babbage, 7 October 1844: Mrs Lyell, ed., *Life letters and journals of Sir Charles Lyell, Bart.* (two vols, London, 1881), vol.2, pp.89–90; See also Lyell to Henry Bence Jones, April 1868: *ibid.*, pp.417–22.

120 See reports in *The Times* for 2, 3, 4, 11 and 14 October 1844. On coroners' inquests see F.A. Barley, 'Coroners' inquests held in the manor of Prescot, 1746–89', *Transactions of the Historical Society of Lancashire and Cheshire*, 86 (1934), 21–39; E.H. East, *A treatise of the pleas of the Crown* (two vols, London, 1803). vol.2, pp.52–70. Berman, *op. cit.* (n.27) is mistaken in characterising the coroner's inquest as a 'trial'.

121 Berman, *op. cit.* (n.27), p.179. Berman incorrectly claims that the explanation of the explosion given by Lyell and Faraday was merely 'a matter of opinion ... [and] had no scientific basis whatsoever'. While it is certainly correct to claim that Faraday and others could only offer opinions, their views were based on various scientific considerations, for example, variations in atmospheric pressure, the role of coaldust, etc.

122 Faraday, 'The ventilation of mines. and the means of preventing explosion from fire damp', *The Civil Engineer and Architect's Journal*, 8 (1845), 115–18, esp. 115.

123 'Report from Messrs. Faraday and Lyell to the Rt. Hon. Sir James Graham, Bart., Secretary of State for the Home Department, on the subject of the explosion at the Haswell Collieries, and on the means of preventing similar accidents', *Philosophical Magazine*, 26 (1845), 17–35, esp. p.18.

124 Faraday, *op. cit.* (n.122).

NOTES FOR CHAPTER 7

1 Especially, D. Gooding, 'Metaphysics versus measurement: The conversion and conservation of force in Faraday's physics', *Annals of Science*, 37 (1980), 1–29; *id.*, 'Empiricism in practice: Teleology, economy, and observation in Faraday's physics', *Isis*, 73 (1982), 46–67.

2 For Faraday on metaphysics see 'On the conservation of force', in *ERCP*, 444–5.

3 L.P. Williams, *Michael Faraday: A biography* (London, 1965), pp.73–89; T.H. Levere, *Affinity and matter. Elements of chemical philosophy 1800–1865* (Oxford, 1971), pp.68–107; P.M. Heimann, 'Faraday's theories of matter and electricity', *British Journal for the History of Science*, 5 (1971), 235–57; F.A.J.L. James, 'Reality or rhetoric? Boscovichianism in Britain: the cases of Davy, Herschel and Faraday' (unpublished).

4 J.H. Brooke, 'Natural theology in Britain from Boyle to Paley', units 9–10 of *Science and belief: From Copernicus to Darwin* (Milton Keynes, 1974).

5 J. Morison, *An introductory to the first four books of Moses* (Perth, 1810); [R. Sandeman?], *The philosophy of the Creation, as narrated in Moses' principia, Gen. chap.I v.1 to 18* (Edinburgh, 1835); J. Hutchinson, *Moses's principia* in R. Spearman and J. Bate, eds, *The philosophical and theological works of the late truly learned John Hutchinson, Esq.* (twelve vols, London, 1748–9), vol.1. For secondary works on Hutchinson and the Hutchinsonians see G.N. Cantor, 'Revelation and the cyclical cosmos of John Hutchinson', in L.J. Jordanova and R.S. Porter, eds., *Images of the earth: essays in the history of the environmental sciences* (Chalfont St Giles, 1979), pp.3–22; C.J. Wilde, 'Hutchinsonianism, natural philosophy and religious controversy in eighteenth century Britain', *History of Science*, 18 (1980), 1–24.

6 S. Pike, *Philosophia sacra: Or, the principles of natural philosophy. Extracted from Divine Revelation* (London, 1753). A second edition appeared in 1815.

7 [Sandeman?], *op. cit.* (n.5), p.vii.

8 *Ibid.*, p.46.

9 [R. Sandeman], *Letters on Theron and Aspasio. Addressed to the author* (4th ed., two vols, Edinburgh, 1803), vol.1, pp.378–81 and vol.2, pp.157–9. Quotations on vol.1, pp.374 and 379–80.

10 J. Glas, 'On Mr Hutchinson's philosophy and divinity', *The works of Mr. John Glas* (2nd ed., five vols, Perth, 1782), vol.2, pp.426–35.

11 Faraday to Ada, Countess of Lovelace, 24 October 1844: *Jones*, 2, 191.

12 S. Turner, *The sacred history of the world, as displayed in the Creation and subsequent events to the Deluge* (three vols, London, 1832–9), vol.1, pp.12, 22 and 32.

13 T.H. Levere, 'Faraday, matter, and natural theology – reflections on an unpublished manuscript', *British Journal for the History of Science*, 4 (1968), 95–107.

14 Faraday, 'A course of lectures on electricity and magnetism', *London Medical Gazette*, 2 (1846), 977 and 3 (1846), 523. The latter part of this quotation may have been a play on *Christianity as old as Creation* (1696), the title of a deist tract by John Toland.

15 Faraday to Benjamin Abbott, 12 July 1812: *Correspondence*, 6.

16 Faraday often used the terms cause and force interchangeably.

17 Faraday, 'On hydrate of chlorine' in *ERCP*, 81–4; Quotation from *ERE*, *1*, 241; F.A.J.L. James, 'Michael Faraday's first law of electrochemistry: How context develops new knowledge', in J.T. Stock and M.V. Orna, eds, *Electrochemistry, past and present* (Washington, 1989), pp.32–49. See also 'Conservation', *op. cit.* (n.2), 454.

18 *Jones*, 2, 47 and 224.

19 'Conservation', *op. cit.* (n.2), 457.

20 *Ibid.*, 454; Levere, *op. cit.* (n.13), 105.

21 Faraday, 'Eight lectures on physics and chemical philosophy', vol.2, f.20: RI MS, F4J7.

22 Faraday, 'On mental education', *ERCP*, 463–91. Quotation on 479, emphasis Faraday's.

23 'Eight lectures', *op. cit.* (n.21), vol.2, f.4.

24 W.R. Grove, *The correlation of physical forces* (London, 1846).

25 'Eight lectures', *op. cit.* (n.21), vol.2, ff.10–11.

26 Levere, *op. cit.* (n.13), p.105.

27 *Jones*, 2, 224–5.

28 *Ibid.*, 239.

29 Faraday, 'On some points of magnetic philosophy', in *ERE*, 3, 528–74, esp.541.

30 Faraday to A.F. Svanberg, 16 August 1850: *Correspondence*, 588; *Jones*, 2, 224.

31 'Lectures', *op. cit.* (n.14), 3 (1846), 529; Faraday to Schoenbein, 9 December 1852: *Schoenbein*, 207–10.

32 A.O. Lovejoy, *The great chain of being* (Cambridge, Mass., 1936).

33 *ERE*, 3, 268.

34 See Gooding, 'Empiricism', *op. cit.* (n.1); *Id.*, 'Final steps to the field theory: Faraday's study of magnetic phenomena, 1845-50', *Historical Studies in the Physical Sciences*, 11 (1981), 231–75.

35 J. Locke, *An essay concerning human understanding*, in *The works of John Locke* (nine vols, London, 1824), vol. 1, pp. 220–74 and 293–6.

36 I. Newton, *Opticks or a treatise of the reflections, refractions, inflections and colours of light* (New York, 1952), esp. query 31.

37 P.M. Heimann and J.E. McGuire, 'Newtonian forces and Lockean powers: Concepts of matter in eighteenth-century thought', *Historical Studies in the Physical Sciences*, 3 (1971), 233–306.

38 See n.3. Also G.N. Cantor, *Optics after Newton. Theories of light in Britain and Ireland, 1704–1840* (Manchester, 1983), pp.71–81.

39 Glas, *op. cit.* (n.10), p.429.

40 J. R[orie], ed., *Selected exhortations delivered to various Churches of Christ by the late Michael Faraday, Wm. Buchanan, John M. Baxter, and Alex. Moir* (Dundee, 1910), p.24.

41 'Lectures', *op. cit.* (n.14), 2 (1846), 977 and 3 (1846), 529.

42 'Conservation', *op. cit.* (n.2), 460. Gooding, 'Metaphysics', *op. cit.* (n.1)) has identified several problems with Faraday's notion of force.

43 'Lectures', *op. cit.* (n.14), 2 (1846), 977 and 3 (1846), 529.

44 *Jones*, 2, 474.

45 *Ibid.*, *1*, 194.

46 F.A.J.L. James, "The optical mode of investigation": Light and matter in Faraday's natural philosophy', in D. Gooding and F.A.J.L. James, eds, *Faraday rediscovered: Essays in the life and work of Michael Faraday, 1791–1867* (Basingstoke and New York, 1985), pp.137–62; Faraday, 'A speculation touching electrical conduction and the nature of matter', *ERE*, 2, 286; *Id.*, *A course of six lectures on the various forces of matter and their relations to each other* (London, 1860).

47 *Jones*, 2, 86.

48 William Rowan Hamilton to his sister Sydney, 30 June 1834: R.P. Graves, *Life of Sir William Rowan Hamilton* (three vols, Dublin and London, 1882–9), vol.2, pp.95–6. Emphasis added. For Hamilton's views see T. Hankins, *Sir William Rowan Hamilton* (Baltimore and London, 1980).

49 Levere, *op. cit.* (n.3), pp.68–75.

50 *ERE*, *1*, 256. First emphasis added.

51 Faraday to F.O. Ward, 16 June 1834: *Correspondence*, 274.

52 'Speculation', *op. cit.* (n.46).

53 *ERE*, *1*, 409–10.

54 See J.E. McGuire, 'Atoms and the "Analogy of nature": Newton's third rule of philosophizing', *Studies in History and Philosophy of Science*, 1 (1970), 3–58.

55 Cantor, *op. cit.* (n.38); J.L. Heilbron, *Electricity in the 17th and 18th centuries: A study of early modern physics* (Berkeley and Los Angeles, 1979).

56 *Jones*, *1*, 268–71.

57 Faraday, 'Thoughts on ray-vibrations', in *ERCP*, 366–72. See also Cantor, *op. cit.* (n.38) and James, *op. cit.* (n.46).

58 Levere, *op. cit.* (n.3), pp.75–9; *Id.*, *op. cit.* (n.13).

59 Faraday's views on time deserve analysis from the perspective of economy.

60 *ERE*, *1*, 352–3 and 514; R. Hare to Faraday, n.d.: *Correspondence*, 350–8.

61 D.C. Gooding, 'Conceptual and experimental bases of Faraday's denial of electrostatic action at a distance', *Studies in History and Philosophy of Science*, 9 (1978), 117–49.

62 Faraday, 'On the limits of vaporization', in *ERCP*, 205–12.

63 'Ray-vibrations', *op. cit.* (n.57), 371.

64 Faraday, 'On some points of magnetic philosophy', in *ERE*, 2, 566–74. See also 'Conservation', *op. cit.* (n.2), 451.

65 *ERE*, 3, 425–6. See also *ibid.*, 194–5.

66 For differing attitudes towards ether theories see G.N. Cantor and M.J.S. Hodge, eds, *Conceptions of ether: Studies in the history of ether theories 1740–1900* (Cambridge, 1981).

67 'Speculation', *op. cit.* (n.46).

68 Faraday, 'Address delivered at the commemoration of the centenary of the birth of Dr Priestley', *Philosophical Magazine*, 2 (1833), 390–1.

69 See Heimann, 'Faraday's theories', *op. cit.* (n.3); Heimann and McGuire, 'Newtonian forces', *op. cit.* (n.37); Levere, *op. cit.* (n.3), pp.23–67.

70 Deuteronomy 33:16 – 'And for the precious things of the earth and the fulness thereof ...'; Isaiah 6:3 – 'And one cried unto another, and said, Holy, holy, holy, is the LORD of hosts: the whole earth is full of his glory'.

71 E.g. T.S. Kuhn, 'Energy conservation as an example of simultaneous discovery', in M. Clagett, ed., *Critical problems in the history of science* (Madison, 1959), pp.321–56.

72 'Lectures', *op. cit.* (n.14), 2 (1846), 977.

73 'Eight lectures', *op. cit.* (n.21), vol.2, f.9.

74 Grove, *op. cit.* (n.24); G.N. Cantor, 'William Robert Grove, the correlation of forces, and the conservation of energy', *Centaurus*, 19 (1976), 273–90.

75 *Course of six lectures, op. cit.* (n.46), pp.130–3 and 154.

76 'Conservation', *op. cit.* (n.2), 447.

77 J.P. Joule to Faraday, 5 June 1849: Royal Institution, Conybeare Papers, f.134.

78 'Conservation', op. cit. (n.2), 453.

79 C.W. Smith, 'Faraday as a referee of Joule's Royal Society paper "On the mechanical equivalent of heat"', *Isis*, 67 (1976), 444–9.

80 *Course of six lectures, op. cit.* (n.46), p.102.

81 J.P. Joule 'On the mechanical equivalent of heat' in *The scientific papers of James Prescott Joule* (two vols, London, 1887), vol.1, pp.298–328, quotation on p.328. Original in italics.

82 *Diary*, *3*, 216.

83 'Lectures', *op. cit.* (n.14), 3 (1846) 95 and 358; *ERE*, *1*, 253.

84 *Diary*, *2*, 246.

85 *ERE*, *1*, 215–17; Gooding, *op. cit.* (n.1).

86 *ERE*, *1*, 76–107; James, *op. cit.* (n.17).

87 E.J. Dijksterhuis, *Mechanization of the world picture* (Oxford, 1961); M. Tamny, 'Atomism and the mechanical philosophy', in R.C. Olby, G.N. Cantor, J.R.R. Christie and M.J.S. Hodge, eds, *Companion to the history of modern science* (London, 1990), pp.597–609.

88 'Eight lectures', *op. cit.* (n.21), vol.1, ff.3–4; 'Conservation', *op. cit.* (n.2), 451–2.

89 *ERE*, *1*, 553–6.

90 J.P. Joule, 'On the changes of temperature produced by the rarefaction and condensation of air', in *The scientific papers of James Prescott Joule* (two vols, London, 1887), vol.1, pp.172–89, esp. p. 189; *Id.*, 'On matter, living force, and heat', *ibid.*, pp.165–76, esp. pp.270 and 274.

91 N.M. Wise and C. Smith, 'Measurement, work and industry in Lord Kelvin's Britain', *Historical Studies in the Physical and Biological Sciences*, 17 (1986), 147–73; C. Smith and M.N. Wise, *Energy and empire: A biographical study of Lord Kelvin* (Cambridge, 1989).

92 Williams, *op. cit.* (n.3), pp.10–94; Levere, *op. cit.* (n.3), pp.68–107.

93 Joule, 'On matter', *op. cit.* (n.90), p.269. See also *Id.*, 'Changes in temperature', *op. cit.* (n.90), p.189.

94 Joule, 'On matter', *op. cit.* (n.90), pp.272–3.

95 C.W. Smith, 'Energy', in *Companion, op. cit.* (n.87), pp.326–41.

96   I. Inkster and J. Morrell, eds, *Metropolis and province: Science in British culture, 1780–1850* (London, 1983); R.H. Kargon, *Science in Victorian Manchester: Enterprise and expertise* (Baltimore and London, 1977); D.S.L. Cardwell, *James Joule: A biography* (Manchester, 1989).

97   M. Berman, *Social change and scientific organization. The Royal Institution, 1799–1844* (London, 1978), pp.156–86.

98   'Lectures', *op. cit.* (n.14), 2 (1846), 978; 'Eight lectures', *op. cit.* (n.21), vol.2, f.21; Faraday, 'On certain metals and metallic properties', *London Medical Gazette*, 1 (1845), 400.

99   Berman, *op. cit.* (n.97), pp.108–15, 125–7 and 185–6.

NOTES FOR CHAPTER 8

1   Faraday to Auguste de la Rive, 2 October 1858: *Correspondence*, 913–4. Emphases added.

2   *Jones*, 2, 304; Faraday, 'Observations on mental education', *ERCP*, 463–91, esp.480.

3   *Jones*, 2, 427.

4   J. Glas, 'Predestination impugned and defended', in *The works of Mr John Glas* (2nd ed., five vols, Perth, 1782), vol.2, pp.395–414, esp. p.398.

5   D.B. Murray, *The social and religious origins of Scottish non-Presbyterian Protestant dissent from 1730–1800* (Unpublished PhD dissertation, University of St Andrews, 1976), pp.112–4. On the similar use of the plain style by Puritan divines see P. Miller, 'The plain style' in S.E. Fish, ed, *Seventeenth century prose* (Oxford, 1971), pp.147–86.

6   J. Glas, 'On Mr Hutchinson's philosophy and divinity', in Glas, *op. cit.* (n.4), vol.2, pp.426–35, esp. p.428.

7   R. Sandeman, *The law of nature defended by Scripture against a learned class of moderns, who think it needful, in order to support the credit of revealed religion against deists, to deny the existence of that law* (1760), in Sandeman, *Discourses on passages of Scripture: With essays and letters* (Dundee, 1857), pp.273–85. Quotation on pp.278–9.

8   'Mental education', *op. cit.* (n.2), pp.464–5.

9   *Jones*, 2, 225.

10   Faraday to Auguste de la Rive, 16 December 1859: *Correspondence*, 943–4.

11   *Jones*, 2, 104 and 239.

12   'Mental education', *op. cit.* (n.2), p.471.

13   William Barrett used a different but not dissimilar metaphor when he claimed that Faraday considered the Bible as 'God's revelation to man of the Divine purpose', and science as 'man's revelation of the Divine handicraft'. William Barrett to J.H. Poynting, 6 March 1911: University of Birmingham, Physics Department.

14   'Mental education', *op. cit.* (n.2), pp.468–9.

15   Faraday to Adolphe Quetelet, 25 February 1850: *Correspondence*, 579–80. Emphasis added.

16   Faraday to William Whewell, 19 September 1835: *ibid.*, 294–6; Faraday to de la Rive: *op. cit.* (n.1). Emphases added. See also *Jones*, *1*, 268 and *2*, 175.

17   Faraday to Schoenbein, 13 November 1858: *Schoenbein*, 314–16.

18   Faraday to Schoenbein, 20 February 1845: *ibid.*, 143–4.

19   'Mental education', *op. cit.* (n.2), p.469; *Jones*, 2, 86.

20   'Mental education', *op. cit.* (n.2); *Jones*, 2, 399.

21   T.H. Levere, 'Faraday, matter, and natural theology – reflections on an unpublished manuscript', *British Journal for the History of Science*, 4 (1968), 95–107. Passages quoted on pp.105 and 107. *Report of H.M. Commissioners appointed to inquire into the revenues and management of certain colleges and schools and the studies pursued and instruction given therein*, in *Parliamentary Papers*, 1864, vol.4, p.381.

22   Faraday to Adolph F. Svanberg, 16 August 1850: *Correspondence*, 588.

23   J. Glas, *Notes on Scriptural texts* in Glas, *op. cit.* (n.4), vol.3, pp.1–344. Quotation on p.274, mispaginated p.472.

24   Sandeman, *op. cit.* (n.7), pp.273–4.

25   A.A. Cooper, Third Earl of Shaftesbury, *An inquiry concerning virtue* (London, 1699); *Id., Characteristicks of men, manners, opinions, times* (three vols, London, 1711).

26   Sandeman, *op. cit.* (n.7); J. Locke, *A paraphrase and notes on the epistle of St. Paul to the Romans*, in *The works of John Locke* (12th ed., ten vols, London, 1824), vol.8, pp.373–427.

27   Sandeman, *op. cit.* (n.7), pp.278–9.

28   J. R[orie], *Selected exhortations delivered to various Churches of Christ by the late Michael Faraday, Wm. Buchanan, John M. Baxter, and Alex. Moir* (Dundee, 1910), pp.15–18.

29   Exhortation notes: RI MS, 8A 101.

30   'Mental education', *op. cit.* (n.2), p.471.

31   Faraday's discussion of prejudice bears a marked resemblance to Francis Bacon's 'Idols', those false images in the mind that vitiate knowledge. Indeed, the three types of prejudice mentioned in the text correspond, respectively, to the 'Idols of the Tribe', the 'Idols of the Market-Place' and the 'Idols of the Theatre'. Bacon, *The great instauration*, in J. Spedding, R.L. Ellis and D.D. Heath, eds, *The works of Francis Bacon* (fourteen vols, London, 1872–4), vol.4, pp.53–65.

32   Faraday to Benjamin Abbott, 12 July 1812: *Correspondence*, 3–6.

33   Faraday, 'Historical sketch of electro-magnetism', *Annals of Philosophy*, 2 (1821), 195–200 and 274–90; 3 (1822), 107–21; L.P. Williams, 'Faraday and Ampère: A critical dialogue', in D. Gooding and F.A.J.L. James, eds, *Faraday rediscovered: Essays on the life and work of Michael Faraday, 1791–1867* (Basingstoke and New York, 1985), pp.83–104, esp. pp.86–90.

34   *Jones*, 2, 321.

35   Faraday to F.O. Ward, 16 June 1834: *Correspondence*, 274.

36   *Diary*, 2, 184.

37   Faraday and P. Reiss, 'On the action of non-conducting bodies in electric induction', *Philosophical Magazine*, 11 (1856), 1–17.

38   Faraday to William Whewell, 19 September 1835: *Correspondence*, 294–6.

39   Faraday to Quetelet: *op. cit.* (n.15).

40   I. Newton, *Mathematical principles of natural philosophy* (two vols, Berkeley, 1934), vol.1, p.xxvii.

41 Faraday, 'On the lines of magnetic force', *ERE*, *3*, 402–6; *Id.*, 'Thoughts on ray-vibrations', *ibid.*, 447–52.

42 D. Gooding, '"In nature's school": Faraday as an experimentalist' in *Faraday rediscovered*, *op. cit.* (n.33), pp.105–35; E. Crawford, 'Learning from experience', *ibid.*, pp.211–27.

43 *Diary*, *5*, 43 and 213.

44 *Jones*, *2*, 175.

45 Faraday to Christian E. Neeff, 24 March 1846: *Correspondence*, 491–2. Emphasis added.

46 Faraday, 'A speculation touching electric conduction and the nature of matter', *ERE*, *2*, 284–93. Quotation on pp.285–6. Emphasis added.

47 *Ibid.*, 290 and 292.

48 L.P. Williams, *Michael Faraday: A biography* (London, 1965), pp.73–80; P.M. Heimann, 'Faraday's theories of matter and electricity', *British Journal for the History of Science*, 5 (1971), 235–57.

49 J. Agassi, *Faraday as a natural philosopher* (Chicago, 1971), p.117.

50 'Ray-vibrations', *op. cit.* (n.41), p.447.

51 Faraday, 'On the physical character of the lines of magnetic force', *ERE*, *3*, 407; *Id.*, *ERE*, *1*, 16–24.

52 *ERE*, *3*, 408.

53 Faraday to Auguste de la Rive, 29 May 1854: *Correspondence*, 737–8.

54 *ERE*, *1*, 1–41, esp. 16–24; Faraday to Whewell: *op. cit.* (n.16).

55 *Jones*, *2*, 44.

56 Chuang Tzu, *Basic writings*, trans. Burton Watson (NY and London, 1964), p.45.

57 Faraday's contribution to 'Addresses delivered at the commemoration of the centenary of the birth of Dr Priestley', *Philosophical Magazine*, 2 (1833), 390–1. The question of who discovered oxygen is now recognised as a complex problem and the palm cannot simply be presented to Priestley.

58 Faraday to Schoenbein, 20 February 1845, 13 November 1858 and 27 March 1860: *Schoenbein*, 143–4, 314–16 and 336–7.

59 Faraday to Schoenbein, 13 November 1845: *Ibid.*, 148.

60 Faraday to Abbott: *op. cit.* (n.32). Cf. J. Locke, *An essay concerning human understanding*, *op. cit.* (n.26), vol. 1, pp. 383–93.

61 'Mental education', *op. cit.* (n.2), pp.478 and 480.

62 Faraday to Robert Hare, 18 April 1840: *Correspondence*, 360–9.

63 Faraday to William Whewell, 29 January 1853: *Correspondence*, 682–3.

64 Faraday to William Whewell, 21 February 1831: *Ibid.*, 190–1; Faraday to Whewell, 24 April 1834: *Ibid.*, 264–5; Whewell to Faraday, 25 April 1834: *Ibid.*, 265–7.

65 Appendix to Senate minutes: University College, London, Ms AM/33.

66 Faraday to Julius Plücker, 23 March 1857: *Correspondence*, 863–4.

67 Augustus de Morgan to William H. Dixon, 17 March 1857: American Philosophical Society.

68 Faraday to William Whewell, 19 September 1835: *Ibid.*, 294–6; Faraday to Charles R. Weld, 6 July 1855: *Ibid.*, 800.

69   'Commonplace book', f.87: Institution of Electrical Engineers, Faraday Papers; *Jones, 1*, 199; S.B. Smith, *The great mental calculators. The psychology, methods, and lives of calculating prodigies, past and present* (New York, 1983), pp.181–210.

70   Faraday to John Tyndall, 19 April 1851: *Correspondence*, 623; Faraday to Peter Riess, 7 April 1855: *Ibid.*, 791–2; J. Tyndall, *Faraday as a discoverer* (5th ed., London, 1894), pp.63–4.

71   Faraday to James Clerk Maxwell, 13 November 1857: *Correspondence*, 884–5.

72   Faraday to Gaspard de la Rive, 9 October 1822: *Ibid.*, 138–9.

73   Faraday to Isambard Kingdom Brunel, 4 May 1835: *Ibid.*, 292.

74   Faraday to Richard Phillips, 29 November 1831: *Ibid.*, 209–11.

75   Faraday, 'On the conservation of force', *ERCP*, 443–63. Quotation on p.458.

76   Quoted in E.C. Patterson, *Mary Somerville and the cultivation of science, 1815–1840* (Boston, 1983), p.135.

77   Faraday to André-Marie Ampère, 3 September 1822: *Correspondence*, 134–5.

78   Faraday to James Clerk Maxwell, 25 March 1857: *Ibid.*, 864–5.

79   J. Hutchinson, *A treatise of power essential and mechanical*, in *The philosophical and theological works of the late truly learned John Hutchinson, Esq* (twelve vols, London, 1748–9), vol.5, pp.222–3. *Moses's principia* occupies the first two volumes. See also C.B. Wilde, 'Hutchinsonianism, natural philosophy and religious controversy in eighteenth century Britain', *History of Science*, 18 (1980), 1–24; *Id.*, 'Matter and spirit as natural symbols in eighteenth-century British natural philosophy', *British Journal for the History of Science*, 15 (1982), 99–131; G.N. Cantor, 'Light and enlightenment: An exploration of mid-eighteenth-century modes of discourse', in D.C. Lindberg and G.N. Cantor, *The discourse of light from the middle ages to the enlightenment* (Los Angeles, 1985), pp.67–106.

80   See Numbers 26:55, 33:54 and 34:13.

81   Anne Connan, a member of the London meeting house, informed me in 1986 that lots had not been used during her long association with the church.

82   Faraday to Tyndall: *op. cit.* (n.70); W. Barratt, review of J.H. Gladstone's *Michael Faraday*, *Nature*, 6 (1872), 410–3, esp. 412.

83   *Jones, 1*, 267.

84   Faraday to Tyndall: *op. cit.* (n.70).

85   *Jones, 2*, 474.

86   Gooding, *op. cit.* (n.42), p.106.

87   Crawford, *op. cit.* (n.42); *Id.*, 'Michael Faraday: Ideas about how he thought', paper delivered to a joint meeting of the British Society for the History of Science and the British Psychological Society, 7 February 1987.

88   Faraday to Schoenbein, 9 December 1852: *Schoenbein*, 207–10; Faraday to Schoenbein, 22 April 1862: *Ibid.*, 355.

89   Faraday to Ernst Becker, 25 October 1860: *Correspondence*, 975–6. Cf. S. Shapin, 'Pump and circumstance: Robert Boyle's literary technology', *Social Studies of Science*, 14 (1984), 481–520.

90   Faraday to Julius Plücker, 23 May 1849: *Ibid.*, 550.

91 J.H. Gladstone, *Michael Faraday* (2nd ed., London, 1873), pp.138–9; Faraday to Becker: *op. cit.* (n.89).

92 *Diary*, 2, 437; Faraday, *ERE*, *1*, 360–416, esp. 366; Gooding, *op. cit.* (n.42), p.127.

93 *Diary*, 5, 152.

94 Faraday to William Whewell, 1 August 1850: *Correspondence*, 586. Emphasis added.

95 Crawford, *op. cit.* (n.87); Tyndall, *op. cit.* (n.70), p.53.

96 'Conservation of force', *op. cit.* (n.75), p.462. Emphasis added.

NOTES FOR CHAPTER 9

1 For example, L.P. Williams, *Michael Faraday: A biography* (London, 1965); several articles in D. Gooding and F.A.J.L. James, eds, *Faraday rediscovered: Essays on the life and work of Michael Faraday, 1791–1867* (Basingstoke and New York, 1985); J. Agassi, *Faraday as a natural philosopher* (Chicago, 1971); W. Berkson, *Fields of force: The development of a world view from Faraday to Einstein* (New York, 1974);D. Gooding, 'Final steps to the field theory: Faraday's study of magnetic phenomena, 1845–1850', *Historical Studies in the Physical Sciences*, 11 (1981), 231–75; S.M. Guralnick, 'The contents of Faraday's electrochemical laws', *Isis*, 70 (1979), 59–75; P.M. Heimann, 'Faraday's theories of matter and electricity', *British Journal for the History of Science*, 5 (1971), 235–57; F.A.J.L. James, 'Michael Faraday's first law of electrochemistry: How context develops new knowledge' in J.T. Stock and M.V. Orna, eds, *Electrochemistry, past and present* (Washington, 1989), pp.32–49; T.H. Levere, *Affinity and matter. Elements of chemical philosophy 1800–1865* (Oxford, 1971).

2 *Diary*, *1*, 45–6.

3 H.C. Oersted, 'Experiments on the effect of a current of electricity on the magnetic needle', *Annals of Philosophy*, 16 (1820), 273–6. See also A.-M. Ampère, 'Conclusions d'un Mémoire sur l'action mutuelle de deux courans électriques, sur celle qui existe entre un courant électrique et un aimant, et celle de deux aimants l'un sur l'autre', *Journal de Physique*, 91 (1820), 76–8.

4 Faraday, 'Historical sketch of electro-magnetism', *Annals of Philosophy*, 2 (1821), 195–200, 274–90 and 3 (1822), 107–21. Quotation on p.193.

5 L.P. Williams, 'Faraday and Ampère: A critical dialogue', in *Faraday rediscovered*, *op. cit.* (n.1), pp.83–104.

6 'Historical sketch', *op. cit.* (n.4), p.111.

7 *Ibid.* and Williams, *op. cit.* (n.5), pp.89–91.

8 B. Gower, 'Speculation in physics: The theory and practice of *Naturphilosophie*', *Studies in History and Philosophy of Science*, 3 (1973), 301–56, esp. p.356; T.H. Levere, *Poetry realized in nature: Samuel Taylor Coleridge and early nineteenth century science* (Cambridge, 1981).

9 'Historical sketch', *op. cit.* (n.4), p.107. See also R.C. Stauffer, 'Speculation and experiment in the background of Oersted's discovery of electromagnetism', *Isis*, 48 (1957), 33–50.

10 Gower, *op. cit.* (n.8).

11 Faraday, 'On some new electro-magnetical motions, and on the theory of magnetism', *ERE*, 2, 127–47; *Id.*, 'Electro-magnetic rotation apparatus', *ibid.*, 147–8.

12 *Diary, 1*, 49–57. See also D. Gooding, '"In nature's school": Faraday as an experimentalist', in *Faraday rediscovered, op. cit.* (n.1), pp.105–36.

13 'On some new electro-magnetical motions', *op. cit.* (n.11), 136.

14 *Ibid.*, 144–5.

15 Faraday to James Stodart, 8 October 1821: *Correspondence*, 125–7.

16 Faraday, 'A history of the condensation of the gases, in reply to Dr Davy, introduced by some remarks on that of electro-magnetic rotation', *ERCP*, 135–141, esp. 140.

17 *Diary, 1*, 71 and 312.

18 *Ibid.*, 91–5, 178 and 310; Faraday, 'Electro-magnetic current', *ERE*, 2, 162–3.

19 *Diary, 1*, 279–80.

20 *Ibid.*, 320.

21 Williams, *op. cit.* (n.1), 142–7; Williams, *op. cit.* (n.5).

22 C. Babbage and J.F.W. Herschel, 'Account of the repetition of M. Arago's experiments on the magnetism manifested by various substances during the act of rotation', *Philosophical Transactions of the Royal Society of London*, 115 (1825), 467–96. Quotation on p.485.

23 *Diary, 1*, 279–80.

24 Williams, *op. cit.* (n.1), pp.174–5.

25 'Electro-magnetic rotation apparatus', *op. cit.* (n.11).

26 *Diary, 1*, 178. Emphasis added.

27 The *Oxford English Dictionary* gives the following as the major uses of the term induction: H. Davy, *Elements of chemical philosophy* (London, 1812), p.132 and J.F.W. Herschel, *Preliminary discourse to the study of natural philosophy* (London, 1830), p.329.

28 See J.L. Heilbron, *Electricity in the 17th and 18th centuries: A study of early modern physics* (Berkeley and Los Angeles, 1979).

29 *Diary, 1*, 279–80. Entries for 28 November and 2 December 1825.

30 *ERE, 1*, 1.

31 'Historical sketch', *op. cit.* (n.4), p.108; J.L. Heilbron, 'The electric field before Faraday', in G.N. Cantor and M.J.S. Hodge, eds, *Conceptions of ether: Studies in the history of ether theories 1740–1900* (Cambridge, 1981), pp.187–214.

32 Williams, *op. cit.* (n.1), p.179; Faraday to Eilhard Mitscherlich, 4 August 1830: *Correspondence*, 181–2.

33 *Diary, 1*, 312.

34 Williams, *op. cit.* (n.1), pp.178–81; *Diary, 1*, 327–59; Faraday, 'On a peculiar class of acoustical figures; and on certain forms assumed by groups of particles upon vibrating elastic surfaces', *ERCP*, 314–58, esp. pp.357–8.

35 *Diary, 1*, 367.

36 Compare the following papers on vaporisation – Faraday, 'On the existence of a limit to vaporization' (1826), *ERCP*, 199–205; *Id.*, 'On the limits of

vaporization' (1830), *ibid.*, 205–12. Unlike the former, the latter shows Faraday's willingness to speculate on how the particles of a vapour behave.

37 Faraday to Richard Phillips, 29 August 1828: *Correspondence*, 172–4; Faraday to Mitscherlich, *op. cit.* (n.32).

38 Faraday, 'On the manufacture of glass for optical purposes' (1829), *ERCP*, 231–91; 'Acoustical figures', *op. cit.* (n.34).

39 *ERE, 1*, 1–7, esp. p.2.

40 *Diary, 1*, 367; *ERE, 1*, 24–40, esp. p.27.

41 *Diary, 1*, 368, 372 and 380–1; *ERE, 1*, 7–16, esp. p.15; *Ibid.*, 38.

42 S. Ross, 'The search for electromagnetic induction', *Notes and Records of the Royal Society of London*, 20 (1965), 184–219.

43 *ERE, 1*, 4; *Diary, 1*, 369.

44 Babbage and Herschel, *op. cit.* (n.22), p.485. *ERE, 1*, 4, 16, 18–9, 24–5 and 36.

45 *ERE, 1*, 4; *Diary, 1*, 369.

46 *Ibid.*, 376.

47 *ERE, 1*, 16–17.

48 *Ibid.*, 20–2. Emphasis added.

49 *Diary, 1*, 371.

50 T. Nickles, 'Discovery', in R.C. Olby, G.N. Cantor, J.R.R. Christie and M.J.S. Hodge, eds, *Companion to the history of modern science* (London and New York, 1990), pp.148–65.

51 Faraday, *A course of six lectures on the various forces of matter and their relations to each other* (London and Glasgow, 1860).

52 Faraday, 'On the probable ultimate analysis of chemical substances', *London Medical Gazette*, 18 (1835–6), 462. See also notes for this lecture: RI MS.

53 *Diary, 3*, 79.

54 Faraday to William Whewell, 13 December 1836: *Correspondence*, 306–7.

55 Faraday, 'On Mossotti's reference of electrical attraction, the attraction of aggregation, and the attraction of gravitation to one cause', *Philosophical Magazine*, 10 (1837), 317–18. O.F. Mossotti, 'On the forces which regulate the internal constitution of bodies', *Scientific Memoirs*, 1 (1837), 448–69. Quotation on p.451.

56 D. Gooding, 'Conceptual and experimental bases of Faraday's denial of electrostatic action at a distance', *Studies in History and Philosophy of Science*, 9 (1978), 117–49. See also *ERE, 1*, 362–3 and 514–5; *ERE, 3*, 79; Faraday, 'An answer to Dr Hare's letter on certain theoretical opinions', *ERE, 2*, 262–74.

57 Gooding, *op. cit.* (n.56), 125; *Diary, 3*, 180; *ERE, 1*, 363.

58 Faraday, 'On the physical character of the lines of magnetic force', *ERE, 3*, 407–37. Quotation on p.409.

59 Faraday, 'A course of lectures on electricity and magnetism', *London Medical Gazette*, 2 (1846), 977–82 and 3 (1846), 1–7, 89–95, 177–82, 265–71, 353–8, 441–7; 523–9. See esp. pp.978–80.

60 Faraday, 'Physico-chemical philosophy', 1846: RI MS, F4J7, f.17.

61 Faraday, 'Observations on the magnetic force', *Philosophical Magazine*, 5 (1853), 218–27.

62 *Diary*, 5, 150.

63 *ERE*, 3, 161–8. Like Einstein, half a century later, Faraday conceived the relevance of relative, rather than absolute, space.

64 *Diary*, 5, 150.

65 *Ibid.*, 152.

66 *Ibid.*, 154–5.

67 *Ibid.*, 156.

68 *Ibid.*, 158.

69 *Ibid.*, 164.

70 *Ibid.*, 167.

71 *Ibid.*, 1, 372.

72 *Ibid.*, 5, 182.

73 *ERE*, 3, 168. Emphasis added.

74 Faraday, 'On the physical lines of magnetic force', *ERE*, 3, 438–43.

75 William Whewell to Faraday, 20 June 1852: *Correspondence*, 660–1.

76 'Physical character', *op. cit.* (n.58); 'Observations', op. cit. (n.61), p.226; *Four letters from Sir Isaac Newton to Doctor Bentley concerning some arguments in proof of a Deity* (London, 1856), reprinted in I.B. Cohen, ed, *Isaac Newton's papers and letters on natural philosophy* (Cambridge, 1958), pp.279–312.

77 Faraday, 'On some points of magnetic philosophy', *ERE*, 3, 566–74.

78 Faraday, 'On the conservation of force', *ERCP*, 443–63, esp. 446–51.

79 James Clerk Maxwell to Faraday, 9 November 1857: *Correspondence*, 881–3.

80 'Conservation', *op. cit.* (n.78), pp.460–3 – dated June 1858. See also Faraday to Edward Jones, 9 June 1857: *Correspondence*, 870–2.

81 W.J.M. Rankine, 'On the conservation of energy', *Philosophical Magazine*, 17 (1859), 250–3 and 347–8.

82 *Diary*, 7, 338.

83 *Ibid.*, 337.

84 *Ibid.*, 380. This experiment discussed on pp.334–54 and 379–81.

85 *Ibid.*, 354–79. Quotation on p.360.

86 George G. Stokes to Faraday, 8 June 1860: *Correspondence*, 956–7; Faraday to Stokes, 11 June 1860: *ibid.*, 957.

87 *Jones*, 2, 412–3.

88 *ERE*, 3, 268. See also Gooding's excellent discussion of the role of the economy of nature in Faraday's researches into diamagnetism: D. Gooding, 'Empiricism in practice: Teleology, economy and observation in Faraday's physics', *Isis*, 73 (1982), 46–67.

89 *Jones*, 2, 239. Emphasis added.

NOTES FOR CHAPTER 10

1 J. Tyndall, *Faraday as a discoverer* (5th ed., London, 1894), p.197.

2 H. von Helmholtz, 'On Faraday', *Nature*, 3 (1870), 51–2. This text is a translation of Helmholtz's preface to the German edition of Tyndall's *Faraday und seine Entdeckungen* (Braunschweig, 1870).

3 J.H. Gladstone, *Michael Faraday* (2nd ed., London, 1873), pp.60–93.
4 D. Gunston, 'He made electricity work', *The Lady*, 9 January 1986, 81.
5 Extract from transcript of interview with Margaret Thatcher broadcast on the BBC 2 programme 'Favourite things', 26 July 1987. Transcript kindly supplied by Margaret Wood, Production Assistant.
6 *Jones*, 2, 478–9.
7 S. Martin, *Michael Faraday: Philosopher and Christian. A lecture* (London, 1867), p.34; J.C. Geikie, *Michael Faraday and David Brewster, philosophers and Christians. Lessons from their lives* (London, 1868); C. Pritchard, *Analogies in the progress of nature and grace. Four sermons preached before the University of Cambridge* (Cambridge, 1868).
8 *Ibid.*, p.120.
9 *Punch*, 53 (1867), 101. Emphasis added. Reproduced in W. Jerrold, *Michael Faraday: Man of science* (London, 1891), p.159.
10 W. James, *The varieties of religious experience. A study of human nature* (Harmondsworth, 1985), pp.78–126.
11 J. Agassi, *Faraday as a natural philosopher* (Chicago, 1971).
12 *Jones*, 2, 126. Extracts from Faraday's 'Swiss Journal' occupy pp.127–60.
13 'Poetry/Carlyle etc./Philosophy etc./Politics': Royal Institution, Tyndall Papers.
14 R. Barton, 'John Tyndall, pantheist: A rereading of the Belfast address', *Osiris*, 3 (1987), 111–34; W.H. Brock, N.D. McMillan and R.C. Mollan, eds, *John Tyndall. Essays on a natural philosopher* (Dublin, 1981).
15 Tyndall, *op. cit.* (n.1), pp. 43–4.
16 *Jones*, 2, 114. Emphasis added.
17 Jerrold, *op. cit.* (n.9), p.113.
18 E. Crawford, 'Michael Faraday: Ideas about how he thought', paper delivered to a joint meeting of the British Society for the History of Science and the British Psychological Society, 7 February 1987; *Id.*, 'Learning from experience', in D. Gooding and F.A.J.L. James, eds, *Faraday rediscovered: Essays on the life and work of Michael Faraday, 1791–1867* (Basingstoke and New York, 1985), pp.211–27; R.D. Tweney, 'Faraday's discovery of induction: A cognitive approach', *ibid.*, pp.189–209.
19 L. Eddar, 'Faraday, The modest genius', *Medical News*, 1 September 1967.
20 James Faraday to Thomas Faraday, 1807: *Jones*, 1, 8.
21 'A list of the London church established at Glovers Hall in the year 1762': EMH MS. Entry for 20 February 1791.
22 *Jones*, 2, 145. See also section 6.1.
23 Gladstone, *op. cit.* (n.3), p.40.
24 F.E. Manuel, *A portrait of Isaac Newton* (Cambridge, Mass., 1968); *Id.*, 'The lad from Lincolnshire', *Texas Quarterly*, 10 (1967), 10–29.
25 D. Knight, 'Davy and Faraday: Fathers and sons', in *Faraday rediscovered, op. cit.* (n.18), pp.32–49.
26 Faraday to Benjamin Abbott, 25 January 1815: *Correspondence*, 84–92. Quotation on p.87.
27 J. Davy, *Memoirs of the life of Sir Humphry Davy* (two vols, London, 1836), vol.2, pp.160–4. Faraday, 'On the history of the condensation of gases, in

reply to Dr Davy, introduced by some remarks on that of electro-magnetic rotation', *ERCP*, 135–141.

28 Faraday, 'On some new electro-magnetical rotations, and on the theory of magnetism', *ERE*, 2, 127–47.

29 Faraday to James Stodart, 8 October 1821: *Correspondence*, 125–7

30 William Hyde Wollaston, 1 November 1821: *ibid.*, 128. See also Faraday to Wollaston, 30 October 1821: *ibid.*, 127–8.

31 D.P. Miller argues that in the early 1820s Davy was trying to reform the Royal Society by reducing the number of new members. In particular he wanted to reduce the large number of antiquarians, natural historians and horticulturalists whom Banks had encouraged to join the Society (Miller, 'Between hostile camps: Sir Humphry Davy's presidency of the Royal Society of London, 1820–1827', *British Journal for the History of Science*, 16 (1983), 1–47.) It might be argued that had he supported Faraday's admission, he would be seen as engaging in the same kind of patronage that Banks had used. However, there is no evidence that his attitude towards Faraday sprang from this concern. Moreover, with his solid scientific background Faraday should have been just the kind of person Davy and the reformers would have wanted in the Society.

32 Certificate book: Royal Society of London. See also correspondence in *Jones*, *1*, 334–40.

33 *Ibid.*, 340; Sarah Faraday to John Tyndall (?), 28 March 1870: RI MS, FII.

34 Faraday to William Buchanan, 26 August 1846: Ferguson MS; Faraday to Sarah, 13 August 1846: *Jones*, 2, 227.

35 Jerrold, *op. cit.* (n.9), pp.96–7.

36 See letters in *Jones*, *1*, 91, 113–7, 130–1, 134–6, 148–50, 173–4, 187–8 and 218–19.

37 Sarah Faraday to Henry Bence Jones, 1 December 1871: RI MS, FII.

38 C.A.H. Crosse, 'Science and society in the fifties', *Temple Bar*, 93 (1891), 33–51.

39 Knight, *op. cit.* (n.25), p.35.

40 See letters in *Jones*, *1*, 323–30, 347–8; 2, 95–6, 244–5, 449, 452 and 453.

41 Crosse, *op. cit.* (n.38); D. Stein, *Ada. A life and a legacy* (Cambridge, Mass. and London, 1985), pp.136–40; M. L[loyd], *Sunny memories, containing personal recollections of some celebrated characters* (two vols, London, 1879–80), vol.1, pp.64–70.

42 Gladstone, *op. cit.* (n.3), p.70.

43 A.F. Yarrow, 'An incident in connection with Faraday's life', *Proceedings of the Royal Institution*, 25 (1926–8), 480.

44 Faraday, *A course of six lectures on the various forces of matter and their relation to each other* (London, 1860), p.2.

45 Jerrold, *op. cit.* (n.9), p.85.

46 E.F. Keller, *Reflections on gender and science* (New Haven and London, 1985).

47 Faraday to William Buchanan, 29 July 1840: Ferguson MS.

48 Faraday to William Buchanan, 15 February 1849: Ferguson MS.

49 Faraday to Caroline Deacon, 12 August 1859: *Correspondence*, 925–6.

50 The words hypocrisy, hypocrite, etc. appear in the Bible on about forty occasions. For example, Jesus accused the Pharisees of hypocrisy in Matthew 23:13. Faraday sometimes used the term against himself (as in his letter of 28 October 1840 to William Buchanan: Ferguson MS) and against others (see Faraday to Edward Magrath, 23 July 1826: *Correspondence*, 165–6).

51 'Many things had arisen to make me anxious but whether I dreamt some of them about Dr B[ence] Jones or whether they are real I am hardly sure.' Faraday to Benjamin Vincent, 12 August 1864: RI MS, F–39.

52 For example, Faraday to Schoenbein, 27 March 1860: *Jones*, 2, 433–4; Faraday to William Whewell, 7 November 1848: *Correspondence*, 528–9.

53 Faraday to William Whewell, 29 January 1853: *ibid.*, 682–3. See also Faraday to André-Marie Ampère, 3 September 1822: *ibid.*, 134–5 and Faraday to James Clerk Maxwell, 13 November 1857: *ibid.*, 884–5.

54 Faraday to William Buchanan, 20 October 1840: Ferguson MS.

55 Faraday to Schoenbein, 12 April 1844: *Schoenbein*, 122–3.

56 Faraday to William Buchanan, 21 September 1844: Ferguson MS.

57 Two letters among the Ferguson MS provide the crucial evidence. Both letters are, however, inadequately dated – one from Sarah to William Buchanan is dated 31 October (n.y.) and the letter from Faraday to Buchanan possesses no date but only 'Sabbath Evg'. According to the usually unreliable George Sims (*The Referee*, 21 June 1891), Faraday experienced a deep internal conflict in 1856. However, since the London and Edinburgh churches ceased communication in 1855, these letters must predate the schism. Since Sarah's second letter is dated 3 November and appears to have been written on a Sunday, 1850 seems the most likely year. Moreover, a further letter of Faraday's to Buchanan, which is dated 6 November 1850, forms part of the same sequence. Hence the full date of Sarah's letter is 31 October 1850 and the undated Faraday letter was probably written on 3 November 1850.

58 Sarah to Buchanan, 31 October 1850: Ferguson MS.

59 Sarah to Buchanan, 3 November 1850: Ferguson MS.

60 Faraday to Buchanan, 3 November 1850: Ferguson MS.

61 Faraday to Buchanan, 6 November 1850: Ferguson MS.

62 Faraday to Buchanan, 3 November 1850: Ferguson MS.

63 *Jones*, 2, 460.

64 Note appended by Archibald Bence Jones to Sarah to Henry Bence Jones, 6 June 1864: RI MS, F11.

65 Sarah to Henry Bence Jones, 31 May 1864: RI MS, F11.

66 See section 5.3.

67 Sarah to Jones: *op. cit.* (n.64).

68 Faraday to Benjamin Vincent, 8 June 1864: RI MS, F–36.

69 One indicator of Faraday's scientific research is the number of pages occupied in his published *Diary*. There is considerable year-to-year fluctuation in this indicator, but his annual average for the period 1832–9 is 131 pages, and 111 pages for the period 1845–58. By contrast, the average for the years 1840–4 is 31 pages, with no contribution for the year 1841.

70 Tyndall, *op. cit.* (n.1), p.89; *Jones*, 2, 126.

71 Faraday to Schoenbein, 16 May 1843: *Schoenbein*, 109.

72 E. Hare, 'Michael Faraday's loss of memory', *Proceedings of the Royal Institution*, 49 (1976), 33–52.

73 Faraday to Benjamin Abbott, 12 July 1812: *Correspondence*, 3–6.

74 Faraday to Schoenbein, 27 March 1841: *Schoenbein*, 85.

75 Faraday to Schoenbein, 20 February 1845: *ibid.*, 143–4; Faraday to Schoenbein, 13 April 1858; *ibid.*, 314.

76 Faraday to Schoenbein, 13 November 1845: *ibid.*, 148.

77 Faraday to Richard Phillips, 29 August 1828: *Correspondence*, 172; Faraday to Charles Babbage, 2 November 1840: *ibid.*, 381.

78 Faraday to Lord Auckland, 29 July 1847: *ibid.*, 508. Emphasis added.

79 Stein, *op. cit.* (n.41), pp.137–9.

80 Hare, *op. cit.* (n.72), p.41, likewise rejects the thesis that Faraday was suffering from mercury poisoning.

81 L.W. Johnson and M.L. Wolbarsht, 'Mercury poisoning: A probable cause of Isaac Newton's physical and mental ills', *Notes and Records of the Royal Society of London*, 34 (1979), 1–10; P.E. Spargo and C.A. Pounds, 'Newton's "derangement of the intellect." New light on an old problem', *ibid.*, 11–32.

82 J.W. Lance, *The mechanism and management of headache* (2nd ed., London, 1973), p.78; N.H. Raskin and O. Appenzeller, *Headache* (Philadelphia, 1980), pp.172–84.

83 Sarah to Bence Jones: *op. cit.* (n.65); Faraday to Whewell: *op. cit.* (n.53).

84 Faraday to Babbage: *op. cit.* (n.77); Faraday to Lord Auckland: *op. cit.* (n.78); Faraday to Edward Magrath, 26 March 1830: *Correspondence*, 179-80; H.P. Laughlin, *The neuroses* (Washington, 1967), pp.81–133.

85 *Ibid.*, pp.227–305, esp. p.239.

86 We also find this emphasis on order exemplified in the work of several other Sandemanians, e.g. Benjamin Vincent's concern with order among books in his superbly organised subject catalogue of the holdings of the Royal Institution Library, or, to take a recent example, Gerard Sandeman's edition of *The Sandeman genealogy* (Edinburgh, 1950).

87 Laughlin, *op. cit.* (n.84), p.225.

88 Gladstone, *op. cit.* (n.3), p.123; Faraday to Schoenbein, 13 November 1845: *Schoenbein*, 148.

89 Crawford, *op. cit.* (n.18), esp. pp.220–2.

90 G.A. Kelly, *The psychology of personal constructs* (two vols, New York, 1955); D. Bannister and F. Fransella, *Inquiring man: The psychology of personal constructs* (Harmondworth, 1980). Quotation on p.37.

91 Laughlin, *op. cit.* (n.84), pp.255–6.

92 Faraday to Julius Plücker, 8 April 1856: *Correspondence*, 833–5.

93 Faraday, 'Observations on mental education', *ERCP*, 463–91.

94 Bannister and Fransella, *op. cit.* (n.90), p.23.

95 For relationship with Davy see section 10.2. Faraday to Schoenbein, 12 April 1844: *Schoenbein*, 122–3; Faraday to William Buchanan, 21 September 1844: Ferguson MS.

NOTES FOR CHAPTER 11

1   James Morison to Alex Hutton, 23 May 1792: Ferguson MS.
2   For analyses of the conflict thesis, see J.H. Brooke, 'Science and religion', in R.C. Olby, G.N. Cantor, J.R.R. Christie and M.J.S. Hodge, eds, *Companion to the history of modern science* (London, 1990), pp.763–82; *Id.*, *Science and religion: Some historical perspectives* (Cambridge, 1991).
3   W.R. Shea, 'Galileo and the church', in D.C. Lindberg and R.L. Numbers, eds, *God and nature: Historical essays on the encounter between Christianity and science* (Berkeley, 1986), pp.114–35; J.L. Langford, *Galileo, science and the church* (Ann Arbor, 1971).
4   J.V. Jensen, 'Return to the Wilberforce–Huxley debate', *British Journal for the History of Science*, 21 (1988), 161–80.
5   J.R. Moore, *The post-Darwinian controversies: A study of the Protestant struggle to come to terms with Darwin in Great Britain and America 1870–1900* (Cambridge, 1979), pp.19–49.
6   For a collection of recent case studies, see Lindberg and Numbers, *op. cit.* (n.3).
7   See R. Gruner, 'Science, nature and Christianity', *Journal of Theological Studies*, 26 (1975), 55–81.
8   See, for example, P. Eichman, 'Michael Faraday: Man of God man of science', *Perspectives on Science and Christian Faith*, 40 (1988), 91–7; C.A. Russell, *Cross-currents: Interactions between science and faith* (Leicester, 1985), pp.256–65. Also such Victorian works as J.C. Geikie, *Michael Faraday and Sir David Brewster; Philosophers and Christians: Lessons from their lives* (London, 1868); S. Martin, *Michael Faraday: Philosopher and Christian* (London, 1867).
9   S.F. Cannon, *Science in culture: The early Victorian period* (New York, 1978); J. Morrell and A. Thackray, *Gentlemen of science: early years of the British Association for the Advancement of Science* (Oxford, 1981).
10 The Astronomer Royal, George Biddell Airy, considered Faraday a mystic. See Cannon, *op. cit.* (n.9), pp.31 and 58.
11 J.H. Brooke, 'Indications of a Creator: Whewell as apologist and priest', in M. Fisch and S. Schaffer, eds, *William Whewell* (Oxford, 1990); P. Corsi, *Science and religion: Baden Powell and the Anglican debate, 1800–1860* (Cambridge, 1988); Moore, *op. cit.* (n.5).
12 [R. Sandeman?], *The philosophy of the Creation, as narrated in Moses' principia, Gen. chap. I v.1 to 18* (Edinburgh, 1835); J. Glas, *The vision of the sealed book, contained in a series of five letters* in *The works of Mr. John Glas* (five vols, Perth, 1782), vol.4, pp.109–80.
13 *Jones*, 2, 191–2.

Index